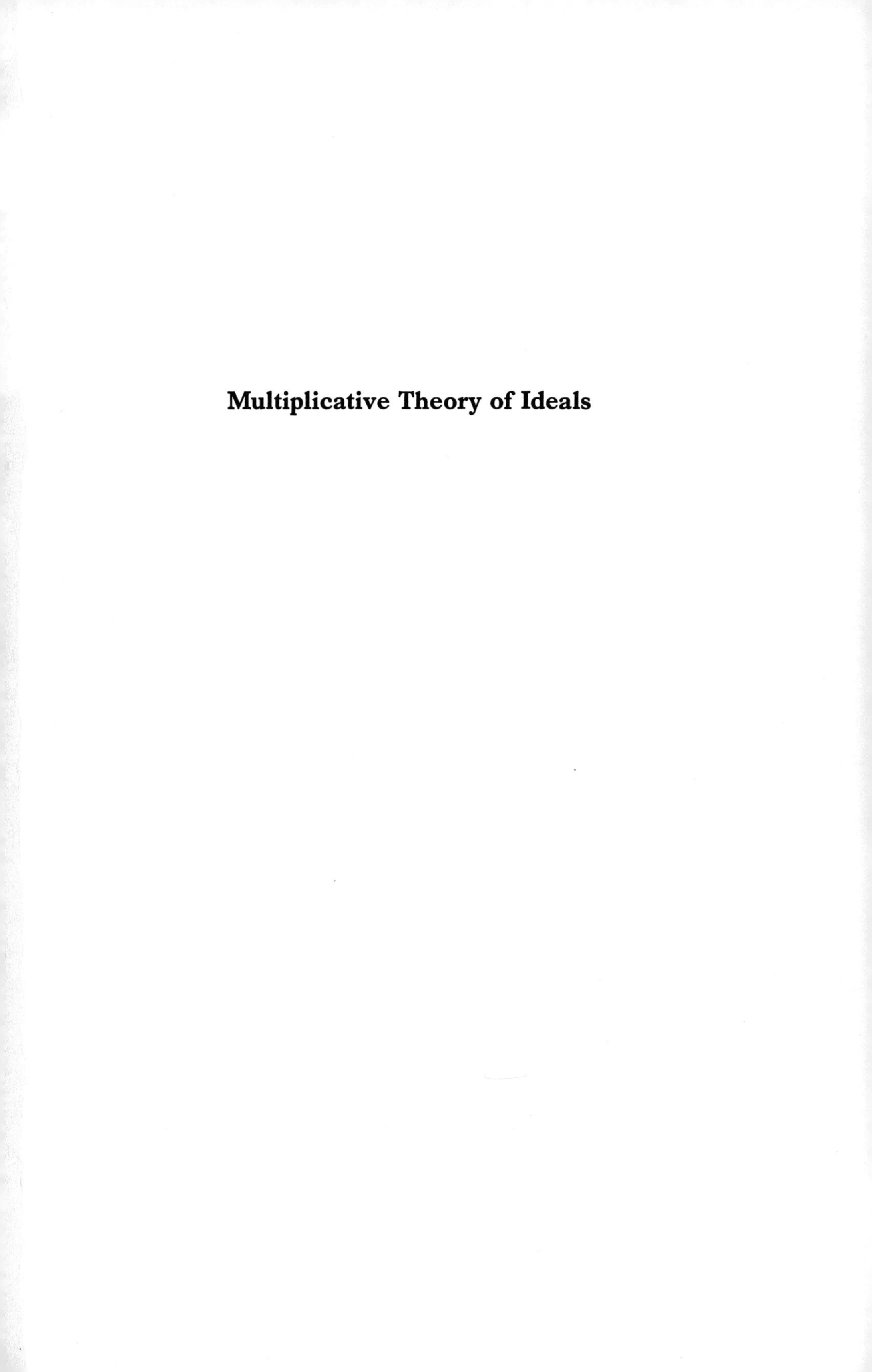

Multiplicative Theory of Ideals

This is Volume 43 in
PURE AND APPLIED MATHEMATICS
A Series of Monographs and Textbooks
Editors: PAUL A. SMITH AND SAMUEL EILENBERG
A complete list of titles in this series appears at the end of this volume

MULTIPLICATIVE THEORY OF IDEALS

MAX D. LARSEN / *PAUL J. McCARTHY*

University of Nebraska
Lincoln, Nebraska

University of Kansas
Lawrence, Kansas

ACADEMIC PRESS New York and London 1971

ACADEMIC PRESS, INC.
111 Fifth Avenue, New York, New York 10003

United Kingdom Edition published by
ACADEMIC PRESS, INC. (LONDON) LTD.
Berkeley Square House, London W1X 6BA

LIBRARY OF CONGRESS CATALOG CARD NUMBER: 72-137621

AMS (MOS)1970 Subject Classification 13F05; 13A05, 13B20,
13C15, 13E05, 13F20

PRINTED IN THE UNITED STATES OF AMERICA

To Lillie and Jean

Contents

Chapter IV. **Integral Dependence**

Chapter V. **Valuation Rings**

Chapter VI. **Prüfer and Dedekind Domains**

Chapter VII. **Dimension of Commutative Rings**

Chapter VIII. **Krull Domains**

Chapter IX. **Generalizations of Dedekind Domains**

Chapter X. **Prüfer Rings**

Preface

The viability of the theory of commutative rings is evident from the many papers on the subject which are published each month. This is not surprising, considering the many problems in algebra and geometry, and indeed in almost every branch of mathematics, which lead naturally to the study of various aspects of commutative rings. In this book we have tried to provide the reader with an introduction to the basic ideas, results, and techniques of one part of the theory of commutative rings, namely, multiplicative ideal theory.

The text may be divided roughly into three parts. In the first part, the basic notions and technical tools are introduced and developed. In the second part, the two great classes of rings, the Prüfer domains and the Krull rings, are studied in some detail. In the final part, a number of generalizations are considered. In the appendix a brief introduction is given to the tertiary decomposition of ideals of noncommutative rings.

The lengthy bibliography begins with a list of books, some on commutative rings and others on related subjects. Then follows a list of papers, all more or less concerned with the subject matter of the text.

This book has been written for those who have completed a course in abstract algebra at the graduate level. Preceding the text there is a discussion of some of the prerequisites which we consider necessary.

At the end of each chapter are a number of exercises. They are of three types. Some require the completion of certain technical details—they might possibly be regarded as busy work. Others contain examples—some of these are messy, but it will be beneficial for the reader to have some experience with examples. Finally, there are exercises which enlarge upon some topic of the text or which contain generalizations of results in the text—the bulk of the exercises are of this type. A number of exercises are referred to in proofs, and those proofs cannot be considered to be complete until the relevant exercises have been done.

We wish to thank those of our colleagues and students who have commented on our efforts over the years. Special thanks goes to Thomas Shores for his careful reading of the entire manuscript, and to our wives for their patience.

Prerequisites

A graduate level course in abstract algebra will provide most of the background knowledge necessary to read this book. In several places we have used a little more field theory than might be given in such a course. The necessary field theory may be found in the first two chapters of "Algebraic Extensions of Fields" by McCarthy, which is listed in the bibliography.

One thing that is certainly required is familiarity with Zorn's lemma. Let S be a set. A **partial ordering** on S is a relation $\leq$ on S such that

(i) $s \leq s$ for all $s \in S$;
(ii) if $s \leq t$ and $t \leq s$, then $s = t$; and
(iii) if $s \leq t$ and $t \leq u$, then $s \leq u$.

The set S, together with a partial ordering on S, is called a **partially ordered set.**

Let S be a partially ordered set. A subset T of S is called **totally ordered** if for all elements s, $t \in T$ either $s \leq t$ or $t \leq s$. Let S' be a subset of S. An element $s \in S$ is called an **upper bound** of S' if $s' \leq s$ for all $s' \in S'$. An element $s \in S$ is called a **maximal element** of S if for an element $t \in S$, $s \leq t$ implies that $t = s$. Note that S may have more than one maximal element.

Zorn's Lemma. *Let S be a nonempty partially ordered set. If every totally ordered subset of S has an upper bound in S, then S has a maximal element.*

If A and B are subsets of some set, then $A \subseteq B$ means that A is a subset of B, and $A \subset B$ means that $A \subseteq B$ but $A \neq B$. If S is a set of subsets of some set, then S is a partially ordered set with $\subseteq$ as the partial ordering. Whenever we refer to a set of subsets as a partially ordered set we mean with this partial ordering.

Let S and T be sets and consider a mapping $f: S \to T$. The mapping can be described explicitly in terms of elements by writing $s \mapsto f(s)$. If A is a subset of S, we write $f(A) = \{f(s) \mid s \in A\}$, and if B is a subset of T, we write $f^{-1}(B) = \{s \mid s \in S \text{ and } f(s) \in B\}$. Thus, f provides us with two mappings, one from the set of subsets of S into the set of subsets of T and another in the opposite direction. We assume that the reader can manipulate with these mappings.

If S, T, and U are sets and $f: S \to T$ and $g: T \to U$ are mappings, their **composition** $gf: S \to U$ is defined by $(gf)\,(s) = g(f(s))$ for all $s \in S$.

On several occasions we shall use the **Kronecker delta** δ_{ij}, which is defined by

$$\delta_{ij} = \begin{cases} 1 & \text{if } \; i = j, \\ 0 & \text{if } \; i \neq j. \end{cases}$$

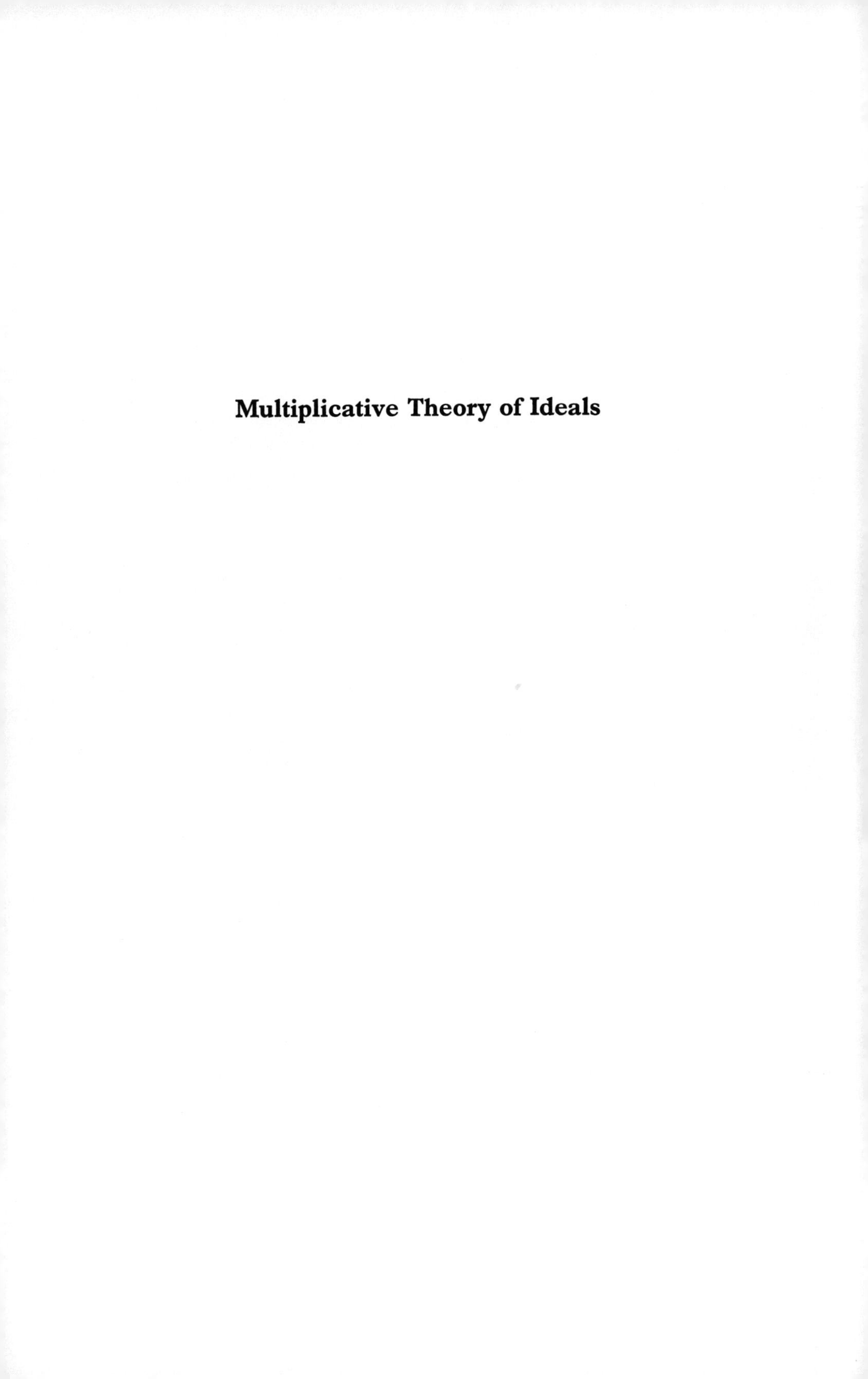

Multiplicative Theory of Ideals

CHAPTER

I

Modules

1 RINGS AND MODULES

We begin by recalling the definition of ring. A **ring** R is a non-empty set, which we also denote by R, together with two binary operations $(a, b) \mapsto a + b$ and $(a, b) \mapsto ab$ (addition and multiplication, respectively), subject to the following conditions:

(i) the set R, together with addition, is an Abelian group;
(ii) $a(bc) = (ab)c$ for all $a, b, c \in R$;
(iii) $a(b + c) = ab + ac$ and $(b + c)a = ba + ca$ for all $a, b, c \in R$.

Let R be a ring. The identity element of the group of (i) will be denoted by 0; the inverse of an element $a \in R$ considered as an element of this group will be denoted by $-a$; $a + (-b)$ will be written $a - b$. The reader may verify for himself such statements as

$$\begin{aligned} 0a = a0 &= 0 && \text{for all } a \in R,\\ a(-b) = (-a)b &= -(ab) && \text{for all } a, b \in R,\\ a(b - c) &= ab - ac && \text{for all } a, b, c \in R. \end{aligned}$$

A ring R is said to be **commutative** if $ab = ba$ for all $a, b \in R$. An element of R is called a **unity**, and is denoted by 1, if $1a = a1 = a$ for all $a \in R$. If R has a unity, then it has exactly one unity. We shall assume throughout this book that all rings under consideration have unities. By a **subring** of a ring R we mean a ring S such that the set S is a subset of the set R and such that the binary operations of R yield the binary operations of S when restricted to $S \times S$. By our assumption concerning unities, both R and S have unities. We shall consider only those subrings of a ring R which have the same unity as R.

By a **left ideal** of a ring R we mean a nonempty subset A of R such that $a - b \in A$ and $ra \in A$ for all $a, b \in A$ and $r \in R$. By a **right ideal** of R we mean a nonempty subset B of R such that $a - b \in B$ and $ar \in B$ for all $a, b \in B$ and $r \in R$. A left ideal of R which is at the same time a right ideal of R is called an **ideal** of R. If R is a commutative ring, then the left ideals, right ideals, and ideals of R coincide. A left ideal, right ideal, or ideal of R is called **proper** if it is different from R. The ideal consisting of the element 0 alone is called the **zero ideal** of R and will be denoted simply by 0; it is a proper ideal if and only if R has more than one element.

Let R be a ring. If $a \in R$, then the set

$$Ra = \{ra \mid r \in R\}$$

is a left ideal of R, while

$$aR = \{ar \mid r \in R\}$$

is a right ideal of R. Since R has a unity, we have $a \in Ra$ and $a \in aR$. The smallest ideal of R containing a, called the **principal ideal generated by** a, is the set of elements of R of the form

$$\sum r_i a s_i,$$

where $r_i, s_i \in R$ and the sum is finite. This ideal will be denoted by (a). If R is commutative, $Ra = aR = (a)$.

1.1. Definition. *Let R be a ring. A* **left R-module** *M is an Abelian group, written additively, together with a mapping $(a, x) \mapsto ax$ from $R \times M$ into M such that*

(1) $a(x + y) = ax + by,$

(2) $(a + b)x = ax + bx,$

(3) $(ab)x = a(bx),$

for all $a, b \in R$ and $x, y \in M$. A **right R-module** *N is an Abelian group, written additively, together with a mapping $(x, a) \mapsto xa$ from $N \times R$ into N such that*

(1′) $(x+y)a = xa + ya$,
(2′) $x(a+b) = xa + xb$,
(3′) $x(ab) = (xa)b$,

for all $a, b \in R$ *and* $x, y \in N$.

We shall adopt the custom of referring to left R-modules simply as R-modules. Many of the results concerning modules, left and right, will be stated for R-modules only. Analogous results hold for right R-modules.

Let R be a commutative ring and let M be an R-module. If $a \in R$ and $x \in M$, we define xa to be ax. This makes M into a right R-module. It is immediate that (1′) and (2′) hold. As for (3′), we note that for all $a, b \in R$ and $x \in M$ we have $x(ab) = (ab)x = (ba)x = b(ax) = (xa)b$. If R is not commutative, then an R-module cannot necessarily be made into a right R-module in this way.

An R-module M is called **unital** if $1x = x$ for all $x \in M$. A similar definition applies to right R-modules. We shall assume throughout this book that all modules considered, whether left or right modules, are unital.

An R-module N is called a **submodule** of an R-module M if N is a subgroup of M and if the module operation $R \times N \to N$ is the restriction of the module operation $R \times M \to M$ to $R \times N$. Suppose that the nonempty set N is a subset of M. Suppose further that $x - y \in N$ and $ax \in N$ for all $x, y \in N$ and $a \in R$. Then N, together with the induced operation of addition, is a subgroup of M. The mapping $(a, x) \mapsto ax$ from $R \times N$ into N makes this subgroup into a submodule of M. We refer simply to the submodule N.

If L and N are submodules of an R-module M, then their intersection $L \cap N$ is also a submodule of M. More generally, if $\{N_\alpha \mid \alpha \in I\}$ is an arbitrary nonempty family of submodules of M, then

$$\bigcap_{\alpha \in I} N_\alpha$$

is a submodule of M. The **sum** of L and N, denoted by $L + N$, is defined by

$$L + N = \{x + y \mid x \in L \quad \text{and} \quad y \in N\}.$$

It is easily seen that $L + N$ is a submodule of M. If $N_1, \ldots, N_k$ are submodules of M, then

$$N_1 + \cdots + N_k = \{x_1 + \cdots + x_k \mid x_i \in N_i \quad \text{for each} \quad i\}$$

is a submodule of M. We denote this submodule by $\sum_{i=1}^{k} N_i$; it is called the sum of $N_1, \ldots, N_k$.

The notion of sum of submodules of M can be extended to an arbitrary nonempty family $\{N_\alpha \mid \alpha \in I\}$ of submodules of M. For each finite subset J of I,

$$\sum_{\alpha \in J} N_\alpha$$

is a submodule of M. In general, the union of submodules of M is not a submodule of M. However,

$$\bigcup_{\substack{J \subseteq I \\ J \text{ finite}}} \sum_{\alpha \in J} N_\alpha$$

is a submodule of M. It is this submodule that we call the **sum** of the family $\{N_\alpha \mid \alpha \in I\}$, and denote by $\sum_{\alpha \in I} N_\alpha$.

Let M be an R-module and let $x \in M$. Then

$$Rx = \{ax \mid a \in R\}$$

is a submodule of M. Since M is unital, Rx contains x. If S is a subset of M, we call the smallest submodule of M which contains S the submodule **generated** by S. This submodule always exists and is unique, for it is precisely the intersection of all of the submodules of M containing S. Note that if S is empty, this submodule consists of the element 0 alone. We call the submodule of M consisting of the element 0 alone the **zero submodule** of M and denote it simply by 0. If S is not empty, then it is clear that the submodule generated by S is precisely

$$\sum_{x \in S} Rx.$$

1.2. Definition. *Let M be an R-module. If there is a finite subset $\{x_1, \ldots, x_n\}$ of M such that $M = Rx_1 + \cdots + Rx_n$, then we say that M is* **finitely generated**.

Example 1. We shall denote the ring of integers by Z. Every Abelian group is a Z-module with respect to the mapping $(n, x) \mapsto nx$ given by

$$nx = \begin{cases} x + \cdots + x \; (n \; x\text{'s}) & \text{if} \quad n > 0 \\ 0 & \text{if} \quad n = 0 \\ -(-n)x & \text{if} \quad n < 0. \end{cases}$$

Example 2. A ring R is an R-module with respect to its own addition and with its own multiplication as the module operation. Similarly R is a right R-module. The submodules of the R-module R are the left ideals of R, while the submodules of the right R-module R are the right ideals of R.

Example 3. Let K be a field. The K-modules are the vector spaces over K. The submodules of a K-module are its vector subspaces.

Example 4. Let M be the set of all ordered n-tuples $(a_1, \ldots, a_n)$ of elements of a ring R. If we set

$$(a_1, \ldots, a_n) + (b_1, \ldots, b_n) = (a_1 + b_1, \ldots, a_n + b_n)$$

and

$$a(a_1, \ldots, a_n) = (aa_1, \ldots, aa_n),$$

then M becomes an R-module. For $i = 1, \ldots, n$, set $x_i = (0, \ldots, 0, 1, 0, \ldots, 0)$, where the 1 appears in the ith place. Then $M = Rx_1 + \cdots + Rx_n$.

Before proceeding with further definitions, we shall give an important relation between the notions of intersection and sum of submodules.

1.3. Theorem (Modular Law for Submodules). *Let K, L, and N be submodules of an R-module M. If $K \subseteq N$, then*

$$K + (L \cap N) = (K + L) \cap N.$$

Proof. We have $K + (L \cap N) \subseteq K + L$. Also, since $K \subseteq N$, we have $K + (L \cap N) \subseteq N$. Hence $K + (L \cap N) \subseteq (K + L) \cap N$. Now suppose $x \in (K + L) \cap N$. Then $x \in N$ and $x = y + z$ where $y \in K$ and $z \in L$. But $K \subseteq N$, so $y \in N$; hence $z = x - y \in L \cap N$. Thus $x \in K + (L \cap N)$, and we have $(K + L) \cap N \subseteq K + (L \cap N)$.

1.4. Definition. *A mapping ϕ from an R-module M into an R-module N is a* **homomorphism** *if $\phi(x + y) = \phi(x) + \phi(y)$ and $\phi(ax) = a\phi(x)$ for all $x, y \in M$ and $a \in R$. A homomorphism $\phi: M \to N$ is* **surjective** *if it maps M onto N,* **injective** *if it is one-to-one, and* **bijective** *if it is both surjective and injective.*

A bijective homomorphism is also called an **isomorphism.** In this case, the inverse mapping exists and is also an isomorphism. If there is an isomorphism from M onto N, we say that M and N are **isomorphic** and write $M \cong N$.

It should be clear how to define homomorphism and the related concepts for right R-modules. Note that the relation of being "isomorphic to" between R-modules is reflexive, symmetric, and transitive. If M is an R-module, then the identity mapping of M is an isomorphism of M onto itself; we denote it by 1_M.

Let M be an R-module and N a submodule of M. Since M is an Abelian group and N is a subgroup of M, we may consider the factor group M/N. We define a mapping from $R \times M/N$ into M/N by $(a, x + N) \mapsto ax + N$; this is a well-defined mapping. For, suppose $x + N = y + N$, then $x - y \in N$ and $ax - ay = a(x - y) \in N$. Hence $ax + N = ay + N$. With respect to this mapping, M/N is an R-module, called a **factor module** of M. The mapping ϕ from M into M/N defined by $\phi(x) = x + N$ is a surjective homomorphism. It is called the **canonical homomorphism** from M onto M/N.

1.5. Theorem. *Let M be an R-module and let ϕ be a homomorphism from M into an R-module N. Then*

$$\operatorname{Ker} \phi = \{x \mid x \in M \quad \text{and} \quad \phi(x) = 0\}$$

is a submodule of M called the **kernel** *of ϕ. Also,*

$$\operatorname{Im} \phi = \{\phi(x) \mid x \in M\}$$

is a submodule of N called the **image** *of ϕ. Furthermore,*

$$\operatorname{Im} \phi \cong M/\operatorname{Ker} \phi.$$

In fact, if η is the canonical homomorphism from M onto M/N, then there is a unique isomorphism ψ from $M/\operatorname{Ker} \phi$ onto $\operatorname{Im} \phi$ such that $\phi = \psi\eta$.

Proof. We leave it to the reader to show that $\operatorname{Ker} \phi$ and $\operatorname{Im} \phi$ are submodules of M and N, respectively. Define a mapping ψ from $M/\operatorname{Ker} \phi$ into $\operatorname{Im} \phi$ by $\psi(x + \operatorname{Ker} \phi) = \phi(x)$. If $x + \operatorname{Ker} \phi = y + \operatorname{Ker} \phi$, then $x - y \in \operatorname{Ker} \phi$; hence $\phi(x) - \phi(y) = \phi(x - y) = 0$. Thus ψ is well-defined and is clearly a surjective homomorphism. If

$\psi(x + \text{Ker}\ \phi) = \psi(x' + \text{Ker}\ \phi)$, then $\phi(x) = \phi(x')$ and $\phi(x - x') = 0$; hence $x - x' \in \text{Ker}\ \phi$ and so $x + \text{Ker}\ \phi = x' + \text{Ker}\ \phi$. Therefore ψ is bijective.

We shall now derive two important corollaries of this theorem. Let L and N be submodules of an R-module M and suppose that $L \subseteq N$. Then L is a submodule of N and the factor module N/L can be imbedded in M/L in the following way. If $x \in N$ we identify $x + L$, as an element of N/L, with $x + L$, as an element of M/L. Clearly this identification is legitimate; each element of N/L is identified with exactly one element of M/L, and distinct elements of N/L are identified with distinct elements of M/L. Furthermore, after this identification has been made, N/L is a submodule of M/L. The elements of N/L are precisely those cosets of L in M which consist entirely of elements of N.

1.6. Corollary. *With M, L, and N as above we have*

$$(M/L)/(N/L) \cong M/N.$$

Proof. Define a mapping from M/L into M/N by $\phi(x + L) = x + N$. Then ϕ is a well-defined surjective homomorphism. Thus, by Theorem 1.5,

$$(M/L)/\text{Ker}\ \phi \cong M/N.$$

However $\phi(x + L) = x + N = 0$ if and only if $x \in N$. Thus $\text{Ker}\ \phi = N/L$.

1.7. Corollary. *If L and N are submodules of an R-module, then*

$$(L + N)/N \cong L/(L \cap N).$$

Proof. Define a mapping from L into $(L + N)/N$ by $\phi(x) = x + N$. It is clear that ϕ is a homomorphism. To show that ϕ is surjective we note that if $(x + y) + N$ is a typical element of $(L + N)/N$, with $x \in L$ and $y \in N$, then $(x + y) + N = x + N = \phi(x)$. Thus, by Theorem 1.5,

$$L/\text{Ker}\ \phi \cong (L + N)/N.$$

However $\phi(x) = 0$ if and only if $x \in N$, that is, if and only if $x \in L \cap N$. Hence $\text{Ker}\ \phi = L \cap N$.

If A is a left ideal of a ring R, we can consider the factor module R/A. If A is, in fact, an ideal, we can make R/A into a ring by defining

$$(a + A)(b + A) = ab + A.$$

We shall show that this is a well-defined binary operation on R/A and leave it to the reader to show that R/A is a ring with this operation as its multiplication. Suppose that $a + A = a' + A$ and $b + A = b' + A$. Then $a - a' \in A$ and $b - b' \in A$, and so $ab - a'b = a(b - b') + (a - a')b' \in A$, since A is both a left ideal and a right ideal of R. Thus $ab + A = a'b' + A$. The ring R/A is called a **residue class ring** of R.

A mapping ϕ from a ring R into a ring R' is called a **homomorphism** if $\phi(a + b) = \phi(a) + \phi(b)$ and $\phi(ab) = \phi(a)\phi(b)$ for all $a, b \in R$, and $\phi(1) = 1$. The modifiers surjective, injective, and bijective retain the same meaning as before when applied to homomorphisms of modules. A bijective homomorphism is called an **isomorphism**, and two rings R and R' are said to be **isomorphic** if there is an isomorphism from R onto R'. In this case we write $R \cong R'$.

Let ϕ be a homomorphism from a ring R into a ring R'. Then $\operatorname{Im} \phi = \{\phi(a) \mid a \in R\}$ is a ring. The unity of $\operatorname{Im} \phi$ is $\phi(1)$, and $\operatorname{Im} \phi$ is a subring of R'. Also $\operatorname{Ker} \phi = \{a \mid a \in R \text{ and } \phi(a) = 0\}$ is an ideal of R, and

$$\operatorname{Im} \phi \cong R/\operatorname{Ker} \phi.$$

This is proved in the same way as Theorem 1.5. Results like those of the corollaries to Theorem 1.5 also hold for rings and ideals.

2 CHAIN CONDITIONS

Let R be a ring and let M be an R-module. By an **ascending chain** of submodules of M we mean an increasing sequence

$$N_1 \subseteq N_2 \subseteq N_3 \subseteq \cdots \tag{1}$$

of submodules of M. By a **descending chain** of submodules of M we mean a decreasing sequence

$$N_1 \supseteq N_2 \supseteq N_3 \supseteq \cdots$$

of submodules of M. Such a chain (ascending or descending) is called **infinite** if for each positive integer n the chain has more than n distinct terms; otherwise the chain is called **finite**. Many of the modules of interest to us satisfy one or both of the following conditions:

Ascending chain condition (ACC): every ascending chain of submodules of M is finite.

Descending chain condition (DCC): every descending chain of submodules of M is finite.

1.8. Theorem. (ACC) *is equivalent to each of the following conditions:*

Finiteness condition (FC): *every submodule of M is finitely generated.*

Maximum condition (MAX): *every nonempty set of submodules of M has a maximal element with respect to set inclusion.*

(DCC) *is equivalent to the following condition:*

Minimum condition (MIN): *every nonempty set of submodules of M has a minimal element with respect to set inclusion.*

Proof. We shall show that (ACC), (FC), and (MAX) are equivalent and leave it to the reader to show that (DCC) and (MIN) are equivalent.

(ACC) $\Rightarrow$ (MAX). Assume that (ACC) holds for M and let $\mathcal{S}$ be a nonempty set of submodules of M. Let $N_1 \in \mathcal{S}$. If N_1 is not a maximal element of $\mathcal{S}$, then there exists $N_2 \in \mathcal{S}$ such that $N_1 \subset N_2$. If N_2 is not a maximal element of $\mathcal{S}$, then there exists $N_3 \in \mathcal{S}$ such that $N_1 \subset N_2 \subset N_3$. If we continue in this way we will arrive at a maximal element of $\mathcal{S}$ in a finite number of steps, for otherwise there exists an infinite ascending chain of submodules of M, contrary to (ACC).

(MAX) $\Rightarrow$ (FC). Assume that (MAX) holds for M and let N be a submodule of M. let $\mathcal{S}$ be the set of all finitely generated submodules of M contained in N. Then $\mathcal{S}$ is not empty since $0 \in \mathcal{S}$. Therefore $\mathcal{S}$ has a maximal element L. If $L \neq N$, choose $x \in N$ with $x \notin L$. Then $L \subset L + Rx \subseteq N$, which contradicts our choice of L since $L + Rx$ is finitely generated. Therefore $L = N$.

(FC) $\Rightarrow$ (ACC). Assume that (FC) holds for M and consider an ascending chain (1) of submodules of M. Let $N = \bigcup_{n=1}^{\infty} N_n$. If $x, y \in N$, then $x \in N_h$ and $y \in N_k$, where we may assume $h \leq k$. Then $x - y \in N_k \subseteq N$. Also, if $a \in R$, then $ax \in N_h \subseteq N$. Thus N is a submodule of M and consequently has a finite set of generators. Suppose $N = Rx_1 + \cdots + Rx_m$ where $x_i \in N_{n_i}$ for $i = 1, \ldots, m$. If $n_0 = \max\{n_1, \ldots, n_m\}$, then $x_i \in N_{n_0}$ for $i = 1, \ldots, m$ and so $N \subseteq N_{n_0}$. Therefore $N = N_{n_0}$ and $N_n = N_{n_0}$ for all $n \geq n_0$. Thus the ascending chain is finite.

1.9. Definition. *A sequence of R-modules and homomorphisms*

$$\cdots \longrightarrow M_{n-1} \xrightarrow{\phi_{n-1}} M_n \xrightarrow{\phi_n} M_{n+1} \longrightarrow \cdots$$

is an **exact sequence** *if*

$$\operatorname{Im} \phi_{n-1} = \operatorname{Ker} \phi_n \qquad \text{for all } n.$$

If, for $n > m$ we have $M_n = 0$, we indicate this by writing

$$\cdots \to M_{m-1} \to M_m \to 0,$$

and in a similar way we indicate that all terms from some point on to the left are zero. Thus a sequence

$$0 \to L \xrightarrow{\phi} M \xrightarrow{\psi} N \to 0 \tag{2}$$

is exact if and only if ϕ is injective, $\operatorname{Im} \phi = \operatorname{Ker} \psi$, and ψ is surjective. An exact sequence of this form is called a **short exact sequence.**

1.10. Theorem. *In the exact sequence* (2), *M satisfies* (ACC) *if and only if both L and N satisfy* (ACC).

Proof. First, assume that M satisfies (ACC). Let $L_1 \subseteq L_2 \subseteq \cdots$ be an ascending chain of submodules of L. Then $\phi(L_1) \subseteq \phi(L_2) \subseteq \cdots$ is an ascending chain of submodules of M, so there is an integer n_0 such that $\phi(L_n) = \phi(L_{n_0})$ for $n \geq n_0$. Since ϕ is injective, we have $L_n = L_{n_0}$ for $n \geq n_0$. Now let $N_1 \subseteq N_2 \subseteq \cdots$ be an ascending chain of submodules of N. Then $\psi^{-1}(N_1) \subseteq \psi^{-1}(N_2) \subseteq \cdots$ is an ascending chain of submodules of M, so there is an integer m_0 such that $\psi^{-1}(N_m) = \psi^{-1}(N_{m_0})$ for $m \geq m_0$. Since ψ is surjective, we have $N_m = \psi(\psi^{-1}(N_m))$

$= \psi(\psi^{-1}(N_{m0})) = N_{m0}$ for $m \geq m_0$. Therefore, both L and N satisfy (ACC).

On the other hand, suppose that both L and N satisfy (ACC) and consider an ascending chain $M_1 \subseteq M_2 \subseteq \cdots$ of submodules of M. Then $\phi^{-1}(\phi(L) \cap M_1) \subseteq \phi^{-1}(\phi(L) \cap M_2) \subseteq \cdots$ is an ascending chain of submodules of L and $\psi(M_1) \subseteq \psi(M_2) \subseteq \cdots$ is an ascending chain of submodules of N. Hence there is an integer n_0 such that $\phi^{-1}(\phi(L) \cap M_n) = \phi^{-1}(\phi(L) \cap M_{n0})$ and $\psi(M_n) = \psi(M_{n0})$ for $n \geq n_0$. Since ϕ maps L onto $\phi(L)$, we have $\phi(L) \cap M_n = \phi(L) \cap M_{n0}$ for $n \geq n_0$. Also $M_n + \operatorname{Ker} \psi = \psi^{-1}(\psi(M_n)) = \psi^{-1}(\psi(M_{n0})) = M_{n0} + \operatorname{Ker} \psi$ for $n \geq n_0$. Therefore, since $\operatorname{Ker} \psi = \operatorname{Im} \phi = \phi(L)$, we have for $n \geq n_0$,

$$
\begin{aligned}
M_n &= M_n \cap (M_n + \phi(L)) \\
&= M_n \cap (M_{n0} + \phi(L)) \\
&= M_{n0} + (M_n \cap \phi(L)) && \text{by Theorem 1.3} \\
&= M_{n0} + (M_{n0} \cap \phi(L)) \\
&= M_{n0}.
\end{aligned}
$$

Thus M satisfies (ACC).

1.11. Definition. *A ring R is* **left Noetherian** *if it satisfies* (ACC) *when considered as an R-module.*

1.12. Theorem. *Let R be a left Noetherian ring. If M is a finitely generated R-module, then M satisfies* (ACC).

Proof. The assertion is certainly true when $M = 0$. Suppose that $M \neq 0$; we shall prove that M satisfies (ACC) by induction on the number of elements required to generate M. Suppose that $M = Rx$. Let N be a submodule of M and set $A = \{a \mid a \in R \text{ and } ax \in N\}$. Then A is a left ideal of R and so A is finitely generated, say, $A = Ra_1 + \cdots + Ra_m$. If $y \in N$, then $y = ax$ for some $a \in A$. If $a = r_1 a_1 + \cdots + r_m a_m$, then $y = r_1 a_1 x + \cdots + r_m a_m x$. Therefore $N = Ra_1 x + \cdots + Ra_m x$, that is, N is finitely generated. Thus, M satisfies (FC) and so satisfies (ACC).

Now suppose that M requires $n > 1$ generators and that every R-module which can be generated by fewer than n of its elements satisfies (ACC). Let $M = Rx_1 + \cdots + Rx_n$ and set $L = Rx_2 + \cdots + Rx_n$. By the induction assumption, L satisfies (ACC). As we have seen, Rx_1 satisfies (ACC); hence every factor module of Rx_1 satisfies (ACC) by Theorem 1.10. Thus, since

$$M/L = (Rx_1 + L)/L \cong Rx_1/(Rx_1 \cap L),$$

by Corollary 1.7, M/L satisfies (ACC). Therefore, making use of Theorem 1.10 again, we conclude that M satisfies (ACC).

3 DIRECT SUMS

Let $\{M_\alpha \mid \alpha \in I\}$ be a family of R-modules indexed by a set I which may be of any nonzero cardinality. Let M be the set of all functions x defined on the set I such that for each $\alpha \in I$ we have $x_\alpha = x(\alpha) \in M_\alpha$, and subject to the restriction that $x_\alpha \neq 0$ for only a finite number of $\alpha \in I$. We make M into an R-module by defining

$$\left.\begin{aligned} (x+y)_\alpha &= x_\alpha + y_\alpha, \\ (ax)_\alpha &= ax_\alpha, \quad a \in R, \end{aligned}\right. \qquad \text{for all} \quad \alpha \in I.$$

1.13. Definition. *The R-module M is the* **(external) direct sum** *of the family $\{M_\alpha \mid \alpha \in I\}$. We write*

$$M = \bigoplus_{\alpha \in I} M_\alpha.$$

If $I = \{1, \ldots, n\}$ we write

$$M = M_1 \oplus \cdots \oplus M_n.$$

Suppose that $M = \bigoplus_{\alpha \in I} M_\alpha$. For each $\alpha \in I$ there is a homomorphism $\phi_\alpha : M \to M_\alpha$ given by

$$\phi_\alpha(x) = x_\alpha$$

and a homomorphism $\psi_\alpha : M_\alpha \to M$ given by

$$(\psi_\alpha(z))_\beta = \begin{cases} z & \text{if} \quad \beta = \alpha \\ 0 & \text{if} \quad \beta \neq \alpha. \end{cases}$$

It is clear that ϕ_α is surjective and ψ_α is injective. Furthermore,

(a) $$\phi_\alpha \psi_\beta = \begin{cases} 1_{M_\alpha} & \text{if} \quad \alpha = \beta \\ 0 & \text{if} \quad \alpha \neq \beta \end{cases}$$

and

(b) $$x = \sum_{\alpha \in I} \psi_\alpha \phi_\alpha(x) \qquad \text{for all} \quad x \in M.$$

Note that the sum in (b) is finite since $\phi_\alpha(x) = 0$ for all but a finite number of $\alpha \in I$.

1.14. Proposition. *Let M be an R-module and let $\{M_\alpha \mid \alpha \in I\}$ be a family of R-modules. Suppose that for each $\alpha \in I$ there exist homomorphisms $\phi_\alpha : M \to M_\alpha$ and $\psi_\alpha : M_\alpha \to M$ such that*

(i) *for each $x \in M$, $\phi_\alpha(x) \neq 0$ for only a finite number of $\alpha \in I$,*
(ii) (a) *and* (b) *hold.*

Then $M \cong \bigoplus_{\alpha \in I} M_\alpha$.

Proof. Define $\Phi : \bigoplus_{\alpha \in I} M_\alpha \to M$ by

$$\Phi(x) = \sum_{\alpha \in I} \psi_\alpha(x_\alpha).$$

Clearly Φ is a homomorphism. It is surjective since for all $x \in M$, if $y_\alpha = \phi_\alpha(x)$ for all $\alpha \in I$, then

$$\Phi(y) = \sum_{\alpha \in I} \psi_\alpha \phi_\alpha(x) = x.$$

If $\Phi(x) = 0$, then for each $\alpha \in I$,

$$0 = \phi_\alpha \Phi(x) = \sum_{\beta \in I} \phi_\alpha \psi_\beta(x_\beta) = x_\alpha,$$

and so $x = 0$. Hence Φ is an isomorphism.

We now consider an R-module M and a family $\{M_\alpha \mid \alpha \in I\}$ of submodules of M. We assume that

(c) $$M = \sum_{\alpha \in I} M_\alpha$$

and

(d) $$M_\alpha \cap \sum_{\beta \in I \setminus \{\alpha\}} M_\beta = 0 \qquad \text{for each} \quad \alpha \in I.$$

Then each element of M can be written in one and only one way in the form $\sum_{\alpha \in I} x_\alpha$, where $x_\alpha \in M$ and $x_\alpha \neq 0$ for only a finite number of $\alpha \in I$. For, each element of M can be written as such a sum in at least one way by (c). Furthermore, if

$$\sum_{\alpha \in I} x_\alpha = \sum_{\alpha \in I} y_\alpha,$$

where $x_\alpha, y_\alpha \in M_\alpha$, then for each $\alpha \in I$,

$$x_\alpha - y_\alpha = \sum_{\beta \in I \setminus \{\alpha\}} (y_\beta - x_\beta) \in M_\alpha \cap \sum_{\beta \in I \setminus \{\alpha\}} M_\beta = 0.$$

so that $x_\alpha = y_\alpha$. We say that M is the **(internal) direct sum** of the family of its submodules $\{M_\alpha \mid \alpha \in I\}$.

1.15. Proposition. *Let $\{M_\alpha \mid \alpha \in I\}$ be a family of submodules of an R-module M. Assume that M is the* (*internal*) *direct sum of this family of submodules; that is, assume* (c) *and* (d) *hold. Then $M \cong \bigoplus_{\alpha \in I} M_\alpha$.*

Proof. For every $\alpha \in I$ we define $\psi_\alpha : M_\alpha \to M$ by $\psi_\alpha(x) = x$ for all $x \in M_\alpha$. If $x \in M$, we write

$$x = \sum_{\alpha \in I} x_\alpha, \quad x_\alpha \in M_\alpha,$$

where $x_\alpha \neq 0$ for only a finite number of $\alpha \in I$. Then set $\phi_\alpha(x) = x_\alpha$; ϕ_α is a well-defined homomorphism from M into M_α. If $x \in M_\beta$, then

$$\phi_\alpha \psi_\beta(x) = \begin{cases} x & \text{if } \beta = \alpha \\ 0 & \text{if } \beta \neq \alpha. \end{cases}$$

Furthermore, for each $x \in M$, if $x = \sum_{\alpha \in I} x_\alpha$, as above, then

$$\sum_{\alpha \in I} \psi_\alpha \phi_\alpha(x) = \sum_{\alpha \in I} \psi_\alpha(x_\alpha) = \sum_{\alpha \in I} x_\alpha = x.$$

Hence the isomorphism exists by Proposition 1.14.

In the light of this result, we shall write

$$M = \bigoplus_{\alpha \in I} M_\alpha$$

whenever $\{M_\alpha \mid \alpha \in I\}$ is a family of submodules of M for which (c) and (d) hold.

Let $\{R_\alpha \mid \alpha \in I\}$ be a family of rings. If we consider only the additive structure of each R, we can form the direct sum of the resulting Abelian groups (as Z-modules),

$$R = \bigoplus_{\alpha \in I} R_\alpha .$$

We define a multiplication on R by

$$(ab)_\alpha = a_\alpha b_\alpha \qquad \text{for all} \quad \alpha \in I.$$

Then R becomes a ring, a fact which is easily verified, called the **direct sum** of the family of rings $\{R_\alpha \mid \alpha \in I\}$. The homomorphisms ϕ_α and ψ_α, as defined in the case of modules, are ring homomorphisms. For each $\alpha \in I$, $\operatorname{Im} \psi_\alpha$ is an ideal of R, and

$$\operatorname{Ker} \phi_\alpha = \sum_{\beta \neq \alpha} \operatorname{Im} \psi_\beta .$$

A result like Proposition 1.14 holds for direct sums of rings. Let R be a ring and let $\{R_\alpha \mid \alpha \in I\}$ be a family of rings. Suppose that for each $\alpha \in I$ there are homomorphisms $\phi_\alpha : R \to R_\alpha$ and $\psi_\alpha : R_\alpha \to R$ such that for each $a \in R$, $\phi_\alpha(a) \neq 0$ for only a finite number of $\alpha \in I$,

$$\phi_\alpha \psi_\beta = \begin{cases} 1_{R_\alpha} & \text{if} \quad \alpha = \beta \\ 0 & \text{if} \quad \alpha \neq \beta \end{cases}$$

and

$$a = \sum_{\alpha \in I} \psi_\alpha \phi_\alpha(a) \qquad \text{for all} \quad a \in R.$$

Then $R \cong \bigoplus_{\alpha \in I} R_\alpha$.

4 TENSOR PRODUCTS

An R-module M is called **free** if there is a nonempty subset S of M such that every element of M can be written uniquely in the form

$$\sum_{x \in S} a_x x,$$

where $a_x \in R$ and $a_x \neq 0$ for only a finite number of elements $x \in S$. In this case, we say that the set S **freely generates** M. Note that

$$M = \bigoplus_{x \in S} Rx.$$

Let S be an arbitrary nonempty set and let M be the set of all formal sums

$$\sum_{x \in S} a_x x,$$

where $a_x \in R$ and $a_x \neq 0$ for only a finite number of elements $x \in S$. Define equality of these sums by

$$\sum_{x \in S} a_x x = \sum_{x \in S} a_x' x$$

if and only if $a_x = a_x'$ for all $x \in S$. We make M into an R-module by defining

$$\sum_{x \in S} a_x x + \sum_{x \in S} b_x x = \sum_{x \in S} (a_x + b_x)x$$

and

$$a \sum_{x \in S} b_x x = \sum_{x \in S} (ab_x)x \qquad \text{for all} \quad a \in R.$$

If we identify $x \in S$ and $1x \in M$, then S is a subset of M. We see immediately that M is a free module freely generated by S. We call M the **free module defined on the set** S.

Let M be the free module defined on a set S. If ϕ is an arbitrary mapping from S into an R-module M', then there is a unique homomorphism $\phi' : M \to M'$ such that $\phi'(x) = \phi(x)$ for all $x \in S$. For, the mapping ϕ' from M into M' defined by

$$\phi'\left(\sum_{x \in S} a_x x\right) = \sum_{x \in S} a_x \phi(x)$$

is a well-defined homomorphism, and is uniquely determined by ϕ. We say that ϕ' is obtained by **extending** ϕ **by linearity**.

Let R be a ring, M a right R-module, and N an R-module. Denote by $Z(M, N)$ the free Z-module defined on the Cartesian product set $M \times N$. Thus the elements of $Z(M, N)$ are formal finite sums

$$n_1(x_1, y_1) + \cdots + n_k(x_k, y_k),$$

where $n_1, \ldots, n_k \in Z$, $x_1, \ldots, x_k \in M$, and $y_1, \ldots, y_k \in N$. Let $Y(M, N)$ be the subgroup of $Z(M, N)$ generated by the set of all elements of the form

$$(x_1 + x_2, y) - (x_1, y) - (x_2, y),$$
$$(x, y_1 + y_2) - (x, y_1) - (x, y_2),$$
$$(xa, y) - (x, ay),$$

where $x, x_1, x_2 \in M$, $y, y_1, y_2 \in N$, and $a \in R$. The factor group $Z(M, N)/Y(M, N)$ is called the **tensor product** of M and N, and is denoted by $M \otimes_R N$. We note that $M \otimes_R N$ is not an R-module; it is simply an Abelian group.

If $x \in M$ and $y \in N$, we denote the element $(x, y) + Y(M, N)$ of $M \otimes_R N$ by $x \otimes y$; we call such an element of $M \otimes_R N$ a **simple tensor**. We have

$$\begin{aligned} (x_1 + x_2) \otimes y &= x_1 \otimes y + x_2 \otimes y, \\ x \otimes (y_1 + y_2) &= x \otimes y_1 + x \otimes y_2, \\ xa \otimes y &= x \otimes ay, \end{aligned}$$

for all $x, x_1, x_2 \in M$, $y, y_1, y_2 \in N$, and $a \in R$. Thus if we fix $y \in N$, the mapping $x \mapsto x \otimes y$ is a (group) homomorphism from M into $M \otimes_R N$. In particular, if n is an integer, $n(x \otimes y) = nx \otimes y$; similarly, $n(x \otimes y) = x \otimes ny$. It follows that

$$0 \otimes y = x \otimes 0 = 0, \qquad (-x) \otimes y = -(x \otimes y) = x \otimes (-y)$$

for all $x \in M$ and $y \in N$.

The set of simple tensors $\{x \otimes y \mid x \in M,\ y \in N\}$ generates the Abelian group $M \otimes_R N$. Indeed, we can make the stronger statement that every element of $M \otimes_R N$ can be written as a sum of simple tensors. For,

$$\sum_{i=1}^{k} n_i(x_i \otimes y_i) = \sum_{i=1}^{k} (n_i x_i) \otimes y_i.$$

1.16. Definition. *Let M be a right R-module and N an R-module. A mapping ϕ from the Cartesian product set $M \times N$ into a group G is a* **bilinear mapping** *if*

$$\begin{aligned} \phi(x_1 + x_2, y) &= \phi(x_1, y) + \phi(x_2, y), \\ \phi(x, y_1 + y_2) &= \phi(x, y_1) + \phi(x, y_2), \\ \phi(xa, y) &= \phi(x, ay), \end{aligned}$$

for all $x, x_1, x_2 \in M$, $y, y_1, y_2 \in N$, and $a \in R$.

1.17. Proposition. *Let M be a right R-module, N an R-module, and ϕ a bilinear mapping from $M \times N$ into a group G. Then there is a unique homomorphism $\psi: M \otimes_R N \to G$ such that $\psi(x \otimes y) = \phi(x, y)$ for all $x \in M$ and $y \in N$.*

Proof. First, we extend ϕ by linearity to a homomorphism ϕ' from $Z(M, N)$ into G. Since ϕ is bilinear, ϕ' maps $Y(M, N)$ onto the identity element of G. Hence there is a homomorphism $\psi: M \otimes_R N \to G$ such that $\psi(x \otimes y) = \phi(x, y)$ for all $x \in M$, $y \in N$. Since every element of $M \otimes_R N$ is a sum of simple tensors, ψ is uniquely determined by how it acts on the simple tensors.

We shall use this proposition a number of times to obtain homomorphisms from tensor products into other groups. The procedure is illustrated in the proof of the following result.

1.18. Proposition. *If N is an R-module, then*

$$R \otimes_R N \cong N,$$

where this is an isomorphism of Abelian groups. Explicitly, there is an isomorphism ψ from $R \otimes_R N$ onto N such that $\psi(a \otimes y) = ay$ for all $a \in R$ and $y \in N$.

Proof. The mapping ϕ from $R \times N$ into N given by $\phi(a, y) = ay$ is bilinear, so there is a unique homomorphism ψ from $R \otimes_R N$ into N such that $\psi(a \otimes y) = ay$ for all $a \in R$, $y \in N$. It is surjective since $\psi(1 \otimes y) = y$. To show it is injective, let $t \in \operatorname{Ker} \psi$. Since

$$t = \sum a_i \otimes y_i = \sum 1a_i \otimes y_i = \sum 1 \otimes a_i y_i = 1 \otimes \sum a_i y_i,$$

if $y = \sum a_i y_i$, we have $0 = \psi(t) = \psi(1 \otimes y) = y$. Hence $t = 1 \otimes 0 = 0$.

Let M, M' be right R-modules, let N, N' be R-modules, and let $f: M \to M'$ and $g: N \to N'$ be homomorphisms. Consider the mapping ϕ from $M \times N$ into $M' \otimes_R N'$ given by

$$\phi(x, y) = f(x) \otimes g(y).$$

This mapping is bilinear. We verify one of the requirements; the other two are verified in a similar manner:

$$\begin{aligned} \phi(x_1 + x_2, y) &= f(x_1 + x_2) \otimes g(y) \\ &= (f(x_1) + f(x_2)) \otimes g(y) \\ &= f(x_1) \otimes g(y) + f(x_2) \otimes g(y) \\ &= \phi(x_1, y) + \phi(x_2, y). \end{aligned}$$

By Proposition 1.17, there is a unique homomorphism from $M \otimes_R N$ into $M' \otimes_R N'$ which maps $x \otimes y$ onto $f(x) \otimes g(y)$ for all $x \in M$ and $y \in N$. We denote this homomorphism by $f \otimes g$. The following result is obvious.

1.19. Proposition. *If* $1_M : M \to M$ *and* $1_N : N \to N$ *are the identity isomorphisms of* M *and* N, *then* $1_M \otimes 1_N$ *is the identity isomorphism of* $M \otimes_R N$. *If* $f: M \to M'$, $f': M' \to M''$, $g: N \to N'$, *and* $g': N' \to N''$ *are homomorphisms, then*

$$(f' \otimes g')(f \otimes g) = f'f \otimes g'g.$$

1.20. Proposition. *If* $f: M \to M'$ *and* $g: N \to N'$ *are surjective homomorphisms, then* $f \otimes g$ *is surjective.*

Proof. Let $\sum x_i' \otimes y_i' \in M' \otimes_R N'$. For each i, let $x_i' = f(x_i)$ and $y_i' = g(y_i)$. Then

$$\begin{aligned}(f \otimes g)(\sum x_i \otimes y_i) &= \sum (f \otimes g)(x_i \otimes y_i) \\ &= \sum f(x_i) \otimes g(y_i) \\ &= \sum x_i' \otimes y_i'.\end{aligned}$$

1.21. Proposition. *If* $f: M \to M'$ *and* $g: N \to N'$ *are surjective homomorphisms, then* $\operatorname{Ker} f \otimes g$ *consists of all finite sums of the form* $\sum x_i \otimes y_i$, *where* $x_i \in \operatorname{Ker} f$ *or* $y_i \in \operatorname{Ker} g$.

Proof. Let K be the set of all these sums. Then K is a subgroup of $M \otimes_R N$, generated by the set of simple tensors $x \otimes y$ where $x \in \operatorname{Ker} f$ or $y \in \operatorname{Ker} g$. For such a simple tensor we have $(f \otimes g)(x \otimes y) = f(x) \otimes g(y) = 0$. Hence $K \subseteq \operatorname{Ker} f \otimes g$. Therefore $f \otimes g$ induces a homomorphism $h: M \otimes_R N/K \to M' \otimes_R N'$ such that

$$h(x \otimes y + K) = f(x) \otimes g(y)$$

for all simple tensors $x \otimes y$ in $M \otimes_R N$. Furthermore,

$$\operatorname{Ker} h = \operatorname{Ker} f \otimes g/K.$$

Define a mapping j from $M' \times N'$ into $M \otimes_R N/K$ by $j(x', y') = x \otimes y + K$, where $f(x) = x'$ and $g(y) = y'$. (We use here the fact that

f and g are surjective.) We must show that j is well-defined. Suppose $f(x_1)=x'$ and $g(y_1)=y'$. Then $x-x_1 \in \operatorname{Ker} f$ and $y-y_1 \in \operatorname{Ker} g$. Hence

$$x \otimes y - x_1 \otimes y_1 = x \otimes (y-y_1) + (x-x_1) \otimes y_1 \in K,$$

and

$$x \otimes y + K = x_1 \otimes y_1 + K.$$

Now, the mapping j is bilinear; we shall verify one of the conditions, again leaving the verification of the other two to the reader. Let $x', x'' \in M'$, $y' \in N'$, and let $f(x_1)=x'$, $f(x_2)=x''$, and $g(y)=y'$. Then $f(x_1+x_2)=x'+x''$ and

$$\begin{aligned} j(x'+x'', y') &= (x_1+x_2) \otimes y + K \\ &= (x_1 \otimes y + K) + (x_2 \otimes y + K) \\ &= j(x', y') + j(x'', y'). \end{aligned}$$

Thus there is a unique homomorphism $k\colon M' \otimes_R N' \to M \otimes_R N/K$ such that $k(x' \otimes y') = x \otimes y + K$, where $f(x)=x'$ and $g(y)=y'$. Then

$$\begin{aligned} hk(x' \otimes y') &= h(x \otimes y + K) \\ &= f(x) \otimes g(y) = x' \otimes y'. \end{aligned}$$

Furthermore

$$\begin{aligned} kh(x \otimes y + K) &= k(f(x) \otimes g(y)) \\ &= x \otimes y + K. \end{aligned}$$

Thus hk and kh are the identity isomorphisms of $M' \otimes_R N'$ and $M \otimes_R N/K$, respectively. It follows that both h and k are isomorphisms. Hence $\operatorname{Ker} h = 0$, that is, $K = \operatorname{Ker} f \otimes g$.

1.22. Theorem. *Let*

$$0 \longrightarrow N' \xrightarrow{\ f\ } N \xrightarrow{\ g\ } N'' \longrightarrow 0$$

be an exact sequence of R-modules and let M be a right R-module. If 1_M *is the identity isomorphism of M, then the sequence*

$$M \otimes_R N' \xrightarrow{1_M \otimes f} M \otimes_R N \xrightarrow{1_M \otimes g} M \otimes_R N'' \longrightarrow 0$$

is exact.

1.23. Theorem. *Let*

$$0 \longrightarrow M' \xrightarrow{f} M \xrightarrow{g} M'' \longrightarrow 0$$

be an exact sequence of right R-modules and let N be an R-module. If 1_N *is the identity isomorphism of N, then the sequence*

$$M' \otimes_R N \xrightarrow{f \otimes 1_N} M \otimes_R N \xrightarrow{g \otimes 1_N} M'' \otimes_R N \longrightarrow 0$$

is exact.

We shall prove Theorem 1.22, the proof of Theorem 1.23 being similar. By Proposition 1.20, $1_M \otimes g$ is surjective. By Proposition 1.21, Ker $1_M \otimes g$ consists of all finite sums $\sum x_i \otimes y_i$ where either $x_i = 0$ or $y_i \in \operatorname{Ker} g = \operatorname{Im} f$. Hence $\operatorname{Ker} 1_M \otimes g = \operatorname{Im} 1_M \otimes f$.

5 FLAT MODULES

1.24. Definition. *A right R-module M is* **flat** *if for each injective homomorphism* $f: N' \to N$ *from one R-module into another, the homomorphism*

$$1_M \otimes f: M \otimes_R N' \to M \otimes_R N$$

is injective, where 1_M *is the identity isomorphism of M.*

Thus M is flat if and only if for every exact sequence of R-modules,

$$0 \longrightarrow N' \xrightarrow{f} N \xrightarrow{g} N'' \longrightarrow 0,$$

the sequence

$$0 \longrightarrow M \otimes_R N' \xrightarrow{1_M \otimes f} M \otimes_R N \xrightarrow{1_M \otimes g} M \otimes_R N'' \longrightarrow 0$$

is exact.

Now we need a technical lemma.

1.25. Lemma. *Let N be an R-module and F a free right R-module, freely generated by a set S. Then every element of* $F \otimes_R N$ *can be written uniquely in the form*

$$\sum_{x \in S} x \otimes y_x,$$

where $y_x \neq 0$ *for only a finite number of* $x \in S$.

Proof. Let $\sum x_i \otimes y_i \in F \otimes_R N$, where $x_i \in F$, $y_i \in N$. Let $x_i = \sum_{x \in S} x a_x^{(i)}$, where $a_x^{(i)} \in R$. Then

$$\sum x_i \otimes y_i = \sum_{x \in S} x \otimes \left(\sum_i a_x^{(i)} y_i\right).$$

Hence each element of $F \otimes_R N$ can be written in the required form in at least one way. Note that the sums on i are finite and that for each i, $a_x^{(i)} \neq 0$ for only a finite number of $x \in S$.

To show the uniqueness, it is sufficient to show that $\sum_{x \in S} x \otimes y_x = 0$ implies $y_x = 0$ for all $x \in S$. For each $x \in S$, define $\phi_x : F \to xR$ as follows: if $x' \in S$, set $\phi_x(x') = x$ when $x' = x$ and $\phi_x(x') = 0$ when $x' \neq x$, and extend ϕ_x to all of F by linearity; that is,

$$\phi_x\left(\sum_{x' \in S} x' a_{x'}\right) = x a_x .$$

Then, if 1_N is the identity isomorphism of N, we have for each $x \in S$,

$$0 = (\phi_x \otimes 1_N)\left(\sum_{x' \in S} x' \otimes y_{x'}\right) = \sum_{x' \in S} \phi_x(x') \otimes y_{x'} = x \otimes y_x .$$

For each $x \in S$ define $\eta : xR \to R$ by $\eta(xa) = a$. It is clear that η is an isomorphism. Then $\eta \otimes 1_N : xR \otimes_R N \to R \otimes_R N$ is an isomorphism. Let $\psi : R \otimes_R N \to N$ be the isomorphism of Proposition 1.18. Then, for all $x \in S$,

$$0 = \psi((\eta \otimes 1_N)(x \otimes y_x)) = \psi(\eta(x) \otimes y_x) = \psi(1 \otimes y_x) = y_x .$$

1.26. Proposition. *A free right R-module is flat.*

Proof. Let N, N' be R-modules and $f : N' \to N$ an injective homomorphism. Let 1_F be the identity isomorphism of the free right R-module F. Let S be a set of free generators of F. Let $t \in F \otimes_R N'$ and suppose that $(1_F \otimes f)(t) = 0$. We can write $t = \sum_{x \in S} x \otimes y_x$ where $y_x \in N'$ and $y_x \neq 0$ for only a finite number of $x \in S$. Then

$$0 = (1_F \otimes f)\left(\sum_{x \in S} x \otimes y_x\right) = \sum_{x \in S} x \otimes f(y_x).$$

By the statement of uniqueness in the lemma we conclude that $f(y_x) = 0$ for all $x \in S$. Since f is injective, this implies that $y_x = 0$ for all $x \in S$. Thus $t = 0$.

We shall have occasion to consider diagrams of R-modules (left or right) and homomorphisms of the type

$$\begin{array}{ccc} M_1 & \xrightarrow{f} & M_2 \\ {\scriptstyle g}\downarrow & & \downarrow{\scriptstyle g'} \\ M_3 & \xrightarrow[f']{} & M_4 \end{array} \qquad \text{or} \qquad \begin{array}{ccc} & M_1 & \\ {\scriptstyle g}\swarrow & & \searrow{\scriptstyle f} \\ M_2 & \xrightarrow[h]{} & M_3 \end{array}$$

Such a diagram is called a **square** or **triangle**, respectively. A square is said to be **commutative** if $g'f = f'g$; a triangle is said to be **commutative** if $hg = f$. We shall consider diagrams of R-modules and homomorphisms which involve squares and triangles, and such a diagram is called **commutative** if every square and triangle in the diagram is commutative.

1.27. Definition. *A right R-module P is* **projective** *if for every diagram*

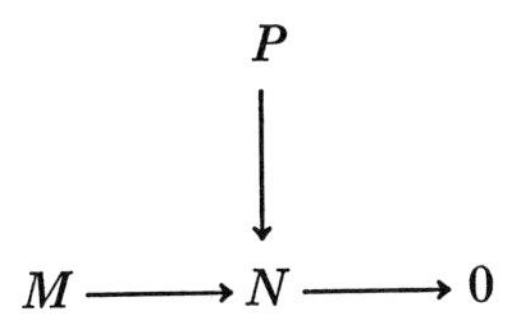

of right R-modules, where the row $M \to N \to 0$ is exact, there is a homomorphism $P \to M$ such that the diagram

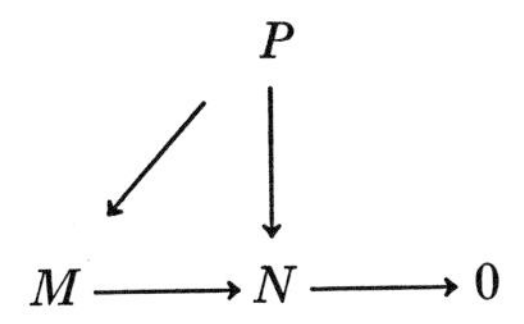

is commutative.

1.28. Theorem. *Let P be a right R-module. Then the following statements are equivalent:*

(1) *P is projective.*
(2) *P is a direct summand of a free right R-module.*

(3) *For every exact sequence of right R-modules of the form*

$$0 \to L \xrightarrow{f} M \to P \to 0,$$

Im f *is a direct summand of* M. (Such a sequence is called a **split** exact sequence.)

Proof. (1) $\Rightarrow$ (2) and (3). Let P be a projective right R-module. Let F be the free right R-module defined on the *set* P. Define $\phi: F \to P$ by setting $\phi(x) = x$ for all $x \in P$ and extending ϕ to all of F by linearity. Then ϕ is a surjective homomorphism and we have an exact sequence

$$0 \longrightarrow L \longrightarrow F \xrightarrow{\phi} P \longrightarrow 0,$$

where $L = \operatorname{Ker} \phi$. We shall show that $F \cong L \oplus P$, and thus prove (2).

More generally, we shall show that if the sequence

$$0 \longrightarrow L \xrightarrow{f} M \xrightarrow{g} P \longrightarrow 0$$

is exact, then $M = (\operatorname{Im} f) \oplus P'$ where $P' \cong P$. This will prove both (2) and (3). Let 1_P be the identity isomorphism of P and consider the diagram

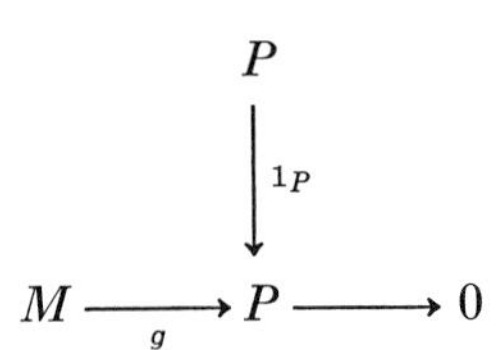

Since P is projective there is a homomorphism $h: P \to M$ such that $1_P = gh$. Let $P' = \operatorname{Im} h$. Since 1_P is injective, so is h, and thus $P' \cong P$. Let $x \in (\operatorname{Im} f) \cap P'$. Then $x \in \operatorname{Ker} g$ and $x = h(y)$, where $y \in P$. Hence $x = h(gh(y)) = hg(x) = 0$. Thus $(\operatorname{Im} f) \cap P' = 0$. Now let $z \in M$; then

$$g(z - hg(z)) = g(z) - 1_P(g(z)) = 0,$$

so $z - hg(z) \in \operatorname{Im} f$. Since $hg(z) \in P'$, we have

$$z = (z - hg(z)) + hg(z) \in (\operatorname{Im} f) + P'.$$

Therefore $M \cong (\operatorname{Im} f) \oplus P'$.

(2) $\Rightarrow$ (1). Assume that P is a direct summand of a free right R-module F. Then there are homomorphisms $\psi: P \to F$ and $\phi: F \to P$ such that $\phi\psi$ is the identity isomorphism of P, ψ being injective and ϕ surjective. Consider a diagram

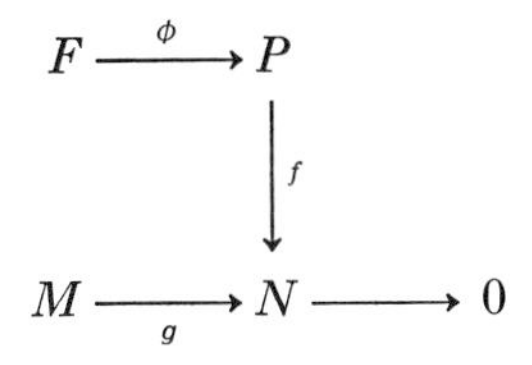

where the row is exact. Let F be freely generated by S. Define $j: F \to M$ by setting $j(x)$ equal to any element of M such that $g(j(x)) = f\phi(x)$ for each $x \in S$ and extending to all of F by linearity. Then the diagram

$$\begin{array}{ccccc} F & \xrightarrow{\phi} & P & & \\ \downarrow{\scriptstyle j} & & \downarrow{\scriptstyle f} & & \\ M & \xrightarrow[g]{} & N & \longrightarrow & 0 \end{array}$$

is commutative. Now consider the homomorphism $j\psi: P \to M$. We have $g(j\psi) = (gj)\psi = (f\phi)\psi = f(\phi\psi) = f$. Therefore P is projective.

(3) $\Rightarrow$ (1). Assume (3) holds and let

$$0 \longrightarrow \mathrm{L} \xrightarrow{\psi} \mathrm{F} \xrightarrow{\phi} P \longrightarrow 0$$

be the exact sequence from the first part of the proof that (1) implies (2) and (3). By hypothesis, $\operatorname{Im} \psi$ is a direct summand of F, so that $F = (\operatorname{Im} \psi) \oplus P'$ where P' is some submodule of F. Define $\eta: P' \to P$ by $\eta(x) = \phi(x)$ for all $x \in P'$. If $y \in P$, then $y = \phi(x)$ for some $x \in F$. Write $x = u + v$ where $u \in \operatorname{Im} \psi = \operatorname{Ker} \phi$ and $v \in P'$. Then

$$y = \phi(u+v) = \phi(u) + \phi(v) = \phi(v) = \eta(v).$$

Hence η is surjective. Suppose $\eta(x) = 0$ where $x \in P'$. Then $x \in \operatorname{Ker} \phi$, so $x \in (\operatorname{Im} \psi) \cap P' = 0$, that is, $x = 0$. Thus η is an isomorphism and P is isomorphic to a direct summand of a free right

R-module. Therefore P is projective by the equivalence of (1) and (2).

1.29. Corollary. *Every free right R-module is projective.*

1.30. Proposition. *Let*

$$0 \longrightarrow M' \xrightarrow{f} M \xrightarrow{g} M'' \longrightarrow 0$$

be a split exact sequence of right R-modules. Let N be an R-module and 1_N the identity isomorphism of N. Then

$$0 \longrightarrow M' \otimes_R N \xrightarrow{f \otimes 1_N} M \otimes_R N \xrightarrow{g \otimes 1_N} M'' \otimes_R N \longrightarrow 0$$

is exact.

Proof. We must show that $f \otimes 1_N$ is injective. By assumption there is a homomorphism $h: M \to M'$ such that hf is the identity isomorphism of M'. Then by Proposition 1.19, $(h \otimes 1_N)(f \otimes 1_N) = hf \otimes 1_N$ is the identity isomorphism of $M' \otimes_R N$. Hence $f \otimes 1_N$ is injective.

1.31. Theorem. *A projective right R-module is flat.*

Proof. Let N and N' be R-modules and $f: N' \to N$ an injective homomorphism. Let 1_P be the identity isomorphism of the projective right R-module P. By Theorem 1.28 there is a free right R-module F such that $F = P \oplus P'$. Then there is an exact sequence $0 \to P \xrightarrow{g} F \to P' \to 0$, where $g(x) = x$ for all $x \in P$. Hence, if 1_N and $1_{N'}$ are the identity isomorphisms of N and N', respectively, both $g \otimes 1_N$ and $g \otimes 1_{N'}$ are injective by Proposition 1.30. If 1_F is the identity isomorphism of F, then the diagram

$$\begin{array}{ccc} P \otimes_R N' & \xrightarrow{1_P \otimes f} & P \otimes_R N \\ \downarrow{\scriptstyle g \otimes 1_{N'}} & & \downarrow{\scriptstyle g \otimes 1_N} \\ F \otimes_R N' & \xrightarrow[1_F \otimes f]{} & F \otimes_R N \end{array}$$

is commutative. By Proposition 1.26, F is flat, so $1_F \otimes f$ is injective. Therefore, $1_P \otimes f$ is injective.

EXERCISES

1. Isomorphisms.
Let M and N be R-modules and suppose there are homomorphisms $\phi: M \to N$ and $\psi: N \to M$ such that $\psi\phi = 1_M$ and $\phi\psi = 1_N$. Show that ϕ and ψ are isomorphisms, inverse to one another.

2. Modules and homomorphisms.
Let M and N be R-modules and let $\phi: M \to N$ be a surjective homomorphism.

(a) Show that there is a one-to-one correspondence between the submodules of N and the submodules of M which contain Ker ϕ. In fact, the correspondence is

$$M_1 \leftrightarrow \phi(M_1) = \{\phi(x) \mid x \in M_1\};$$

the inverse correspondence is

$$N_1 \leftrightarrow \phi^{-1}(N_1) = \{x \mid x \in M \quad \text{and} \quad \phi(x) \in N_1\}.$$

(b) Let M_1 be an arbitrary submodule of M. Show that

$$\phi^{-1}(\phi(M_1)) = M_1 + \operatorname{Ker} \phi.$$

(c) Drop the requirement that ϕ be surjective, and let M_1 and N_1 be submodules of M and N, respectively. Show that

$$M/(M_1 + \phi^{-1}(N_1)) \cong (\phi(M) + N_1)/(\phi(M_1) + N_1).$$

3. Chain conditions.

(a) Prove the equivalence of (DCC) and (MIN).

(b) Show that the additive group of integers satisfies (ACC) but not (DCC).

(c) Let p be a prime. Let G be the additive group of rational numbers and H the subgroup of G consisting of those rational numbers which can be written with denominator prime to p. Show that G/H satisfies (DCC) but not (ACC).

(d) Let $M_1, \ldots, M_n$ be submodules of an R-module M such that $M = M_1 + \cdots + M_n$. Show that M satisfies (ACC) [or (DCC)] if and only if M_i does for each i.

4. Composition series.
Let M be an R-module. By a **normal series** of M we mean a finite chain of submodules of M,

$$M = M_0 \supseteq M_1 \supseteq \cdots \supseteq M_r = 0.$$

The integer r is called the **length** of the normal series. If $M_{i-1} \neq M_i$ for $i = 1, \ldots, r$, the normal series is said to be **without repetition**. One normal series is called a **refinement** of a second normal series if every term of the second series is a term of the first series; it is called a **proper refinement** if it has some terms not occurring in the second series. A normal series of M is called a **composition series** of M if it is without repetition and has no proper refinement. The factor modules M_{i-1}/M_i, $i = 1, \ldots, r$, are called the **factors** of the normal series. Two normal series are called **equivalent** if they have the same length and if the factors of the two series can be put in one-to-one correspondence in such a way that corresponding factors are isomorphic.

(a) (Zassenhaus' lemma) Let L, L', N, and N' be submodules of M such that $L' \subseteq L$ and $N' \subseteq N$. Show that

$$(L' + (L \cap N))/(L' + (L \cap N'))$$
$$\cong (N' + (N \cap L))/(N' + (N \cap L')).$$

(b) Show that any two normal series of M have equivalent refinements.

(c) Assume that M has a composition series. Show that any two composition series of M are equivalent, and that every normal series without repetition of M has a refinement which is equivalent to a composition series.

(d) Show that M has a composition series if and only if it satisfies both (ACC) and (DCC).

5. The length of a module.
If an R-module M has a composition series of length r, we call r the **length** of M and write $L_R(M) = r$. If M has no composition series we write $L_R(M) = \infty$.

(a) Show that if $0 \to K \to M \to N \to 0$ is an exact sequence of R-modules then $L_R(M) = L_R(K) + L_R(N)$ (we assume ∞ added to ∞ or to a nonnegative integer gives ∞).

(b) Show that if K and N are submodules of an R-module then

$$L_R(K) + L_R(N) = L_R(K + N) + L_R(K \cap N).$$

(c) Let V be a vector space over a field F. Show that $L_F(V) = \dim_F V$. Show that V satisfies (ACC) if and only if it satisfies (DCC).

(d) Let $0 \to M_1 \to M_2 \to \cdots \to M_n \to 0$ be an exact sequence of R-modules with $L_R(M_i)$ finite for $i = 1, \ldots, n$. Show that

$$\sum_{i=1}^{n} (-1)^i L_R(M_i) = 0.$$

(e) Let M be an R-module and let $\mathscr{S}$ be the family of all finitely generated submodules of M. Show that

$$L_R(M) = \sup_{N \in \mathscr{S}} L_R(N).$$

(f) Let R be a commutative ring and M an R-module. Let $a \in R$ be such that if $x \in M$, then $ax = 0$ implies $x = 0$. Show that $aM = \{ax \mid x \in M\}$ is a submodule of M isomorphic to M. If N is a submodule of M show that $M/N \cong aM/aN$. Finally, show that if n is positive integer, then

$$L_R(M/a^nM) = nL_R(M/aM).$$

6. Direct sums.

(a) Let $\{M_\alpha \mid \alpha \in I\}$ be a family of R-modules and let $M = \bigoplus_{\alpha \in I} M_\alpha$. Define $\psi_\alpha : M_\alpha \to M$ as in Section 3. Let N be an R-module and for each $\alpha \in I$, let $\eta_\alpha : M_\alpha \to N$ be a homomorphism. Show that there is a unique homomorphism $\eta : M \to N$ such that $\eta\psi_\alpha = \eta_\alpha$ for all $\alpha \in I$.

(b) Let M' be an R-module and, for each $\alpha \in I$, let $\psi_\alpha' : M_\alpha \to M'$ be a homomorphism. Suppose that for each R-module N and for each family of homomorphisms $\eta_\alpha : M_\alpha \to N$, one for each $\alpha \in I$, there is a unique homomorphism $\eta' : M' \to N$ such that $\eta'\psi_\alpha' = \eta_\alpha$ for all $\alpha \in I$. Show that $M' \cong M$.

(c) Let $\{M_\alpha \mid \alpha \in I\}$ and $\{N_\alpha \mid \alpha \in I\}$ be families of R-modules and let $M = \bigoplus_{\alpha \in I} M_\alpha$ and $N = \bigoplus_{\alpha \in I} N_\alpha$. For each $\alpha \in I$, let $f_\alpha : M_\alpha \to N_\alpha$ be a homomorphism. Define $f : M \to N$ by $f(x)_\alpha = f_\alpha(x_\alpha)$ for all $\alpha \in I$; f is called the **direct sum of the family of homomorphisms** $\{f_\alpha \mid \alpha \in I\}$. Show that

$$\operatorname{Ker} f = \bigoplus_{\alpha \in I} \operatorname{Ker} f_\alpha,$$

$$\operatorname{Im} f = \bigoplus_{\alpha \in I} \operatorname{Im} f_\alpha,$$

and

$$\operatorname{Im} f \cong \bigoplus_{\alpha \in I} M_\alpha / \operatorname{Ker} f_\alpha .$$

(d) Continue the notation of (c). Let $\{L_\alpha \mid \alpha \in I\}$ be another family of R-modules and for each $\alpha \in I$, let $g_\alpha : L_\alpha \to M_\alpha$ be a homomorphism. Let $L = \bigoplus_{\alpha \in I} L_\alpha$. Let g be the direct sum of the family $\{g_\alpha \mid \alpha \in I\}$. Show that $L \xrightarrow{g} M \xrightarrow{f} N$ is exact if and only if $L_\alpha \xrightarrow{g_\alpha} M_\alpha \xrightarrow{f_\alpha} N_\alpha$ is exact for each $\alpha \in I$.

(e) Let π be a one-to-one mapping of the set I onto itself. Show that

$$\bigoplus_{\alpha \in I} M_\alpha \cong \bigoplus_{\alpha \in I} M_{\pi(\alpha)} .$$

(f) Let J and K be nonempty subsets of I such that $I = J \cup K$ and $J \cap K$ is empty. Show that

$$\bigoplus_{\alpha \in I} M_\alpha \cong \Big(\bigoplus_{\alpha \in J} M_\alpha\Big) \oplus \Big(\bigoplus_{\alpha \in K} M_\alpha\Big).$$

7. Tensor products.

(a) Let M be a right R-module and N an R-module. Let $\eta(x, y) = x \otimes y$ for all $x \in M$ and $y \in N$. Show that if ϕ is bilinear mapping from $M \times N$ into an Abelian group G, then there is a unique homomorphism $\psi : M \otimes_R N \to G$ such that $\psi\eta = \phi$. For $i = 1, 2$, let (T_i, η_i) be a pair consisting of an Abelian group T_i and a bilinear mapping η_i from $M \times N$ into T_i. Assume that for each bilinear mapping ϕ from $M \times N$ into an Abelian group G there is a unique homomorphism $\psi_i : T_i \to G$ such that $\psi_i \eta_i = \phi$. Show that there is a unique isomorphism $\sigma : T_1 \to T_2$ such that $\sigma\eta_1 = \eta_2$.

(b) Let R and R' be rings and let M be a right R-module and an R'-module such that $(a'x)a = a'(xa)$ for all $a' \in R'$, $x \in M$, and $a \in R$. Then M is called an R'-R-**bimodule**.

Let N be an R-module and define a mapping from $R' \times (M \otimes_R N)$ into $M \otimes_R N$ by

$$(a', \sum x_i \otimes y_i) \mapsto \sum a'x_i \otimes y_i.$$

Show that this mapping is well-defined and that it makes $M \otimes_R N$ into an R'-module.

(c) Let L be a right R'-module, M an R'-R-bimodule, and N an R-module. Show that there is a unique isomorphism $L \otimes_{R'} (M \otimes_R N) \to (L \otimes_{R'} M) \otimes_R N$ such that $x \otimes (y \otimes z) \mapsto (x \otimes y) \otimes z$ for all $x \in L$, $y \in M$, and $z \in N$.

(d) Let R be a commutative ring and let M and N be R-modules (we may consider them as R-R-bimodules). Show that there is a unique isomorphism $M \otimes_R N \to N \otimes_R M$ such that $x \otimes y \mapsto y \otimes x$ for all $x \in M$ and $y \in N$.

(e) Let R be a commutative ring and let M and N be R-modules. Then, according to part (b), $M \otimes_R N$ is an R-module. Let $Z'(M, N)$ be the free R-module defined on the set $M \times N$ and let $Y'(M, N)$ be the submodule of $Z'(M, N)$ generated by the set of all elements of the form $(x, y_1 + y_2) - (x, y_1) - (x, y_2)$, $(x_1 + x_2, y) - (x_1, y) - (x_2, y)$, $(xa, y) - a(x, y)$, and $(x, ay) - a(x, y)$, where $x, x_1, x_2 \in M$, $y, y_1, y_2 \in N$, $a \in R$. Show that as R-modules we have

$$M \otimes_R N \cong Z'(M, N)/Y'(M, N).$$

(f) Let $\{M_\alpha \mid \alpha \in I\}$ be a family of right R-modules and let $\{N_\beta \mid \beta \in J\}$ be a family of R-modules. Show that

$$\Big(\bigoplus_{\alpha \in I} M_\alpha\Big) \otimes_R \Big(\bigoplus_{\beta \in J} N_\beta\Big) \cong \bigoplus_{(\alpha, \beta) \in I \times J} (M_\alpha \otimes_R N_\beta).$$

8. Some special tensor products.

(a) Let G be an Abelian group and let n be a positive integer. Let $nG = \{nx \mid x \in G\}$ and let Z_n be the additive group of integers modulo n. Show that $Z_n \otimes_Z G \cong G/nG$.

(b) Let k be the gcd of the positive integers m and n. Show that $Z_m \otimes_Z Z_n \cong Z_k$.

(c) Let U and V be vector spaces over a field K. Show that $U \otimes_K V$ is a vector space over K and that

$$\dim_K U \otimes_K V = (\dim_K U)(\dim_K V).$$

(d) An Abelian group G is called a **torsion group** if every element of G has finite order. Show that if either G or H is a torsion group, then $G \otimes_Z H$ is a torsion group.

(e) An Abelian group G is said to be **divisible** if $nG = G$ for all positive integers n. Show that if either G or H is divisible, then $G \otimes_Z H$ is divisible.

(f) Show that if G is a torsion group and H is a divisible group, then $G \otimes_Z H = 0$.

9. Flat modules.

(a) Let M be a right R-module. Show that M is flat if and only if for every injective homomorphism $f: N' \to N$ of a finitely generated R-module N' into an R-module N, $1_M \otimes f$ is injective.

(b) Let $\{M_\alpha \mid \alpha \in I\}$ be a family of right R-modules. Show that $\bigoplus_{\alpha \in I} M_\alpha$ is flat if and only if M_α is flat for each $\alpha \in I$.

(c) Let M be a right R-module and N and R-R'-bimodule. Show that if M is flat and if N is flat as a right R'-module, then $M \otimes_R N$ is a flat right R'-module.

(d) Let R be a subring of a ring R'; then, in a natural way, R' is an R-R-bimodule. Every R'-module (right or left) is also an R-module. Show that if M is a flat right R'-module, and if R' is a flat right R-module, then M is a flat right R-module.

10. Flat modules and isomorphisms.

Let M be a flat right R-module and N an R-module, and let N_1 and N_2 be submodules of N. If N' is a submodule of N, and if $f: N' \to N$ is given by $f(y) = y$ for all $y \in N'$, then $1_M \otimes f$ is injective. We shall identify each element of $M \otimes_R N'$ with its image under $1_M \otimes f$, and thus consider $M \otimes_R N'$ as a subgroup of $M \otimes_R N$. Show that the following isomorphism and equalities hold:

(a) $M \otimes_R N / M \otimes_R N_1 \cong M \otimes_R (N/N_1)$

(b) $M \otimes_R (N_1 + N_2) = (M \otimes_R N_1) + (M \otimes_R N_2)$

(c) $M \otimes_R (N_1 \cap N_2) = (M \otimes_R N_1) \cap (M \otimes_R N_2)$.

11. A criterion for flatness.

Let M be a right R-module such that if A is a left ideal of R and $f: A \to R$ is given by $f(a) = a$ for all $a \in A$, then $1_M \otimes f$ is injective.

(a) Show that if F is a free R-module, and if $g: E \to F$ is injective, then $1_M \otimes g$ is injective.
(b) Show that M is flat.

12. Flatness and exact sequences.
Let

$$0 \longrightarrow L \xrightarrow{f} M \xrightarrow{g} N \longrightarrow 0$$

be an exact sequence of right R-modules and assume that N is flat.
(a) Prove that

$$0 \longrightarrow L \otimes_R K \xrightarrow{f \otimes 1_K} M \otimes_R K \xrightarrow{g \otimes 1_K} N \otimes_R K \longrightarrow 0$$

is exact for every R-module K.
(b) Show that M is flat if and only if L is flat.

13. Elementwise criterion for flatness.
Let M be a right R-module.
(a) Show that M is flat if and only if for every R-module N and for every pair of finite subsets $\{x_1, \ldots, x_n\}$ and $\{y_1, \ldots, y_n\}$ of M and N, respectively, such that $\sum_{i=1}^{n} x_i \otimes y_i = 0$, there are elements $z_1, \ldots, z_k \in M$ and elements $b_{ji} \in R$, $i = 1, \ldots, n$, $j = 1, \ldots, k$, such that

$$\sum_{i=1}^{n} b_{ji} y_i = 0, \qquad j = 1, \ldots, k,$$

and

$$x_i = \sum_{j=1}^{k} z_j b_{ji}, \qquad i = 1, \ldots, n,$$

(b) Show that M is flat if and only if for every pair of finite subsets $\{x_1, \ldots, x_n\}$ and $\{a_1, \ldots, a_n\}$ of M and R, respectively, such that $\sum_{i=1}^{n} x_i a_i = 0$, there exist elements $z_1, \ldots, z_k \in M$ and elements $b_{ji} \in R$, $i = 1, \ldots, n$, $j = 1, \ldots, k$, such that

$$\sum_{i=1}^{n} b_{ji} a_i = 0, \qquad j = 1, \ldots, k,$$

and

$$x_i = \sum_{j=1}^{k} z_j b_{ji}, \qquad i = 1, \ldots, n.$$

(c) Let R be a subring of a ring R'. Show that R' is a flat right R-module if and only if for every solution $c_1, \ldots, c_n$ in R' of a system of equations

$$\sum_{i=1}^{n} x_i a_{ih} = 0, \qquad h = 1, \ldots, r,$$

where $a_{ih} \in R$ for each i and h, we have

$$c_i = \sum_{j=1}^{k} d_j b_{ji}, \qquad i = 1, \ldots, n,$$

where $d_1, \ldots, d_k \in R'$, $b_{ji} \in R$ for each j and i, and

$$\sum_{i=1}^{n} b_{ji} a_{ih} = 0, \qquad j = 1, \ldots, k, \quad h = 1, \ldots, r.$$

14. Split exact sequences.

(a) Let $0 \to L \xrightarrow{f} M \xrightarrow{g} N \to 0$ be an exact sequence of R-modules. Show that this is a split exact sequence if and only if there is a homomorphism $h: N \to M$ such that $gh = 1_N$.

(b) Show that the exact sequence in part (a) is a split exact sequence if and only if there is a homomorphism $k: M \to L$ such that $kf = 1_L$.

(c) Let L and N be R-modules. Show that there is a split exact sequence $0 \to L \to M \to N \to 0$.

(d) If n is a positive integer, let Z_n be the additive group of integers modulo n. Construct an exact sequence $0 \to Z_2 \to Z_4 \to Z_2 \to 0$ which is not a split exact sequence. Construct other examples of this same type.

15. Projective modules.

(a) Let P be an R-module. Show that P is projective if and only if there is a subset S of P and for each $x \in S$ there exists a homomorphism $\phi_x: P \to R$ such that for all $y \in P$, $\phi_x(y) \neq 0$ for only a finite number of $x \in S$, and $y = \sum_{x \in S} \phi_x(y)x$.

(b) Show that a direct summand of a projective module is projective.

(c) Let $\{P_\alpha \mid \alpha \in I\}$ be a family of R-modules. Show that

$P = \bigoplus_{\alpha \in I} P_\alpha$ is projective if and only if each P_α is projective.

(d) Let P be a projective R-module. Show that if, in the diagram

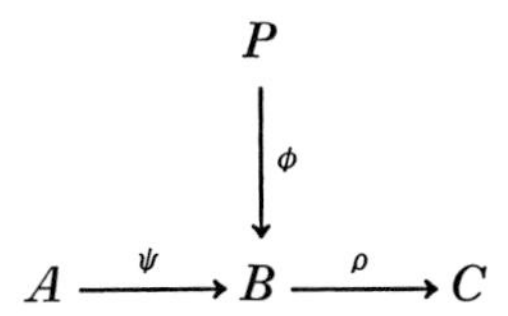

of R-modules, the row is exact and $\rho\phi = 0$, then there is a homomorphism $\eta : P \to A$ such that $\psi\eta = \phi$.

(e) Let P be a projective R-module. Show that if, in the diagram

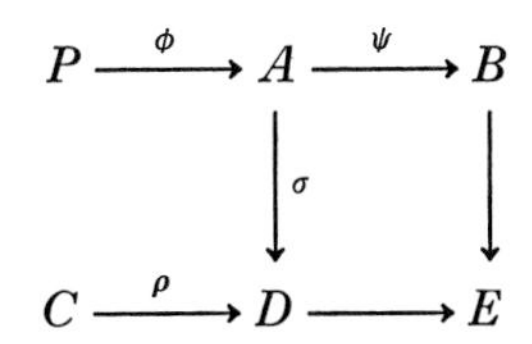

of R-modules, the square is commutative, the bottom row is exact, and $\psi\phi = 0$, then there is a homomorphism $\eta : P \to C$ such that $\rho\eta = \sigma\phi$.

CHAPTER

II

Primary Decompositions and Noetherian Rings

1 OPERATIONS ON IDEALS AND SUBMODULES

Throughout the remainder of this book, with the exception of the appendix, all rings which are considered will be assumed to be commutative. Thus it will no longer be necessary to distinguish between left and right modules. We shall think of all modules as left modules and use the appropriate notation.

Let R be a ring. If A and B are ideals of R we set

$$AB = \{\text{finite sums } \sum a_i b_i \mid a_i \in A,\ b_i \in B\}.$$

Then AB is an ideal of R. For, if $\sum a_i b_i$ and $\sum c_j d_j$ are in AB, then

$$\sum a_i b_i - \sum c_j d_j = \sum a_i b_i + \sum(-c_j)d_j \in AB,$$

and if $r \in R$, then

$$r(\sum a_i b_i) = \sum(ra_i)b_i \in AB.$$

The ideal AB is called the **product** of A and B. If $a \in R$ we write aB or Ba for $(a)B$; we note that $aB = \{ab \mid b \in B\}$. If A and B are ideals of R, then their **sum** $A + B$ as submodules of the R-module R is an ideal of R. If A is generated by $a_1, \ldots, a_n$, we will usually write $A = (a_1, \ldots, a_n)$ rather than $Ra_1 + \cdots + Ra_n$.

2.1. Proposition. *Let A, B, and C be ideals of R. Then*

(1) $A(BC) = (AB)C$,
(2) $AB = BA$,
(3) $AB \subseteq A \cap B$,
(4) $A(B + C) = AB + AC$,
(5) $A \subseteq B$ *implies that* $AC \subseteq BC$,
(6) $A(B \cap C) \subseteq AB \cap AC$.

Proof. (1), (2), and (3) are clear. If $a \in A$, $b \in B$, and $c \in C$ then $a(b + c) = ab + ac \in AB + AC$, and so sums of elements of this type are in $AB + AC$. Hence $A(B + C) \subseteq AB + AC$. It follows from (5), which is clearly true, that $AB \subseteq A(B + C)$ and $AC \subseteq A(B + C)$. Thus $AB + AC \subseteq A(B + C)$. Finally, $A(B \cap C) \subseteq AB$ and $A(B \cap C) \subseteq AC$, so that $A(B \cap C) \subseteq AB \cap AC$.

Having defined the product of two ideals of R we can define now the ***n*th power** of an ideal A of R, where n is a positive integer, in the usual way:

$$A^1 = A,$$
$$A^n = AA^{n-1} \qquad \text{if} \quad n > 1.$$

The rules of exponents hold:

$$A^m A^n = A^{m+n}, \qquad (A^m)^n = A^{mn}, \qquad (AB)^n = A^n B^n.$$

It is customary to set $A^0 = R$.

Let M be an R-module. If A is an ideal of R and N is a submodule of M, we set

$$AN = \{\text{finite sums } \textstyle\sum a_i x_i \mid a_i \in A, \quad x_i \in N\}.$$

Then AN is a submodule of M, and the proof of the following proposition is like that of Proposition 2.1.

2.2. Proposition. *Let A and B be ideals of R and let L and N be submodules of M. Then*

(1) $A(BN) = (AB)N$,
(2) $AN \subseteq N$,
(3) $A(L + N) = AL + AN$ *and* $(A + B)N = AN + BN$,

(4) $A \subseteq B$ *implies that* $AN \subseteq BN$,
(5) $L \subseteq N$ *implies that* $AL \subseteq AN$,
(6) $A(L \cap N) \subseteq AL \cap AN$ *and* $(A \cap B)N \subseteq AN \cap BN$.

If $a \in R$ we write aN for $(a)N$, and if $x \in M$ we write Ax for $A(Rx)$.

Again, let M be an R-module, A an ideal of R, and N a submodule of M. Set

$$N: A = \{x \mid x \in M \quad \text{and} \quad Ax \subseteq N\}.$$

If $x, y \in N: A$ then $a(x-y) = ax - ay \in N$ for all $a \in A$, that is, $x - y \in N: A$. Also, $a(rx) = (ar)x \in N$ for all $a \in A$, where r is any element of R, that is, $rx \in N: A$. Thus $N: A$ is a submodule of M. If L is any submodule of M, then $AL \subseteq N$ if and only if $L \subseteq N: A$.

If L and N are submodules of M, we set

$$L: N = \{a \mid a \in R \quad \text{and} \quad aN \subseteq L\}.$$

Then $L: N$ is an ideal of R, and if A is any ideal of R, then $AN \subseteq L$ if and only if $A \subseteq L: N$.

The submodule $N: A$ and the ideal $L: N$ are called the **residual of** N **by** A and **the residual of** L **by** N, respectively.

2.3. Proposition. *Let A and B be ideals of R and let K, L, and N be submodules of M. Then*

(1) $A \subseteq B$ *implies* $N: A \supseteq N: B$,
(2) $(N: A): B = N: AB$,
(3) $(L \cap N): A = (L: A) \cap (N: A)$,
(4) $N: (A + B) = (N: A) \cap (N: B)$,
(5) $L \subseteq N$ *implies that* $L: A \subseteq N: A$ *and* $L: K \subseteq N: K$,
(6) $L \subseteq N$ *implies that* $K: L \supseteq K: N$,
(7) $(L \cap N): K = (L: K) \cap (N: K)$,
(8) $K: (L + N) = (K: L) \cap (K: N)$.

Proof. We shall prove (1), (2), (3), and (8) and leave the others for the reader.

(1) We have $A(N: B) \subseteq B(N: B) \subseteq N$ and so $N: B \subseteq N: A$.
(2) We have $AB((N: A): B) \subseteq A(N: A) \subseteq N$ so that $(N: A): B$

$\subseteq N: AB$. Also, $AB(N: AB) \subseteq N$ and so $B(N: AB) \subseteq N: A$. Hence $N: AB \subseteq (N: A): B$.

(3) We have $A((L: A) \cap (N: A)) \subseteq A(L: A) \cap A(N: A) \subseteq L \cap N$ and consequently $(L: A) \cap (N: A) \subseteq (L \cap N): A$. Also $A((L \cap N): A) \subseteq L \cap N \subseteq L$ so that $(L \cap N): A \subseteq L: A$; likewise $(L \cap N): A \subseteq N: A$. Thus $(L \cap N): A \subseteq L: A \cap N: A$.

(8) We have $L \subseteq L+N$ and so by (6), $K: (L+N) \subseteq K: L$; likewise $K: (L+N) \subseteq K: N$. Hence $K: (L+N) \subseteq (K: L) \cap (K: N)$. Let $C = (K: L) \cap (K: N)$. Then $C \subseteq K: L$ so $CL \subseteq K$. Likewise $CN \subseteq K$, and therefore $C(L+N) = CL + CN \subseteq K$. Hence $C \subseteq K: (L+N)$.

If A and B are ideals of R then the residual $A: B$ is an ideal of R. If C is another ideal of R, then by (2) of the above proposition, $(A: B): C = A: BC$.

2 PRIMARY SUBMODULES

2.4. Definition. (1) *Let M be an R-module. A submodule Q of M is* **primary** *if for all $a \in R$ and $x \in M$, $ax \in Q$ and $x \notin Q$ imply that $a^n M \subseteq Q$ for some positive integer n.*

(2) *A submodule N of an R-module M is* **irreducible** *if, for submodules L_1 and L_2 of M, $N = L_1 \cap L_2$ implies that either $L_1 = N$ or $L_2 = N$.*

2.5. Proposition. *If M satisfies* (ACC) *then every irreducible submodule of M is primary.*

Proof. Let Q be an irreducible submodule of M. Let $a \in R$ and $x \in M$ be such that $ax \in Q$ but $x \notin Q$. By (ACC) there is a positive integer n such that $Q: (a^n) = Q: (a^{n+1}) = \cdots$. We have $Q \subseteq (Q + a^n M) \cap (Q + Rx)$. Let $y \in (Q + a^n M) \cap (Q + Rx)$; then $y = z + a^n u = z' + bx$, where $z, z' \in Q$, $u \in M$, and $b \in R$. Since $ay = az' + abx \in Q$, it follows that $a^{n+1} u = ay - az \in Q$; hence $u \in Q: (a^{n+1}) = Q: (a^n)$. Thus $a^n u \in Q$ and consequently $y \in Q$. Therefore $Q = (Q + a^n M) \cap (Q + Rx)$. Since Q is irreducible and $Q \neq Q + Rx$, we must have $Q = Q + a^n M$. Thus $a^n M \subseteq Q$. We conclude that Q is primary.

2.6. Proposition. *If M satisfies* (ACC), *then every submodule of M can be written as an intersection of a finite number of irreducible submodules of M.*

Proof. Let $\mathscr{S}$ be the set of all submodules of M which cannot be written as an intersection of a finite number of irreducible submodules of M. If $\mathscr{S}$ is empty, we have finished. Suppose that $\mathscr{S}$ is not empty; then $\mathscr{S}$ has a maximal element N. Since N is not irreducible there are submodules L_1 and L_2 of M such that $N = L_1 \cap L_2$, $N \subset L_1$, and $N \subset L_2$. Then $L_1 \notin \mathscr{S}$ and $L_2 \notin \mathscr{S}$ and so there are irreducible submodules $K_1, \ldots, K_m$ and $K_1', \ldots, K_n'$ of M such that $L_1 = K_1 \cap \cdots \cap K_m$ and $L_2 = K_1' \cap \cdots \cap K_n'$. But then $N = K_1 \cap \cdots \cap K_m \cap K_1' \cap \cdots \cap K_n'$, which is contrary to the fact that $N \in \mathscr{S}$. Thus, $\mathscr{S}$ must be empty.

Thus, if M satisfies (ACC), it contains many primary submodules. If we combine Propositions 2.5 and 2.6, we obtain the following result.

2.7. Theorem. *If the R-module M satisfies* (ACC), *then every submodule of M can be written as an intersection of a finite number of primary submodules.*

An ideal Q of a ring R is called **primary** if it is a primary submodule of R when considered as an R-module. Since R has a unity, Q is a primary ideal if and only if for all $a, b \in R$, $ab \in Q$ and $b \notin Q$ imply that $a^n \in Q$ for some positive integer n.

Since R is commutative we shall call it simply **Noetherian** if it is left Noetherian. It follows from Theorem 2.7 that if R is Noetherian, then every ideal of R can be written as an intersection of a finite number of primary ideals.

2.8. Definition. *An ideal P of a ring R is* **prime** *if for all elements $a, b \in R$, $ab \in P$ implies that either $a \in P$ or $b \in P$.*

2.9. Proposition. *Let P be an ideal of a ring R. Then the following statements are equivalent:*

(1) *P is a prime ideal.*
(2) *If $a, b \in R$ and $(a)(b) \subseteq P$, then either $a \in P$ or $b \in P$.*
(3) *If A and B are ideals of R and if $AB \subseteq P$, then either $A \subseteq P$ or $B \subseteq P$.*

Proof. Clearly (3) implies (2) and (2) implies (1). Now suppose that P is prime and that $AB \subseteq P$, where A and B are ideals of R. If $A \nsubseteq P$ then there is an element $a \in A$ with $a \notin P$. For every $b \in B$, $ab \in AB \subseteq P$, and so $b \in P$. Thus $B \subseteq P$.

2.10. Definition. (1) *Let A be an ideal of a ring R. Set*

$$\text{Rad}(A) = \{a \mid a \in R \quad \text{and} \quad a^n \in A \quad \text{for some positive integer} \quad n\},$$

and call Rad(A) *the* **radical** *of A.*

(2) *Let N be a submodule of an R-module M. Set*

$$\text{Rad}(N) = \text{Rad}(N{:}M),$$

and call Rad(N) *the* **radical** *of N.*

Let $a, b \in \text{Rad}(A)$ and $r \in R$; let $a^m \in A$ and $b^n \in A$, where m and n are positive integers. Then $(ra)^m = r^m a^m \in A$ and

$$(a-b)^{m+n} = \sum_{k=0}^{m+n} \binom{m+n}{k} (-1)^k a^{m+n-k} b^k \in A,$$

since either $m + n - k \geq m$ or $k \geq n$. Hence ra, $a - b \in \text{Rad}(A)$, and we conclude that Rad(A) is an ideal of R. A number of properties of the mapping $A \mapsto \text{Rad}(A)$ are given in Exercise 1. We note that a submodule Q of M is primary and if and only if for all $a \in R$ and $x \in M$, $ax \in Q$ and $x \notin Q$ imply that $a \in \text{Rad}(Q)$.

2.11. Proposition. *If Q is a primary submodule of M then* Rad(Q) *is a prime ideal of R. If A is an ideal of R and if $A \subseteq P$, where P is a prime ideal of R, then* $\text{Rad}(A) \subseteq P$.

Proof. Let $ab \in \text{Rad}(Q)$; then for some positive integer n we have $a^n b^n M = (ab)^n M \subseteq Q$. If $b \notin \text{Rad}(Q)$, then $b^n x \notin Q$ for some $x \in M$. Since $a^n b^n x \in Q$ but $b^n x \notin Q$ we have $a^{nk} M \subseteq Q$ for some positive integer k. Thus $a \in \text{Rad}(Q)$.

If $a \in \text{Rad}(A)$, then $a^n \in A$; hence a^n, and therefore a itself, belongs to every prime ideal of R which contains A.

If Q is a primary submodule of M and $\text{Rad}(Q) = P$, we say that Q is P-**primary**.

A nonempty subset S of R is said to be **multiplicatively closed** if $ab \in S$ whenever $a, b \in S$. It is clear that a proper ideal P of R is prime if and only if $R \backslash P$ is multiplicatively closed.

2.12. Proposition. *Let A be an ideal of R and let S be a multiplicatively closed set in R such that $A \cap S$ is empty. Then there is an ideal P of R which is maximal with respect to the properties that $A \subseteq P$ and $P \cap S$ are empty. Furthermore, P is a prime ideal.*

Proof. Let $\mathscr{X}$ be the set of all ideals B of R such that $A \subseteq B$ and $B \cap S$ is empty; $\mathscr{X}$ is not empty since $A \in \mathscr{X}$. By Zorn's lemma, $\mathscr{X}$ has a maximal element P. To show that P is prime, suppose that it is not and let $ab \in P$, $a \notin P$, $b \notin P$. Then $P \subset P + (a)$ and $P \subset P + (b)$, and so there are elements $s, t \in S$ such that $s \in P + (a)$ and $t \in P + (b)$. Hence $s = p_1 + r_1 a$, $t = p_2 + r_2 b$, where $p_1, p_2 \in P$ and $r_1, r_2 \in R$. Then $st = p_1 p_2 + p_1 r_2 b + r_1 a p_2 + r_1 r_2 ab \in P \cap S$, which contradicts the fact that $P \cap S$ is empty.

2.13. Definition. *An ideal P of R is* **maximal** *if $P \neq R$ and if there is no ideal A of R such that $P \subset A \subset R$.*

If we take $S = \{1\}$ in Proposition 2.12, we see that each proper ideal of R is contained in at least one maximal ideal of R.

An element $a \in R$ is called a **zero-divisor** if there is an element $b \in R$, $b \neq 0$, such that $ab = 0$. Thus, if R has more than one element, 0 is a zero-divisor. A zero-divisor other than 0 is referred to as **proper**. An element of R which is not a zero-divisor is called **regular**. If R has more than one element, then 1 is regular. An ideal of R is called **regular** if it contains a regular element. An element $u \in R$ is called a **unit** if there is an element $v \in R$ such that $uv = 1$; there is only one such element v; it is denoted by u^{-1}, and it is called the **inverse** of u. Clearly, 1 is a unit and if R has more than one element then every unit is regular. An element $a \in R$ is called **nilpotent** if

$a \in \text{Rad}(0)$, that is, if $a^n = 0$ for some positive integer n. If $a \in R$ and if a is not nilpotent, then 0 does not belong to the multiplicatively closed set $S = \{a^n \mid n \text{ is a positive integer}\}$. Hence there is a prime ideal P of R such that $a \notin P$.

2.14. Proposition. *If A is an ideal of R, then*

$$\text{Rad}(A) = \bigcap P,$$

where the intersection is over all prime ideals of R containing A.

Proof. By Proposition 2.11, $\text{Rad}(A)$ is contained in the ideal on the right. Let $a \notin \text{Rad}(A)$. If $S = \{a^n \mid n \text{ is a positive integer}\}$ then $A \cap S$ is empty; hence there is a prime ideal P of R such that $A \subseteq P$ and $a \notin P$. Thus a does not belong to the ideal on the right.

Let A be an ideal of R. A prime ideal P of R is called a **minimal prime divisor of** A if $A \subseteq P$ and if there is no prime ideal P' of R such that $A \subseteq P' \subset P$. If N is a submodule of an R-module M, then the minimal prime divisors of $N : M$ are called the **minimal prime divisors of** N.

2.15. Lemma. *Let A be an ideal of R and suppose that $A \subseteq P$ where P is a prime ideal of R. Then P contains a minimal prime divisor of A.*

Proof. Let $\mathscr{X}$ be the set of all prime ideals P' of R such that $A \subseteq P' \subseteq P$. Then $\mathscr{X}$ is not empty since $P \in \mathscr{X}$. Partially order $\mathscr{X}$ by the reverse of inclusion. Then any maximal element of $\mathscr{X}$ is a minimal prime divisor of A contained in P. The existence of a maximal element of $\mathscr{X}$ will follow from Zorn's lemma if we show that the intersection P' of the ideals in a totally ordered subset $\{P_\alpha\}$ of $\mathscr{X}$ is prime. Let $ab \in P'$ and suppose that $a \notin P'$. Then $a \notin P_\alpha$ for some α and so $b \in P_\alpha$. If $\beta \neq \alpha$, then either $P_\beta \subseteq P_\alpha$ or $P_\alpha \subseteq P_\beta$. If $P_\beta \subseteq P_\alpha$, then $a \notin P_\beta$ and so $b \in P_\beta$. If $P_\alpha \subseteq P_\beta$, then $b \in P_\beta$. Hence $b \in \cap P_\alpha = P'$.

From this lemma and Proposition 2.14 we obtain the following proposition.

2.16. Proposition. *If N is a submodule of M, then*

$$\operatorname{Rad}(N) = \bigcap P,$$

where the intersection is over all minimal prime divisors of N.

3 NOETHERIAN RINGS

In this section we shall prove several assertions which are of considerable importance in the theory of commutative rings. One consequence of the first of these theorems is that there is an abundance of Noetherian rings.

2.17. Theorem (Hilbert Basis Theorem). *If R is a Noetherian ring, then the polynomial ring $R[X]$ is a Noetherian ring.*

Proof. Let A' be an ideal of $R[X]$ and let A be the set of all $a \in R$ such that there is an element of A' of the form $aX^s + c_{s-1}X^{s-1} + \cdots + c_0$. Then A is an ideal of R and so $A = (a_1, \ldots, a_k)$. For each i, let $a_i X^{s_i} + \cdots \in A'$. If $s = \max\{s_1, \ldots, s_k\}$, then $a_i X^s + \cdots \in A'$; we denote this polynomial by f_i. Let $A'' = (f_1, \ldots, f_k)$. If we consider $R[X]$ as an R-module, then A' and A'' are submodules. Let N be the submodule of $R[X]$ consisting of all polynomials of degree at most $s - 1$. We shall show that

$$A' = (A' \cap N) + A''.$$

Once this has been done the assertion will follow. For N is finitely generated and so, by Theorem 1.12, $A' \cap N$ is a finitely generated R-module. Hence there are polynomials $f_{k+1}, \ldots, f_n \in A' \cap N$ such that $A' = (f_1, \ldots, f_n)$.

It remains to verify the equality stated above. Clearly, the right-hand side is contained in the left-hand side. Let

$$g = b_t X^t + \cdots + b_1 X + b_0 \in A'.$$

Then $b_t \in A$, so

$$b_t = r_1 a_1 + \cdots + r_k a_k$$

where $r_1, \ldots, r_k \in R$. If $t < s$, then $g \in A' \cap N$. Suppose that $t \geq s$.

Then

$$\begin{aligned} g &= (r_1 a_1 + \cdots + r_k a_k)X^t + b_{t-1}X^{t-1} + \cdots + b_0 \\ &= r_1 X^{t-s}(f_1 - f_1') + \cdots + r_k X^{t-s}(f_k - f_k') \\ &\quad + b_{t-1}X^{t-1} + \cdots + b_0, \end{aligned}$$

where $f_i' = f_i - a_i X^s \in N$. Thus $g = g_1' + g_1$, where $g_1' \in A''$ and

$$g_1 = c_{t-1}X^{t-1} + \cdots + c_1 X + c_0 \in A'.$$

If we repeat this argument several times, we will obtain finally

$$g = g_1' + \cdots + g_{t-s+1}' + g_{t-s+1},$$

where $g_i' \in A''$ for $i = 1, \ldots, t-s+1$ and $g_{t-s+1} \in A' \cap N$. This completes the proof.

2.18. Corollary. *If R is a Noetherian ring, then any polynomial ring $R[X_1, \ldots, X_n]$ is a Noetherian ring.*

Thus, if K is a field, then any polynomial ring $K[X_1, \ldots, X_n]$ is a Noetherian ring. The ring Z of integers is a Noetherian ring (in fact, each of its ideals is principal) so the same is true of any polynomial ring $Z[X_1, \ldots, X_n]$.

2.19. Theorem (Artin–Rees Lemma). *Let R be a Noetherian ring and let A, B, and C be ideals of R. Then there is a positive integer r such that*

$$A^n B \cap C = A^{n-r}(A^r B \cap C) \qquad \textit{for all} \quad n > r.$$

Proof. Let $A = (a_1, \ldots, a_k)$. Let $X_1, \ldots, X_k$ be indeterminates and let S_n be the set of all homogeneous polynomials $f \in R[X_1, \ldots, X_k]$ of degree n such that $f(a_1, \ldots, a_k) \in A^n B \cap C$. (All terms of f have degree n.) Let $S = \bigcup_{n=0}^{\infty} S_n$ and let A' be the ideal of $R[X_1, \ldots, X_k]$ generated by S. By Theorem 2.17, A' is finitely generated, say $A' = (f_1, \ldots, f_t)$. It is clear that we may assume that $f_i \in S$ for $i = 1, \ldots, t$. Let d_i be the degree of f_i and let $r = \max\{d_1, \ldots, d_t\}$. Let $n > r$. If $a \in A^n B \cap C$, then, since $a \in A^n$, we have $f(a_1, \ldots, a_k) = a$ for some $f \in S$ of degree n. If $f = \sum f_i g_i$, where $g_i \in R[X_1, \ldots, X_k]$, then we may assume that for each i, g_i is of degree $n-d_i$ and is homogeneous. Then

$$\begin{aligned} a=f(a_1,\ldots,a_k)&=\sum f_i(a_1,\ldots,a_k)g_i(a_1,\ldots,a_k)\\ &\in \sum A^{n-d_i}(A^{d_i}B\cap C)\\ &=\sum A^{n-r}A^{r-d_i}(A^{d_i}B\cap C)\\ &\subseteq A^{n-r}\sum(A^rB\cap A^{r-d_i}C)\\ &\subseteq A^{n-r}(A^rB\cap C),\end{aligned}$$

and consequently $A^nB\cap C\subseteq A^{n-r}(A^rB\cap C)$ when $n>r$. On the other hand, $A^{n-r}(A^rB\cap C)\subseteq A^nB\cap A^{n-r}C\subseteq A^nB\cap C$ for all $n>r$, so all is proved.

Now we shall extend the Artin–Rees lemma to modules. Let M be an R-module and let $R^*=R\oplus M$, the direct sum of R and M as R-modules. Each element of R^* can be written uniquely in the form $a+x$, where $a\in R$ and $x\in M$. Define a multiplication in R^* by

$$(a+x)(a'+x')=aa'+a'x+ax'.$$

Then R^* is a commutative ring with unity. If R is Noetherian and M is a finitely generated R-module, then every submodule of the R-module R^* is finitely generated; hence R^* is a Noetherian ring. The subring $\{a+0\mid a\in R\}$ of R^* is identified with R, and $\{0+x\mid x\in M\}$ is an ideal of R^* which we identify with M. We have $M^2=0$ and every submodule of M is an ideal of R^*. Furthermore $R^*/M\cong R$ (as rings).

2.20. Theorem. *Let R be a Noetherian ring and M a finitely generated R-module. Let A be an ideal of R and L and N submodules of M. Then there is a positive integer r such that*

$$A^nL\cap N=A^{n-r}(A^rL\cap N)\qquad \textit{for all}\quad n>r.$$

Proof. Since $A+M$, L, and N are ideals of R^*, it follows from Theorem 2.19 that there is a positive integer r such that for all $n>r$,

$$\begin{aligned} A^nL\cap N\subseteq(A+M)^nL\cap N&=(A+M)^{n-r}((A+M)^rL\cap N)\\ &\subseteq(A^{n-r}+M)((A^r+M)L\cap N)\\ &=A^{n-r}(A^rL\cap N),\end{aligned}$$

since $M^2 = 0$. On the other hand,

$$A^{n-r}(A^rL \cap N) \subseteq A^nL \cap A^{n-r}N \subseteq A^nL \cap N \qquad \text{for all} \quad n > r,$$

so the equality is proved.

2.21. Corollary. *Let R, A, and M be as in Theorem 2.20. If $N = \bigcap_{n=1}^{\infty} A^nM$, then $AN = N$.*

Proof. If r is as in Theorem 2.20, with $L = M$, then for all $n > r$,

$$A^nM \cap N = A^{n-r}(A^rM \cap N) \subseteq AN.$$

Since $N \subseteq A^nM$ this implies that $N \subseteq AN$. But $AN \subseteq N$, so $A = AN$.

Now a technical lemma!

2.22. Lemma. *Let $a_{ij}, b_j \in R$ for $i, j = 1, \ldots, k$, and let $d = \det[a_{ij}]$. If $\sum_{j=1}^{k} a_{ij} b_j = 0$ for $i = 1, \ldots, k$, then $db_j = 0$ for $j = 1, \ldots, k$.*

Proof. Let d_{ij} be the cofactor of a_{ij} in the matrix $[a_{ij}]$. Then

$$\sum_{i=1}^{k} d_{ij} a_{ih} = \begin{cases} d & \text{if } j = h \\ 0 & \text{if } j \neq h. \end{cases}$$

Hence

$$0 = \sum_{i=1}^{k} d_{ij} \sum_{h=1}^{k} a_{ih} b_h = \sum_{h=1}^{k} \left(\sum_{i=1}^{k} d_{ij} a_{ih} \right) b_h = db_j \qquad \text{for} \quad j = 1, \ldots, k.$$

2.23. Theorem (Krull Intersection Theorem). *Let R be a Noetherian ring and M a finitely generated R-module. Let A be an ideal of R. Then*

$$\bigcap_{n=1}^{\infty} A^nM = 0$$

if and only if $1 - a \in A$ and $ax = 0$, where $a \in R$ and $x \in M$, imply that $x = 0$.

Proof. Suppose that $1 - a \in A$ and $ax = 0$ imply that $x = 0$. Let $N = \bigcap_{n=1}^{\infty} A^nM$; then $AN = N$. Let $N = Rx_1 + \cdots + Rx_k$. For

$i = 1, \ldots, k$, $x_i = \sum_{j=1}^{k} a_{ij} x_j$, with $a_{ij} \in A$, that is $\sum_{j=1}^{k} (\delta_{ij} - a_{ij})x_j = 0$. Let $d = \det[\delta_{ij} - a_{ij}]$. Then $1 - d \in A$, and by Lemma 2.22, applied to R^*, we have $dx_j = 0$ for $j = 1, \ldots, k$. Hence $x_j = 0$ for $j = 1, \ldots, k$, and so $N = 0$.

Conversely, suppose that for some $a \in R$ and nonzero $x \in M$ we have $1 - a \in A$ and $ax = 0$. Then, for each positive integer n we have $x = (1 - a)^n x \in A^n M$. Hence $\bigcap_{n=1}^{\infty} A^n M \neq 0$.

2.24. Corollary. *Let R, A, and M be as in Theorem 2.23. If A is contained in every maximal ideal of R, then*

$$\bigcap_{n=1}^{\infty} A^n M = 0.$$

Proof. Let $a \in R$ and $x \in M$ be such that $1 - a \in A$ and $ax = 0$. If a is not a unit in R then (a) is a proper ideal of R and so is contained in some maximal ideal of R by Proposition 2.12; this implies that 1 is contained in some maximal ideal of R, which is not true. Hence, a is a unit in R, and it follows that $x = 0$.

4 UNIQUENESS RESULTS FOR PRIMARY DECOMPOSITIONS

Let R be a ring and let M be an R-module. Let N be a submodule of M, and suppose that

$$N = Q_1 \cap \cdots \cap Q_n,$$

where $Q_1, \ldots, Q_n$ are primary submodules of M; this is called a **primary decomposition** of N. Such a decomposition is called **reduced** if:

(i) No Q_i contains the intersection of the remaining Q_j; and
(ii) $\mathrm{Rad}(Q_i) \neq \mathrm{Rad}(Q_j)$ for $i \neq j$.

2.25. Proposition. *If a submodule N of M has a primary decomposition, then N has a reduced primary decomposition.*

Proof. Let $N = Q_1 \cap \cdots \cap Q_n$ where Q_i is primary for $i = 1, \ldots, n$. If some Q_i contains the intersection of the remaining Q_j, simply delete it. Hence we may assume (i) holds. Suppose $Q_{i_1}, \ldots, Q_{i_k}$ have the same radical P. By Exercise 6(f), $Q = Q_{i_1} \cap \cdots \cap Q_{i_k}$ is P-primary. In the decomposition for N replace $Q_{i_1} \cap \cdots \cap Q_{i_k}$ by Q. Proceeding in this manner we finally obtain a reduced primary decomposition for N.

It can happen that N has no primary decomposition whatsoever [Exercise 13(d)], but if it does have such a decomposition, it may have more than one reduced decomposition [Exercise 11(c)]. In this section we shall determine certain features of primary decompositions which are unique.

Let

$$N = Q_1 \cap \cdots \cap Q_k$$

be a reduced primary decomposition of N and let $P_i = \text{Rad}(Q_i)$ for $i = 1, \ldots, k$. The prime ideals $P_1, \ldots, P_k$ are called **prime divisors** of N. We have

$$\text{Rad}(N) = \text{Rad}(Q_1) \cap \cdots \cap \text{Rad}(Q_k) = P_1 \cap \cdots \cap P_k.$$

Hence, if $a \in P_1 \cdots P_k$, then some power of a is in $N : M$. Consequently, if P is any prime ideal of R which contains $N : M$, then $P_1 \cdots P_k \subseteq P$ and so $P_i \subseteq P$ for some i. It follows that every minimal prime divisor of N is a prime divisor of N and is minimal in the set of prime divisors of N. The next theorem characterizes the prime divisors of N in terms of N itself. We assume throughout this discussion that $N \neq M$.

2.26. Theorem. *Let N be a submodule of M and assume that N has a primary decomposition. Let $N = Q_1 \cap \cdots \cap Q_k$ be a reduced primary decomposition of N. Let P be a prime ideal of R. Then $P = \text{Rad}(Q_i)$ for some i if and only if $N : Rx$ is a P-primary ideal of R for some $x \notin N$.*

Proof. Let $P_i = \text{Rad}(Q_i)$ for $i = 1, \ldots, k$ and suppose $P = P_1$. Since the decomposition is reduced there is an element $x \in Q_2 \cap \cdots \cap Q_k$ with $x \notin Q_1$ (simply choose $x \notin N$ if $k = 1$). By Exercise 6(h), $Q_1 : Rx$ is a P-primary ideal of R. Furthermore, $Q_i : Rx = R$ for $i = 2, \ldots, k$.

Hence $N: Rx = Q_1 : Rx$ and so $N: Rx$ is P-primary; note that $x \notin N$. Conversely, suppose that $N: Rx$ is P-primary for some $x \notin N$. Then

$$P = \mathrm{Rad}(N: Rx) = \mathrm{Rad}(Q_1 : Rx) \cap \cdots \cap \mathrm{Rad}(Q_k : Rx).$$

For each i, $\mathrm{Rad}(Q_i : Rx) = P_i$ or R, and is equal to P_i for at least one i, since $x \notin N$. Thus P is the intersection of some of the prime ideals $P_1, \ldots, P_k$. By Exercise 5(b), P is equal to P_i for some i.

2.27. Corollary. *Let N be a submodule of M and assume that N has a primary decomposition. If*

$$N = Q_1 \cap \cdots \cap Q_m = Q_1' \cap \cdots \cap Q_n'$$

are two reduced primary decompositions of N, then $m = n$ and the Q_i and Q_i' can be so numbered that $\mathrm{Rad}(Q_i) = \mathrm{Rad}(Q_i')$ *for $i = 1, \ldots, n$.*

2.28. Corollary. *Let R be a Noetherian ring and let $P_1, \ldots, P_k$ be the prime divisors of the ideal 0. Then the set of zero-divisors of R is precisely $P_1 \cup \cdots \cup P_k$.*

Proof. Let $a \in P_i$; then there exists $b \in R$, $b \neq 0$, such that $a^n \in 0: (b)$ for some smallest positive integer n. Hence $a(a^{n-1}b) = 0$ and a is a zero-divisor. Conversely, suppose a is a proper zero-divisor and let $ab = 0$ where $b \neq 0$. Then for some i, $b \notin Q_i$ where $0 = Q_1 \cap \cdots \cap Q_k$ is a reduced primary decomposition of 0. It follows that $a \in \mathrm{Rad}(Q_i) = P_i$.

2.29. Definition. *Let S be a multiplicatively closed set in R. If N is a submodule of M, we set*

$$N_S = \{x \mid x \in M \quad \text{and} \quad sx \in N \quad \text{for some} \quad s \in S\}.$$

Note that $N_S = M$ if $(N: M) \cap S$ is not empty and, in particular, if $0 \in S$. Let $x_1, x_2 \in N_S$ and $a \in R$. If $s_1 x_1 \in N$ and $s_2 x_2 \in N$, where $s_1, s_2 \in S$, then $s_1 s_2 \in S$ and $s_1 s_2(x_1 - x_2) = s_2(s_1 x_1) - s_1(s_2 x_2) \in N$; also $s_1(ax_1) = a(s_1 x_1) \in N$. Hence $x_1 - x_2$, $ax_1 \in N_S$ and we conclude that N_S is a submodule of M; certainly $N \subseteq N_S$. The submodule N_S of M is called the **component of N determined by** S, or simply the S-**component** of N.

2.30. Proposition. *Let N be a submodule of M which has a primary decomposition, $N = Q_1 \cap \cdots \cap Q_k$. Let S be a multiplicatively closed set in R. Let Q_i be P_i-primary and assume that $P_i \cap S$ is empty for $i = 1, \ldots, h$ and that $P_i \cap S$ is not empty for the remaining i. Then $N_S = Q_1 \cap \cdots \cap Q_h$.*

Proof. Let $x \in N_S$. Then $sx \in N = Q_1 \cap \cdots \cap Q_k$ for some $s \in S$. For $i = 1, \ldots, h$, $sx \in Q_i$ but $s \notin P_i$. Hence $x \in Q_i$. Therefore $N_S \subseteq Q_1 \cap \cdots \cap Q_h$. Now let $x \in Q_1 \cap \cdots \cap Q_h$. If $h = k$, then $x \in N$. If $h \neq k$, choose $s_j \in P_j \cap S$ for $j = h + 1, \ldots, k$. For large enough n we will have

$$(s_{h+1} \cdots s_k)^n M \subseteq Q_{h+1} \cap \cdots \cap Q_k,$$

and so

$$(s_{h+1} \cdots s_k)^n x \in Q_1 \cap \cdots \cap Q_k = N.$$

Since $(s_{h+1} \cdots s_k)^n \in S$, we have $x \in N_S$. Therefore $N_S = Q_1 \cap \cdots \cap Q_h$.

Again, let N be a submodule of M which has a primary decomposition

$$N = Q_1 \cap \cdots \cap Q_k,$$

which we assume to be reduced. Let $P_i = \mathrm{Rad}(Q_i)$, $i = 1, \ldots, k$. A set of these prime ideals, $\{P_{i_1}, \ldots, P_{i_r}\}$, is called an **isolated set of prime divisors** of N if each P_i which is contained in at least one of the ideals in this set is itself in this set. If P_i is a minimal prime divisor of N, then $\{P_i\}$ is an isolated set of prime divisors of N.

2.31. Proposition. *If $\{P_{i_1}, \ldots, P_{i_r}\}$ is an isolated set of prime divisors of N, then $Q_{i_1} \cap \cdots \cap Q_{i_r}$ depends only on this set and not on the particular reduced primary decomposition of N.*

Proof. Let $S = R \backslash (P_{i_1} \cup \cdots \cup P_{i_r})$; then S is a multiplicatively closed set in R. The proposition will follow if we show that $N_S = Q_{i_1} \cap \cdots \cap Q_{i_r}$, and to do this it suffices to show that $S \cap P_i$ is empty when $i = i_j$ for some j, and $S \cap P_i$ is not empty when $i \neq i_1, \ldots, i_r$. The former is certainly true. Suppose $i \neq i_1, \ldots, i_r$. We have $P_i \nsubseteq P_{i_j}$ for $j = 1, \ldots, r$, and so $P_i \nsubseteq \bigcup_{j=1}^{r} P_{i_j}$ by Exercise 5(c). Hence $P_i \cap S$ is not empty.

In the notation of this proposition, $Q_{i_1} \cap \cdots \cap Q_{i_r}$ is called an **isolated component of** N. The proposition asserts that the isolated components of N are uniquely determined by N.

Our principal uniqueness theorem is an immediate corollary to Proposition 2.31 and is stated now.

2.32. Theorem. *Let P be a minimal prime divisor of a submodule N of M. Suppose that N has a primary decomposition. If the P-primary submodule Q occurs in a reduced primary decomposition of N, then Q occurs in every reduced primary decomposition of N.*

We close this section with a result related to the Krull intersection theorem (Theorem 2.23).

2.33. Proposition. *Let R be a Noetherian ring and let M be a finitely generated R-module. Let A be an ideal of R. Then $\bigcap_{n=1}^{\infty} A^n M$ is a component of the ideal* 0.

Proof. Let $S = \{1 - a \mid a \in A\}$. If $a, b \in A$, then $(1-a)(1-b) = 1 - (a + b - ab) \in S$, so S is a multiplicatively closed set in R. Let $N = \bigcap_{n=1}^{\infty} A^n M$; by Corollary 2.21, we have $N = AN$. Let $x \in N$ and let $Ax = Q_1 \cap \cdots \cap Q_k$, where Q_i is primary and $\operatorname{Rad}(Q_i) = P_i$ for $i = 1, \ldots, k$. For each i, $Ax \subseteq Q_i$ so that either $A \subseteq P_i$ or $x \in Q_i$ [see Exercise 6(c)]. Suppose that $A \subseteq P_j$ for some j. Then there is a positive integer m such that $A^m \subseteq P_j^m \subseteq Q_j : M$ [see Exercise 6(a)]; that is, $A^m M \subseteq Q_j$. Hence $N \subseteq Q_j$, and consequently $x \in Q_j$. Thus $x \in Q_i$ for $i = 1, \ldots, k$ and therefore we have $x \in Ax$. Hence $x = ax$ for some $a \in A$. Conversely, if $x \in M$ and $x = ax$ for some $a \in A$, then $x = a^n x \in A^n M$ for all n; hence $x \in N$. Thus, $x \in N$ if and only if $sx = 0$ for some $s \in S$; that is, if and only if $x \in 0_S$.

EXERCISES

1. The radical of an ideal.

Let R be a ring and let A, B, and C be ideals of R.

(a) Show that $\operatorname{Rad}(AB) = \operatorname{Rad}(A \cap B) = \operatorname{Rad}(A) \cap \operatorname{Rad}(B)$.

(b) Show that $\operatorname{Rad}(A + B) = \operatorname{Rad}(\operatorname{Rad}(A) + \operatorname{Rad}(B))$.

(c) Show that

$$\begin{aligned}\operatorname{Rad}(A+BC) &= \operatorname{Rad}(A+(B\cap C))\\ &= \operatorname{Rad}(A+B)\cap \operatorname{Rad}(A+C).\end{aligned}$$

(d) Show that $\operatorname{Rad}(\operatorname{Rad}(A)) = \operatorname{Rad}(A)$.

(e) Show that $A+B=R$ if and only if $\operatorname{Rad}(A)+\operatorname{Rad}(B)=R$.

2. Comaximal ideals.
Two ideals A, B of a ring R are said to be **comaximal** if $A+B=R$. A finite collection $A_1, \ldots, A_k$ of ideals of R is comaximal if each pair is comaximal. Let $A_1, \ldots, A_k$ be comaximal ideals of R and let $n_1, \ldots, n_k$ be positive integers. Prove the following:

(a) $A_1^{n_1}, \ldots, A_k^{n_k}$ are comaximal.

(b) $A_1 \cap \cdots \cap A_k = A_1 \cdots A_k$.

(c) For $a_1, \ldots, a_k \in R$, there exists $x \in R$ such that $x - a_i \in A_i$ for $i = 1, \ldots, k$.

(d) $R/(A_1^{n_1} \cap \cdots \cap A_k^{n_k}) \cong R/A_1^{n_1} \oplus \cdots \oplus R/A_k^{n_k}$.

(e) The ideals $B_1, \ldots, B_n$ of R are comaximal if and only if $\operatorname{Rad}(B_1), \ldots, \operatorname{Rad}(B_n)$ are comaximal.

3. Maximal, prime, and primary ideals.
Let R be a ring.

(a) Show that an ideal P of R is maximal if and only if R/P is a field.

(b) Show that an ideal P of R is prime if and only if R/P has no proper zero-divisors.

(c) Show that an ideal Q of R is primary if and only if every zero-divisor of R/Q is nilpotent.

(d) Show that if $R \neq 0$ and if R and 0 are the only ideals of R, then R is a field.

4. Ideals in a direct sum of rings.
Let $R_1, \ldots, R_n$ be rings and let $R = R_1 \oplus \cdots \oplus R_n$.

(a) Show that if A is an ideal of R, then A can be written uniquely in the form $A_1 \oplus \cdots \oplus A_n$, where A_i is an ideal of R_i for $i = 1, \ldots, n$.

(b) Show that A is a maximal ideal of R if and only if for some i, A_i is a maximal ideal of R_i and $A_j = R_j$ for $j \neq i$.

(c) Show that the statement of part (b) is true when "maximal" is replaced by "prime" or by "primary."

5. Intersections and unions of primary ideals.

(a) Let R be a ring, P a prime ideal of R, and $Q_1, \ldots, Q_k$ primary ideals of R. Show that if $Q_1 \cap \cdots \cap Q_k \subseteq P$, then $\text{Rad}(Q_i) \subseteq P$ for some i.

(b) Let $P_1, \ldots, P_k$ be prime ideals of R. Show that if $P = P_1 \cap \cdots \cap P_k$ is a prime ideal, then $P = P_i$ for some i and $P \subseteq P_j$ for $j = 1, \ldots, k$.

(c) Let $P_1, \ldots, P_k$ be prime ideals of R and let A be an ideal of R such that $A \subseteq P_1 \cup \cdots \cup P_k$. Show that $A \subseteq P_i$ for some i.

6. Primary submodules.

(a) Let R be a ring and let A be an ideal of R such that $\text{Rad}(A)$ is finitely generated. Show that there is a positive integer n such that $(\text{Rad}(A))^n \subseteq A$.

(b) Assume that R is Noetherian and let M be an R-module. Let Q be a P-primary submodule of M. Show that for some positive integer n we have $P^n \subseteq Q : M$.

(c) If R is Noetherian, show that a submodule Q of an R-module M is primary if and only if for all ideals A of R and submodules N of M, $AN \subseteq Q$ and $N \nsubseteq Q$ imply that $A^n M \subseteq Q$ for some positive integer n.

(d) Let M be an R-module, P an ideal of R, and Q a submodule of M. Suppose that:

(i) $Q : M \subseteq P \subseteq \text{Rad}(Q)$; and

(ii) If $ax \in Q$ and $x \notin Q$, where $a \in R$ and $x \in M$, then $a \in P$.

Show that P is a prime ideal and Q is P-primary.

(e) Show that a submodule Q of M if primary if $\text{Rad}(Q)$ is a maximal ideal of R.

(f) Let $Q_1, \ldots, Q_k$ be P-primary submodules of M. Show that $Q_1 \cap \cdots \cap Q_k$ is P-primary.

(g) Let ϕ be a homomorphism from R into R'. Let Q' be a P'-primary ideal of R'. Show that $P = \phi^{-1}(P')$ is prime and that $Q = \phi^{-1}(Q')$ is P-primary.

(h) Let M be an R-module, A an ideal of R, P a prime ideal of R, Q a P-primary submodule of M, and N a submodule of M. Show that if $A \nsubseteq P$ then $Q: A = Q$ and that if $N \nsubseteq Q$ then the ideal $Q: N$ is P-primary.

(i) Assume R is Noetherian and M is finitely generated. If N is a submodule of M, show that $N: A = N$ if and only if A is contained in no prime divisor of N.

(j) Show that $N: A = N$ if and only if no prime divisor of A is contained in any prime divisor of N.

7. The Jacobson radical.
Let $J(R)$ denote the intersection of the maximal ideals of the ring R. Then $J(R)$ is an ideal called the **Jacobson radical** of R.

(a) Show that $J(R)$ is the set

$$\{a \mid a \in R \quad \text{and} \quad 1 + ra \quad \text{is a unit in} \quad R \quad \text{for all} \quad r \in R\}.$$

(b) Show that $J(R/J(R)) = 0$.

(c) (Nakayama's lemma) Let A be an ideal of R. Show that the following statements are equivalent:

(1) $A \subseteq J(R)$.

(2) If M is a finitely generated R-module and $AM = M$, then $M = 0$.

(3) If M is a finitely generated R-module and N is a submodule of M such that $M = AM + N$, then $N = M$.

(4) If $a \in A$, then $1 + a$ is a unit in R.

(d) Let R be a Noetherian ring and M a finitely generated R-module. Let N be a submodule of M and A an ideal of R with $A \subseteq J(R)$. Show that $\bigcap_{n=1}^{\infty} (N + A^n M) = N$.

8. Corollaries to the Artin–Rees lemma.

(a) Let R be a Noetherian ring and M a finitely generated R-module. Let A be an ideal of R and let $a \in R$. Show that for every positive integer n we have $a(A^n M : (a)) = A^n M \cap aM$.

(b) Show that there is a positive integer r such that

$$A^n M : (a) \subseteq 0 : (a) + A^{n-r} M \quad \text{for all} \quad n > r.$$

(c) Let N be a submodule of M. Show that there is a positive integer r such that

$$(N + A^n M) : (a) \subseteq N : (a) + A^{n-r} M \quad \text{for all} \quad n > r.$$

(d) Let $x \in M$. Show that there is a positive integer r such that

$$(N + A^n M) : Rx \subseteq N : Rx + A^{n-r} \quad \text{for all} \quad n > r.$$

9. *A* component of 0.
Let R be a Noetherian ring and A an ideal of R.

(a) Let $B = \{b \mid b \in R \text{ and } bA^t = 0 \text{ for some positive integer } t\}$. Show that B is an ideal of R.

(b) Let $0 = Q_1 \cap \cdots \cap Q_k$ be a reduced primary decomposition of 0 and let $P_i = \text{Rad}(Q_i)$ for $i = 1, \ldots, k$. Show that

$$B = \bigcap_{A \not\subseteq P_i} Q_i .$$

(c) Show that there is an integer s such that $A^s \cap B = 0$.

(d) Show that $B : A = B$.

10. The ring of integers.

(a) Show that every ideal in the ring of integers is a principal ideal.

(b) Determine all of the prime ideals and primary ideals of the ring of integers.

(c) Prove the unique factorization theorem for integers as a corollary to results concerning primary decomposition.

11. Ideals in $K[X, Y]$.
Let K be a field and consider the ring $R = K[X, Y]$ of polynomials in two indeterminates X and Y over K.

(a) Show that (X, Y^2) is a primary ideal of R with radical (X, Y). Show that (X, Y^2) is not a power of (X, Y). This shows that *a primary ideal need not be a power of a prime ideal* (even in a Noetherian ring).

(b) Show that $\text{Rad}((X^2, XY)) = (X)$. Show that (X^2, XY) is not primary. This shows that *an ideal with prime radical need not be primary.*

(c) Show that for each $c \in K$, $(Y - cX, X^2)$ is a primary ideal of R, and that for $c, d \in K$ with $c \neq d$ we have $(Y - cX, X^2) \neq (Y - dX, X^2)$. Show that for each $c \in K$, $(X^2, XY) = (X) \cap (Y - cX, X^2)$ is a reduced primary decomposition of (X^2, XY). This shows that *an ideal in a Noetherian ring may have infinitely many distinct reduced primary decompositions*. What are the prime divisors of (X^2, XY)?

12. Ideals in $Z[X]$.

Let $R = Z[X]$, the ring of polynomials with coefficients in Z.

(a) Show that $(4, X)$ is primary and $\text{Rad}((4, X)) = (2, X)$, but that $(4, X)$ is not a power of $(2, X)$.

(b) Show that $(X^2, 2X)$ is not primary. Show that (X) is prime and $(X)^2 \subseteq (X^2, 2X)$.

(c) Show that $(4, 2X, X^2)$ is primary and that $(4, 2X, X^2) = (4, X) \cap (2, X^2)$. This shows that *a primary ideal need not be irreducible.*

(d) Let S be the subring of R consisting of all elements of R having the coefficient of X divisible by 3. Let P be the ideal $(3X, X^2, X^3)$ of S. Show that P is a prime ideal but P^2 is not a primary ideal. This shows that *a power of a prime ideal need not be primary.*

13. A non-Noetherian ring.

Let R be the set of all sequences $\{a_n\}$ $(n \geq 1)$ of elements of the field having two elements, such that for some m_0, depending on the sequence, $a_m = a_{m0}$ for all $m \geq m_0$. If we define operations on R by

$$\{a_n\} + \{b_n\} = \{a_n + b_n\} \qquad \text{and} \qquad \{a_n\}\{b_n\} = \{a_n b_n\},$$

then R is a commutative ring with unity. For each $i > 0$, let $P_i = \{\{a_n\} \mid a_i = 0\}$, and let $P_0 = \{\{a_n\} \mid$ for some m_0, depending on the sequence, $a_m = 0$ for $m \geq m_0\}$.

(a) Show that for $i \geq 0$, P_i is a prime ideal of R.

(b) Show that $\{P_i \mid i \geq 0\}$ is the set of proper prime ideals of R.

(c) Show that an ideal of R is primary if and only if it is prime.

(d) Show that $\bigcap_{i=1}^{\infty} P_i = 0$, but for every $j > 0$,

$$P_0 \cap \left(\bigcap_{\substack{i=1 \\ i \neq j}}^{\infty} P_i \right) \neq 0.$$

Therefore, 0 is not an intersection of a finite number of primary ideals of R.

(e) By part (d) R is not Noetherian. Construct an infinite ascending chain of ideals of R, a nonempty set of ideals of R without a maximal element, and an ideal of R not finitely generated.

14. A theorem of Cohen.

(a) Let R be a ring and let A be an ideal of R. Show that if $b \in R$ and if $A+(b)$ and $A:(b)$ are finitely generated, then A is finitely generated.

(b) Let $\mathscr{S}$ be the set of all ideals of R which are not finitely generated. Show that if $\mathscr{S}$ is not empty then $\mathscr{S}$ has a maximal element.

(c) Assume that $\mathscr{S}$ is not empty and let P be a maximal element of $\mathscr{S}$. Show that P is a prime ideal. Arguing from this, complete the proof of the following assertion: *R is Noetherian if and only if every prime ideal of R is finitely generated.*

15. Irreducible ideals.

Let R be a ring and A an ideal of R. Let $A = B_1 \cap \cdots \cap B_k$, where each B_i is irreducible and no B_i contains the intersection of the remaining B_j.

(a) Show that if $A = B_1' \cap \cdots \cap B_k'$ where $B_i \subseteq B_i'$ for $i = 1, \ldots, k$, then $B_i = B_i'$ for $i = 1, \ldots, k$.

(b) Suppose that we also have $A = C_1 \cap \cdots \cap C_n$ where each C_i is irreducible and no C_i contains the intersection of the remaining C_j. For $i = 1, \ldots, k$, let $D_i = B_1 \cap \cdots \cap B_{i-1} \cap B_{i+1} \cap \cdots \cap B_k$. Show that for each i, $A = D_i \cap C_j$ for some j.

(c) Show that $k = n$.

16. Primal ideals.

Let R be a ring. For each ideal A of R, let

$$U(A) = \{b \mid b \in R \quad \text{and} \quad b + A \quad \text{is a regular element of} \quad R/A\}.$$

An ideal A of R is said to be **primal** if $R \backslash U(A)$ is an ideal of R. If A is a primal ideal of R, then $R \backslash U(A)$ is called its **adjoint ideal**. The adjoint ideal P of a primal ideal A is the unique ideal maximal with respect to the properties $A \subseteq P$ and $U(A) \cap P$ is empty.

(a) Show that every irreducible ideal of R is primal.

(b) Let K be a field and let $R = K[X, Y]$. Show that (X^2, XY) is primal with adjoint ideal (X, Y). Recall that the radical of (X^2, XY) is (X) [Exercise 11(b)]. By symmetry,

(Y^2, XY) is primal with adjoint ideal (X, Y). Show that $(XY) = (X^2, XY) \cap (Y^2, XY)$ is not primal.

(c) For $i = 1, \ldots, k$, let A_i be a primal ideal with adjoint ideal P_i. Let $B = A_1 \cap \cdots \cap A_k$ and assume that no A_i can be replaced by a strictly larger ideal without changing the intersection. Show that B is primal if and only if some P_j contains every P_i, and that in this case P_j is the adjoint ideal of B.

(d) Show that every ideal of R is the intersection of all of the primal ideals containing it.

(e) A primal decomposition $B = A_1 \cap \cdots \cap A_k$ is called **reduced** if no A_i contains the intersection of the remaining A_j, if no A_i can be replaced by a strictly larger ideal without changing the intersection, and if no intersection of two or more of the A_i is primal. Show that if R is Noetherian then every ideal of R has a reduced primal decomposition.

(f) Let $B = A_1 \cap \cdots \cap A_k = A_1' \cap \cdots \cap A_n'$ be reduced primal decompositions of B. Show that $k = n$ and that the A_i and A_i' can be so numbered that A_i and A_i' have the same adjoint ideal for each i.

17. Prime ideals associated with a module.

Let R be a Noetherian ring and M a finitely generated R-module. For $x \in M$ set $\text{Ann}(x) = \{a \mid a \in R \text{ and } ax = 0\}$, the **annihilator** of x. If P is a prime ideal of R such that $P = \text{Ann}(x)$ for some $x \in M$, we say that P is **associated** with M. The set of all prime ideals of R associated with M is denoted by $\text{Ass}(M)$.

(a) Show that if P is a prime ideal of R then $P \in \text{Ass}(M)$ if and only if M has a submodule isomorphic to R/P.

(b) Show that if $M = \bigcup_{\alpha \in I} M_\alpha$, where each M_α is a submodule of M, then $\text{Ass}(M) = \bigcup_{\alpha \in I} \text{Ass}(M_\alpha)$.

(c) Let P be a prime ideal of R and M a nonzero submodule of R/P. Show that $\text{Ass}(M) = \{P\}$.

(d) Show that every maximal element (with respect to set inclusion) of the set $\{\text{Ann}(x) \mid x \in M \text{ and } x \neq 0\}$ belongs to $\text{Ass}(M)$.

(e) Show that $M \neq 0$ if and only if $\text{Ass}(M)$ is not empty.

(f) Let $a \in R$. Show that $ax = 0$, for $x \in M$, implies $x = 0$ if and only if a belongs to no element of $\text{Ass}(M)$.

18. Properties of $\mathrm{Ass}(M)$.

Let R be a Noetherian ring and M a finitely generated R-module.

(a) Show that if N is a submodule of M then

$$\mathrm{Ass}(N) \subseteq \mathrm{Ass}(M) \subseteq \mathrm{Ass}(N) \cup \mathrm{Ass}(M/N).$$

(b) Show that if $M = \bigoplus_{\alpha \in I} M_\alpha$ then $\mathrm{Ass}(M) = \bigcup_{\alpha \in I} \mathrm{Ass}(M_\alpha)$.

(c) Show that if $N_1 \cap \cdots \cap N_k = 0$, where the N_i are submodules of M, then $\mathrm{Ass}(M) \subseteq \bigcup_{i=1}^{k} \mathrm{Ass}(M/N_i)$.

(d) Let T be a subset of $\mathrm{Ass}(M)$. Show that there is a submodule N of M such that $\mathrm{Ass}(M/N) = T$ and $\mathrm{Ass}(N) = \mathrm{Ass}(M) \backslash T$.

19. Polynomial rings.

Let R be a ring and let X be an indeterminate.

(a) Show that $R[X]$ is a flat R-module.

(b) Show that $a_0 + a_1 X + \cdots + a_n X^n \in R[X]$ is a zero-divisor in $R[X]$ if and only if there is an element $b \in R$, $b \neq 0$, such that $ba_i = 0$ for $i = 0, \ldots, n$.

(c) Let Q be a P-primary ideal of R. Show that $QR[X]$ is a $PR[X]$-primary ideal of $R[X]$, and that $QR[X] \cap R = Q$.

(d) Show that if $A_1, \ldots, A_k$ are ideals of R then

$$(A_1 \cap \cdots \cap A_k)R[X] = A_1 R[X] \cap \cdots \cap A_k R[X].$$

CHAPTER

III

Rings and Modules of Quotients

1 DEFINITION

Let R be a ring and let M be an R-module. Let S be a multiplicatively closed set in R. Let T be the set of all ordered pairs (x, s) where $x \in M$ and $s \in S$. Define a relation on T by

$$(x, s) \sim (x', s')$$

if there exists $t \in S$ such that $t(sx' - s'x) = 0$. This is an equivalence relation on T, and we denote the equivalence class of (x, s) by x/s. Let $S^{-1}M$ denote the set of equivalence classes of T with respect to this relation. We can make $S^{-1}M$ into an R-module by setting

$$\begin{aligned} x/s + y/t &= (tx + sy)/st, \\ a(x/s) &= ax/s, \qquad a \in R. \end{aligned}$$

We must show that these are well-defined operations; once this has been done it is easily verified that $S^{-1}M$ is an R-module with respect to these operations.

Suppose $x/s = x'/s'$ and $y/t = y'/t'$. Then there are elements $u, v \in S$ such that $u(s'x - sx') = 0$ and $v(t'y - yt') = 0$. It follows that

$$\begin{aligned} &uv(s't'(tx + sy) - st(t'x' + s'y')) \\ &\quad = t'tvu(s'x - sx') + s'suv(t'y - ty') = 0. \end{aligned}$$

Hence

$$(tx + sy)/st = (t'x' + s'y')/s't'.$$

Also

$$u(s'ax - sax') = au(s'x - sx') = 0,$$

so that

$$ax/s = ax'/s'.$$

The R-module $S^{-1}M$ is called a **quotient module**, or a **module of quotients**. Note that if $0 \in S$, then $S^{-1}M = 0$. In our discussion of quotient modules we shall always assume that $0 \notin S$. In fact, we will call S a **multiplicative system in** R if S is a multiplicatively closed subset of R with $0 \notin S$.

Since R may be considered as an R-module we can form the quotient module $S^{-1}R$. An element of $S^{-1}R$ has the form a/s, where $a \in R$ and $s \in S$. We can make $S^{-1}R$ into a ring by setting

$$(a/s)(b/t) = ab/st.$$

If $a/s = a'/s'$ and $b/t = b'/t'$, and if $u(s'a - sa') = 0$ and $v(t'b - tb') = 0$, where $u, v \in S$, then

$$uv(s't'ab - sta'b') = vt'bu(s'a - sa') + usa'v(t'b - tb') = 0.$$

Consequently,

$$ab/st = a'b'/s't',$$

and thus we have a well-defined multiplication in $S^{-1}R$. It is clear that with this operation, $S^{-1}R$ is a ring. The ring $S^{-1}R$ is called a **quotient ring** of R, or a **ring of quotients** of R.

3.1. Proposition. *Let R and S be as above.*

(a) *If $a \in R$, then as/s is independent of the element $s \in S$, and the mapping $\eta: R \to S^{-1}R$ given by $\eta(a) = as/s$ is a homomorphism.*

(b) Ker $\eta = 0_S$, *the S-component of* 0.

(c) *Every element of $\eta(S)$ is a unit in $S^{-1}R$.*

(d) *If $\phi: R \to R'$ is a homomorphism from R into a ring R' such that every element of $\phi(S)$ is a unit in R', then there is a unique homomorphism $\psi: S^{-1}R \to R'$ such that*

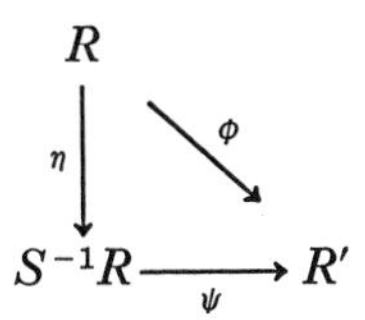

is commutative.

Proof. (a) If $a \in R$ and $s, t \in S$, then $u(t(as) - s(at)) = 0$ for all $u \in S$. Hence $as/s = at/t$. Thus $\eta(a) = as/s$ depends only on a. We have

$$\begin{aligned}\eta(a+b) &= (a+b)s^2/s^2 = as/s + bs/s \\ &= \eta(a) + \eta(b)\end{aligned}$$

and

$$\eta(ab) = abs^2/s^2 = (as/s)(bs/s) = \eta(a)\eta(b).$$

Thus η is a homomorphism.

(b) We have $a \in \operatorname{Ker} \eta$ if and only if $as/s = 0$; this is equivalent to $ts^2 a = 0$ for some $t \in S$. Since $ts^2 \in S$ this implies $a \in 0_S$. Conversely, if $va = 0$, where $v \in S$, then $av/v = 0/v = 0$, so that $a \in \operatorname{Ker} \eta$.

(c) Note that $s/s = t/t$ for all $s, t \in S$, and that this element is the unity of $S^{-1}R$. If $s \in S$ then $\eta(s) = s^2/s$, and $(s^2/s)(s/s^2) = 1$, so $\eta(s)$ is a unit in $S^{-1}R$.

(d) Define ψ: $S^{-1}R \to R'$ by $\psi(a/s) = \phi(a)\phi(s)^{-1}$. Suppose $a/s = b/t$ and that $u(ta - sb) = 0$ where $u \in S$. Then $0 = \phi(u)(\phi(t)\phi(a) - \phi(s)\phi(b))$ and since $\phi(u)$ is a unit in R', $\phi(t)\phi(a) - \phi(s)\phi(b) = 0$. Thus, $\phi(a)\phi(s)^{-1} = \phi(b)\phi(t)^{-1}$. Hence ψ is well-defined. It is a homomorphism since

$$\begin{aligned}\psi(a/s + b/t) &= \psi((ta+sb)/st) = \phi(ta+sb)\phi(st)^{-1} \\ &= (\phi(t)\phi(a) + \phi(s)\phi(b))\phi(s)^{-1}\phi(t)^{-1} \\ &= \phi(a)\phi(s)^{-1} + \phi(b)\phi(t)^{-1} \\ &= \psi(a/s) + \psi(b/t)\end{aligned}$$

and

$$\begin{aligned}\psi((a/s)(b/t)) &= \psi(ab/st) = \phi(ab)\phi(st)^{-1} \\ &= \phi(a)\phi(s)^{-1}\phi(b)\phi(t)^{-1} \\ &= \psi(a/s)\psi(b/t).\end{aligned}$$

Furthermore for all $a \in R$, $\psi\eta(a) = \psi(as/s) = \phi(a)\phi(s)\phi(s)^{-1} = \phi(a)$. Clearly ψ is the only homomorphism from $S^{-1}R$ into R' which will make the triangle commutative.

Let R, S, and M be as above. We can make $M \otimes_R S^{-1}R$ into an R-module by the method described in Exercise 7(b) of Chapter I. Then $a(x \otimes b/s) = x \otimes ab/s = ax \otimes b/s$.

3.2. Proposition. *$S^{-1}M \cong M \otimes_R S^{-1}R$, as R-modules.*

Proof. Define $\phi: S^{-1}M \to M \otimes_R S^{-1}R$ by

$$\phi(x/s) = x \otimes 1/s.$$

If $x/s = x'/s'$ and if $u(s'x - sx') = 0$, where $u \in S$, then

$$\begin{aligned} x \otimes 1/s &= x \otimes us'/sus' = us'x \otimes 1/sus' = usx' \otimes 1/sus' \\ &= x' \otimes us/sus' = x' \otimes 1/s'. \end{aligned}$$

Therefore ϕ is well-defined. Since

$$\begin{aligned} \phi(x/s + y/t) &= \phi((tx + sy)/st) = (tx + sy) \otimes 1/st \\ &= tx \otimes 1/st + sy \otimes 1/st \\ &= x \otimes 1/s + y \otimes 1/t = \phi(x/s) + \phi(y/t) \end{aligned}$$

and

$$\phi(a(x/s)) = \phi(ax/s) = ax \otimes 1/s = a(x \otimes 1/s) = a\phi(x/s),$$

ϕ is a homomorphism (of R-modules). Now define a mapping ψ' from the Cartesian product $M \times S^{-1}R$ into $S^{-1}M$ by $\psi'(x, a/s) = ax/s$. If $a/s = b/t$ and if $v(ta - sb) = 0$, where $v \in S$, then $v(tax - sbx) = 0$, and so $ax/s = bx/t$. Hence ψ' is well-defined. After some verification, it follows from Proposition 1.17 that there is a homomorphism (of Abelian groups) $\psi: M \otimes_R S^{-1}R \to S^{-1}M$ such that $\psi(x \otimes a/s) = ax/s$. Furthermore,

$$\begin{aligned} \psi(a(x \otimes b/t)) &= \psi(x \otimes ab/t) = abx/t \\ &= a(bx/t) = a\psi(x \otimes b/t), \end{aligned}$$

so that ψ is, in fact, a homomorphism of R-modules. Since

$$\phi\psi(x \otimes a/s) = \phi(ax/s) = ax \otimes 1/s = x \otimes a/s$$

and

$$\psi\phi(x/s) = \psi(x \otimes 1/s) = x/s,$$

ϕ and ψ are isomorphisms, inverse to one another.

3.3. Theorem. *$S^{-1}R$ is a flat R-module.*

Proof. Let N and L be R-modules and let $\rho: L \to N$ be an injective homomorphism. Let $\phi_1: S^{-1}L \to L \otimes_R S^{-1}R$ and $\phi_2: S^{-1}N \to N \otimes_R S^{-1}R$ be defined in the same manner as ϕ in the proof of Proposition 3.2. Define $\sigma: S^{-1}L \to S^{-1}N$ by $\sigma(x/s) = \rho(x)/s$; we verify as usual that σ is a well-defined homomorphism. If $i = 1_{S^{-1}R}$, then the diagram

$$\begin{array}{ccc} S^{-1}L & \xrightarrow{\sigma} & S^{-1}N \\ \downarrow{\scriptstyle \phi_1} & & \downarrow{\scriptstyle \phi_2} \\ L \otimes_R S^{-1}R & \xrightarrow[\rho \otimes i]{} & N \otimes_R S^{-1}R \end{array}$$

is commutative, for

$$\begin{aligned}(\rho \otimes i)\phi_1(x/s) &= (\rho \otimes i)(x \otimes 1/s) = \rho(x) \otimes 1/s \\ &= \phi_2(\rho(x)/s) = \phi_2\,\sigma(x/s).\end{aligned}$$

Now suppose that $\alpha \in L \otimes_R S^{-1}R$ and that $(\rho \otimes i)(\alpha) = 0$. Since ϕ_1 is surjective, $\alpha = \phi_1(x/s)$ for some $x/s \in S^{-1}L$, and $\phi_2\,\sigma(x/s) = 0$. Since ϕ_2 is injective, $\rho(x)/s = \sigma(x/s) = 0$. Hence, for some $u \in S$, $\rho(ux) = u\rho(x) = 0$. Since ρ is injective, $ux = 0$. Thus $x/s = ux/us = 0$. Therefore, $\alpha = \phi_1(x/s) = 0$, and we conclude that $\rho \otimes i$ is injective.

If S is the set of all regular elements of R, then S is a multiplicative system in R. The ring $S^{-1}R$ is called the **total quotient ring** of R. In this case, the homomorphism η of Proposition 3.1 is injective, and we shall identify each element of R with its image in $S^{-1}R$ under η. Thus R is a subring of its total quotient ring. Let T be a multiplicatively closed subset of S and let ϕ be the homomorphism from R into $T^{-1}R$ given by $\phi(a) = at/t$, $t \in T$. Since $\eta(t)$ is a unit in $S^{-1}R$ for all $t \in T$, it follows from Proposition 3.1 that there is a unique homomorphism $\psi: T^{-1}R \to S^{-1}R$ such that the diagram

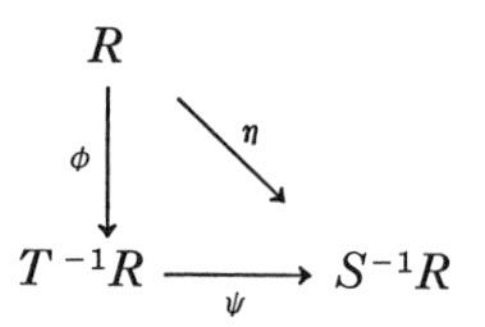

is commutative; in fact $\psi(a/t) = a/t$. It follows that ψ is injective, and we shall identify each element of $T^{-1}R$ with its image in $S^{-1}R$ under ψ.

If R has more than one element and no proper zero-divisors, then R is called an **integral domain**. In this case, the total quotient ring of R is a field, called the **quotient field** of R.

2 EXTENSION AND CONTRACTION OF IDEALS

In this section we develop the machinery by which we can compare the ideal structures of a ring R and one of its rings of quotients $S^{-1}R$.

3.4. Definition. *Let R be a ring and let S be a multiplicative system in R. Let η be the homomorphism $\eta: R \to S^{-1}R$ of Proposition* 3.1.

(1) *If A is an ideal of R, denote by $S^{-1}A$ the ideal of $S^{-1}R$ generated by $\eta(A)$. The ideal $S^{-1}A$ is the* **extension** *of A to $S^{-1}R$.*

(2) *If A' is an ideal of $S^{-1}R$, denote by $A' \cap R$ the complete inverse image of A' under η. The ideal $A' \cap R$ is the* **contraction** *of A' to R.*

To reiterate, if A is an ideal of R, $S^{-1}A$ is the set of all finite sums $\sum \eta(a_i)r_i$ where $a_i \in A$, $r_i \in S^{-1}R$. In fact, one can easily show that

$$S^{-1}A = \{a/s \mid a \in A, s \in S\}.$$

On the other hand, if A' is an ideal of $S^{-1}R$, then $A' \cap R = \eta^{-1}(A' \cap \eta(R))$. Beware of the notation $A' \cap R$ for the contraction of A'; for unless η is injective, R is not a subset of $S^{-1}R$, so interpreting $A' \cap R$ as the intersection of two sets is nonsense.

3.5. Proposition. *If A is an ideal of R, then $S^{-1}A \neq S^{-1}R$ if and only if $A \cap S$ is empty.*

Proof. If $s \in A \cap S$, then $\eta(s) \in S^{-1}A$. Hence since $\eta(s)$ is a unit in $S^{-1}R$ this means that $S^{-1}A = S^{-1}R$. Conversely, suppose $S^{-1}A = S^{-1}R$. Then $1 \in S^{-1}A$, so there exist $a \in A$ and $s \in \mathrm{S}$ such that $1 = a/s$. Hence since $1 = t/t$ for some $t \in S$, there exist $u \in S$ such that $u(st - ta) = 0$; that is, $ust = uta$ which is an element of $A \cap S$. Thus $A \cap N$ is not empty.

3.6. Proposition. *If A' is an ideal of $S^{-1}R$, then*

$$S^{-1}(A' \cap R) = A'.$$

Proof. The ideal $S^{-1}(A' \cap R)$ is the ideal of $S^{-1}R$ generated by $\eta(A' \cap R) = \eta[\eta^{-1}(A' \cap \eta(R))] \subseteq A'$. Thus $S^{-1}(A' \cap R) \subseteq A'$. On the other hand, if $a/s \in A'$ for some $a \in R$, $s \in S$, then $as/s = (s^2/s)(a/s) \in A' \cap \eta(R)$ and consequently $a \in \eta^{-1}(A' \cap \eta(R))$. Therefore $a/s = (as/s)(1/s)$ is in the ideal of $S^{-1}R$ generated by $\eta[\eta^{-1}(A' \cap \eta(R))] = \eta(A' \cap R)$; that is, $a/s \in S^{-1}(A' \cap R)$.

3.7. Proposition. *If R is Noetherian, then $S^{-1}R$ is Noetherian.*

Proof. Let $A_1' \subseteq A_2' \subseteq A_3' \subseteq \cdots$ be an ascending chain of ideals of $S^{-1}R$. Then $A_1' \cap R \subseteq A_2' \cap R \subseteq A_3' \cap R \subseteq \cdots$ and so there is an n_0 such that $A_n' \cap R = A'_{n0} \cap R$ for all $n \geq n_0$. Then, by Proposition 3.6, $A_n' = A'_{n0}$ for all $n \geq n_0$.

3.8. Proposition. *Let $\{A_\alpha' \mid \alpha \in I\}$ be an arbitrary family of ideals of $S^{-1}R$. Then*

$$\Big(\bigcap_{\alpha \in I} A_\alpha'\Big) \cap R = \bigcap_{\alpha \in I} (A_\alpha' \cap R).$$

Proof. We have

$$\begin{aligned}\Big(\bigcap_{\alpha \in I} A_\alpha'\Big) \cap R &= \eta^{-1}\Big(\Big(\bigcap_{\alpha \in I} A_\alpha'\Big) \cap \eta(R)\Big) \\ &= \eta^{-1}\Big(\bigcap_{\alpha \in I} (A_\alpha' \cap \eta(R))\Big) \\ &= \bigcap_{\alpha \in I} \eta^{-1}(A' \cap \eta(R)) = \bigcap_{\alpha \in I} (A' \cap R).\end{aligned}$$

In contrast to the result of Proposition 3.6, if A is an ideal of R, it can happen that $A \subset S^{-1}A \cap R$, even when $S^{-1}A \neq S^{-1}R$. We shall now examine the process of extending and contracting ideals, in that order. First we consider prime and primary ideals. Recall that an ideal Q of R is primary if and only if $ab \in Q$ and $b \notin Q$, where $a, b \in R$, imply that $a^n \in Q$ for some positive integer n.

3.9. Proposition. *Let Q be a primary ideal of R and let $P = \text{Rad}(Q)$. Assume $Q \cap S$ is empty. Then $P \cap S$ is empty, $S^{-1}P$ is a prime ideal of $S^{-1}R$, and $S^{-1}Q$ is $S^{-1}P$-primary. Furthermore $S^{-1}P \cap R = P$ and $S^{-1}Q \cap R = Q$.*

Proof. If $P \cap S$ is not empty, and if $s \in P \cap S$, then for some positive integer n we have $s^n \in Q \cap S$, contrary to assumption. To show that $S^{-1}Q$ is $S^{-1}P$-primary, we shall verify that the conditions of Exercise 6(d) of Chapter II hold; this will have as a consequence that $S^{-1}P$ is prime. We have $S^{-1}Q \subseteq S^{-1}P$. Let $a/s \in S^{-1}P$ where $a \in P$ and $s \in S$. There is a positive integer n such that $a^n \in Q$. Then $(a/s)^n = a^n/s^n \in S^{-1}Q$. Hence $S^{-1}P \subseteq \text{Rad}(S^{-1}Q)$. Now consider elements $a/s,\ b/t \in S^{-1}R$ such that $(a/s)(b/t) = ab/st \in S^{-1}Q$, and suppose $a/s \notin S^{-1}Q$. For some $c \in Q$ and $u \in S$ we have $ab/st = c/u$, so there is an element $v \in S$ such that $v(uab - stc) = 0$. Then $avub = vstc \in Q$, but $a \notin Q$, and consequently $vub \in P$. Thus $b/t = vub/vut \in S^{-1}P$. Therefore $S^{-1}Q$ is $S^{-1}P$-primary.

We certainly have $Q \subseteq S^{-1}Q \cap R$. Let $a \in S^{-1}Q \cap R$. Then $\eta(a) = as/s \in S^{-1}Q$ and so $as/s = b/t$ where $b \in Q$ and $t \in S$. For some $u \in S$, $u(ast - bs) = 0$. Thus $aust = ubs \in Q$, but $ust \notin P$ since $ust \in S$. Hence $a \in Q$; thus we have $Q = S^{-1}Q \cap R$. Since P is P-primary, we conclude also that $S^{-1}P \cap R = P$.

3.10. Corollary. *Let P be a prime ideal of R such that $P \cap S$ is empty. Then there is a one-to-one order preserving correspondence between the P-primary ideals of R and the $S^{-1}P$-primary ideals of $S^{-1}R$. This correspondence is $Q \leftrightarrow S^{-1}Q$; the inverse correspondence is $Q' \leftrightarrow Q' \cap R$.*

Proof. By Proposition 3.9, the mapping $Q \mapsto S^{-1}Q$ is one-to-one from the set of P-primary ideals of R into the set of $S^{-1}P$-primary ideals of $S^{-1}R$. Let Q' be a $S^{-1}P$-primary ideal of $S^{-1}R$. By Proposition 3.6, $Q' = S^{-1}(Q' \cap R)$, and $Q' \cap R$ is a P-primary ideal of R by Proposition 3.9 and Exercise 6(g) of Chapter II.

3.11. Corollary. *There is a one-to-one correspondence, $P \leftrightarrow S^{-1}P$, between the prime ideals of R which do not meet S and the proper prime ideals of $S^{-1}R$. This correspondence is order preserving.*

The proof of this corollary is similar to that of Corollary 3.10 and is left to the reader.

3.12. Proposition. *If A is an ideal of R and A_S is the S-component of A, then*

$$S^{-1}A \cap R = A_S.$$

Proof. Suppose $a \in A_S$; then $as \in A$ for some $s \in S$. Then $\eta(a)\eta(s) \in \eta(A)$ and since $\eta(s)$ is a unit in $S^{-1}R$, $\eta(a)$ is in the ideal of $S^{-1}R$ generated by $\eta(A)$; that is, $\eta(a) \in S^{-1}A$. Therefore $a \in S^{-1}A \cap R$. Conversely, suppose $a \in S^{-1}A \cap R$. Then $\eta(a) \in S^{-1}A$ and so $as/s = b/s$ for some $b \in A$, $s \in S$. Hence there exists $u \in S$ such that $u(as^2 - bs) = 0$. From this we see that $uas^2 = ubs \in A$. Since $us^2 \in S$ this proves that $a \in A_S$.

If P is a proper prime ideal of R, then $S = R\backslash P$ is a multiplicative system in R. In this case we denote $S^{-1}R$ by R_P and call it the **quotient ring of R with respect to P**. If A is an ideal of R, we denote the extension of A to R_P by AR_P rather than $S^{-1}A$. However for an ideal A' of R_P, the contraction of A' to R will still be denoted by $A' \cap R$. The ideal PR_P of R_P is a proper prime ideal of R_P. If P' is a proper prime ideal of R_P, then by Corollary 3.11, $P' = P_1R_P$ for some prime ideal P_1 of P with $P_1 \cap S$ empty. Then $P_1 \subseteq P$ and consequently $P' \subseteq PR_P$. It follows that *PR_P is the unique maximal ideal of R_P*. Since every nonunit of R_P is contained in some maximal ideal, *every element of R_P not in PR_P is a unit in R_P*. A ring in which there is only one maximal ideal is called a **local ring**. Hence R_P is often called the **localization of R at P**.

3.13. Proposition. *Let R be a ring and let A and B be ideals of R. Then $A = B$ if and only if $AR_P = BR_P$ for every maximal ideal P of R.*

Proof. We only need to prove the sufficiency of the condition. Let $a \in A$. Then $a/1 \in BR_P$ for every maximal ideal P of R. Hence, by the usual argument, for every maximal ideal P of R, there is an element $r_P \in R \backslash P$ such that $r_P a \in B$. Let C be the ideal of R generated by $\{r_P \mid P \text{ a maximal ideal of } R\}$. If $C \neq R$, then C is contained in some maximal ideal of R (Proposition 2.12). Since this is not true, we must have $C = R$. Thus $1 \in C$ and so there are maximal ideals $P_1, \ldots, P_k$ of R such that $1 = x_1 r_{P_1} + \cdots + x_k r_{P_k}$, where $x_1, \ldots, x_k \in R$. Then $a = x_1 r_{P_1} a + \cdots + x_k r_{P_k} a \in B$. Thus $A \subseteq B$, and we show in a similar manner that $B \subseteq A$. Therefore $A = B$.

Let P be a proper prime ideal of R. If n is a positive integer then P^n is not necessarily P-primary [see Exercise 12(d) of Chapter II]. However, every minimal prime divisor of P^n must contain P, so that P is the only minimal prime divisor of P^n. Denote by $P^{(n)}$ the component of P^n determined by $R \backslash P$. The ideal $P^{(n)}$ is called the ***n*th symbolic power** of P. By Proposition 3.12, $P^{(n)} = P^n R_P \cap R$. If P^n has a primary decomposition, then it follows from Proposition 2.30 and Theorem 2.32 that $P^{(n)}$ is the primary ideal with radical P which occurs in every reduced primary decomposition of P^n.

3.14. Proposition. *Let P be a proper prime ideal of a Noetherian ring R and let $S = R \backslash P$. Then*

$$\bigcap_{n=1}^{\infty} P^{(n)} = 0_S.$$

Proof. By Exercise 4(a), $P^n R_P = (PR_P)^n$ for each positive integer n, and

$$\bigcap_{n=1}^{\infty} (PR_P)^n = 0$$

by Corollary 2.24. Therefore

$$\left(\bigcap_{n=1}^{\infty} P^n R_P\right) \cap R = 0_S$$

by Proposition 3.1(b). Hence, by Proposition 3.8,

$$\bigcap_{n=1}^{\infty} P^{(n)} = \bigcap_{n=1}^{\infty} (P^n R_P \cap R) \\ = 0_S.$$

3 PROPERTIES OF RINGS OF QUOTIENTS

In this section we shall obtain several properties of rings of quotients which will prove to be useful. In particular, we shall show that the formation of rings of quotients is transitive and that the formation of rings of quotients commutes with the formation of residue class rings. These facts are consequences of the following universal property of rings of quotients. Let R be a ring and S a multiplicative system in R; let $\eta: R \to S^{-1}R$ be the homomorphism of Proposition 3.1.

3.15. Proposition. *Let R' be a ring and let $\mu: R \to R'$ be a homomorphism such that:*

(i) *elements of $\mu(S)$ are units in R';*
(ii) *the kernel of μ is 0_S; and*
(iii) *every element of R' can be expressed in the form $\mu(x)\mu(s)^{-1}$ for some $x \in R$, $s \in S$.*

Then there is a unique isomorphism $\psi: S^{-1}R \to R'$ such that the diagram

$$\begin{array}{ccc} R & & \\ {\scriptstyle\eta}\downarrow & \searrow{\scriptstyle\mu} & \\ S^{-1}R & \xrightarrow[\psi]{} & R' \end{array}$$

is commutative.

Proof. By Proposition 3.1(d), there is a unique homomorphism $\psi: S^{-1}R \to R'$ such that the diagram is commutative; explicitly, $\psi(a/s) = \mu(a)\mu(s)^{-1}$. It follows from (ii) and (iii) that ψ is injective and surjective.

3.16. Theorem. *Let T be a multiplicative system in R with $S \subseteq T$. Then*

$$\eta(T)^{-1}S^{-1}R \cong T^{-1}R.$$

Proof. Let $\eta' : R \to T^{-1}R$ be the homomorphism of Proposition 3.1. Since every element of $\eta'(S)$ is a unit in $T^{-1}R$ there is, by Proposition 3.1(d), a homomorphism $\mu: S^{-1}R \to T^{-1}R$ such that the diagram

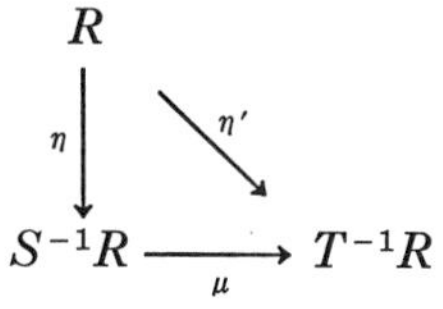

$$\begin{array}{ccc} R & & \\ \downarrow{\scriptstyle \eta} & \searrow{\scriptstyle \eta'} & \\ S^{-1}R & \xrightarrow{\mu} & T^{-1}R \end{array}$$

is commutative. We shall show that $\eta(T)$ and μ satisfy the conditions of Proposition 3.15.

Clearly, elements of $\mu(\eta(T)) = \eta'(T)$ are units in $T^{-1}R$. Suppose $\mu(a/s) = 0$ for some $a \in R$ and $s \in S$, that is, $\eta'(a)\eta'(s)^{-1} = 0$. Then there exists an element $t \in T$ such that $at = 0$; hence

$$(a/s)\eta(t) = \eta(a)\eta(s)^{-1}\eta(t) = \eta(at)\eta(s)^{-1} = 0.$$

Thus $a/s \in 0_{\eta(T)}$. Conversely, if $a/s \in 0_{\eta(T)}$, then $\eta(at) = 0$ for some $t \in T$; hence there is an element $u \in S$ such that $aut = 0$. This implies that $a \in 0_T$, so that $\mu(a/s) = \eta'(a)\eta'(s)^{-1} = 0$. Therefore, the kernel of μ is $0_{\eta(T)}$.

Finally, if $a/t \in T^{-1}R$, where $a \in R$ and $t \in T$, then $a/t = \eta'(a)\eta'(t)^{-1} = \mu(\eta(a))\mu(\eta(t))^{-1}$, which is the form required in (iii) of Proposition 3.15.

3.17. Theorem. *Let A be an ideal of R such that $A \cap S$ is empty. Let $\phi: R \to R/A$ be the canonical homomorphism. Then*

$$\phi(S)^{-1}(R/A) \cong S^{-1}R/S^{-1}A$$

Proof. Note that $\phi(S)$ is a multiplicative system in R/A. Let $\phi' : S^{-1}R \to S^{-1}R/S^{-1}A$ be the canonical homomorphism. Define a mapping $\mu: R/A \to S^{-1}R/S^{-1}A$ by $\mu(a + A) = \eta(a) + S^{-1}A$; η is well-defined since $A \subseteq S^{-1}A \cap R$, it is a homomorphism, and the diagram

$$\begin{array}{ccc} R & \xrightarrow{\eta} & S^{-1}R \\ \downarrow{\scriptstyle \phi} & & \downarrow{\scriptstyle \phi'} \\ R/A & \xrightarrow[\mu]{} & S^{-1}R/S^{-1}A \end{array}$$

is commutative. We shall show that $\phi(S)$ and μ satisfy the conditions of Proposition 3.15.

Elements of $\eta(S)$ are units in $S^{-1}R$, and since $\eta(S) \cap S^{-1}A$ is empty, the elements of $\phi'\eta(S)$ are units in $S^{-1}R/S^{-1}A$. Hence, the elements of $\mu(\phi(S))$ are units in $S^{-1}R/S^{-1}A$.

Now we determine the kernel of μ. The kernel of ϕ' is $S^{-1}A$ and hence the kernel of $\mu\phi = \phi'\eta$ is $\eta^{-1}(S^{-1}A \cap \eta(R)) = S^{-1}A \cap R = A_S$, by Proposition 3.12. Hence the kernel of μ is A_S/A. But A_S/A is precisely the $\phi(S)$-component of the zero ideal in R/A.

Finally, every element of $S^{-1}R/S^{-1}A$ can be written in the form

$$\phi'(\eta(a)\eta(s)^{-1}) = \phi'\eta(a)/\phi'\eta(s)^{-1} = \mu\phi(a)\mu\phi(s)^{-1} = \mu(a+A)\mu(s+A)^{-1},$$

so that (iii) of Proposition 3.15 holds.

3.18. Definition. *The set*

$$\bar{S} = \{x \mid x \in R \quad \text{and} \quad \eta(x) \quad \text{is a unit in} \quad S^{-1}R\}$$

is the **saturation** *of* S.

It is clear that $\bar{S}$ is a multiplicative system in R, and that $S \subseteq \bar{S}$.

3.19. Proposition.

$$\bar{S} = \{x \mid x \in R \quad \text{and} \quad xy \in S \quad \text{for some} \quad y \in R\}.$$

Proof. If $xy = s \in S$ then $\eta(x)\eta(y) = \eta(s)$ is a unit in $S^{-1}R$, and so $\eta(x)$ is a unit in $S^{-1}R$. Conversely, suppose $\eta(x) = xs/s$ is a unit in $S^{-1}R$; here $s \in S$. If the inverse of xs/s is z/t, where $z \in R$ and $t \in S$, then $xsz/st = (xs/s)(z/t) = t/t$, so that for some $u \in S$ we have $x(sztu) = ust^2 \in S$.

3.20. Theorem. $S^{-1}R \simeq \bar{S}^{-1}R$; *explicitly, there is an isomorphism from* $S^{-1}R$ *onto* $\bar{S}^{-1}R$ *which maps* $a/s \in S^{-1}R$, *where* $s \in S$, *onto* $a/s \in \bar{S}^{-1}R$.

Proof. Let $\mu: R \to \bar{S}^{-1}R$ be the homomorphism defined by $\mu(a) = at/t$, where $t \in \bar{S}$. Since $S \subseteq \bar{S}$, every element of $\mu(S)$ is a unit in $\bar{S}^{-1}R$. The kernel of μ is $0_{\bar{S}}$, which contains 0_S. If $a \in 0_{\bar{S}}$ then $ax = 0$ for some $x \in \bar{S}$. Then, if $xy \in S$, we have $a(xy) = 0$. Hence $a \in 0_S$, and we conclude that Ker $\mu = 0_S$. Consider an element $a/t \in \bar{S}^{-1}R$, where $a \in R$ and $t \in \bar{S}$. If $tz \in S$, then $a/t = az/tz = \mu(az)\mu(tz)^{-1}$. We have verified that S and μ satisfy (i)–(iii) of Proposition 3.15. Therefore, there is a unique isomorphism $\psi: S^{-1}R \to \bar{S}^{-1}R$ such that the diagram

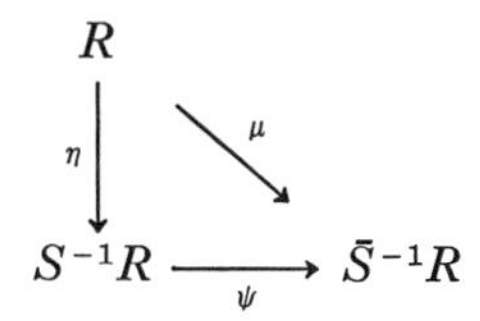

is commutative. For $a \in R$ and $s \in S$ we have

$$\begin{aligned}\psi(a/s) &= \psi(as/s)\psi(s/s^2)\\ &= \psi(as/s)\psi(s^2/s)^{-1}\\ &= \psi\eta(a)\psi\eta(s)^{-1}\\ &= \mu(a)\mu(s)^{-1}\\ &= a/s.\end{aligned}$$

If S consists entirely of regular elements of R, then we have agreed to regard $S^{-1}R$ as a subring of the total quotient ring of R (see the end of Section 1). If, in Theorem 3.16, T also consists entirely of regular elements, it follows that we actually have $\eta(T)^{-1}S^{-1}R = T^{-1}R$. In fact, $\eta(T) = T$, so that

$$T^{-1}S^{-1}R = T^{-1}R.$$

Furthermore, the saturation $\bar{S}$ of S consists entirely of regular elements of R, and

$$S^{-1}R = \bar{S}^{-1}R.$$

EXERCISES

1. Quotient modules.
Let R be a ring, S a multiplicative system in R, and M an R-module.

(a) Show that if we set $(a/s)(x/t) = ax/st$, where $a/s \in S^{-1}R$ and $x/t \in S^{-1}M$, then $S^{-1}M$ is an $S^{-1}R$-module.
(b) Show that if $M \otimes_R S^{-1}R$ is made into an $S^{-1}R$-module by the method of Exercise 7(b) of Chapter I, then the isomorphism of Proposition 3.2, is an isomorphism of $S^{-1}R$-modules.

2. Isomorphisms of modules of quotients.
Let M and N be R-modules and let S be a multiplicative system in R. Verify the following isomorphisms of $S^{-1}R$-modules.
(a) $S^{-1}(M \oplus N) \cong S^{-1}M \oplus S^{-1}N$.
(b) $S^{-1}M \otimes_R N \cong S^{-1}(M \otimes_R N) \cong S^{-1}M \otimes_{S^{-1}R} S^{-1}N \cong M \otimes_R S^{-1}N$.

3. Quotient module mappings.
Let M and N be R-modules and let $f: M \to N$ be a homomorphism. Let S be a multiplicative system in R. Define $S^{-1}f: S^{-1}M \to S^{-1}N$ by $S^{-1}f(x/s) = f(x)/s$ for all $x \in M$, $s \in S$.
(a) Prove that $S^{-1}f$ is a $S^{-1}R$-module homomorphism.
(b) If L is another R-module such that $0 \to L \xrightarrow{g} M \xrightarrow{f} N \to 0$ is an exact sequence of R-modules, prove that the sequence

$$0 \longrightarrow S^{-1}L \xrightarrow{S^{-1}g} S^{-1}M \xrightarrow{S^{-1}f} S^{-1}N \longrightarrow 0$$

is exact.
(c) Show that if M is a flat R-module, then $S^{-1}M$ is a flat $S^{-1}R$-module.

4. Properties of extension of ideals.
Let A, B be ideals of R and let S be a multiplicative system in R. Prove the following formulas regarding extension of ideals.
(a) $S^{-1}(AB) = (S^{-1}A)(S^{-1}B)$.
(b) $S^{-1}(A \cap B) = S^{-1}A \cap S^{-1}B$.
(c) $S^{-1}(A + B) = S^{-1}A + S^{-1}B$.
(d) $S^{-1}(A : B) = S^{-1}A : S^{-1}B$ if B is finitely generated, or if A and B are contractions of ideals of $S^{-1}R$ [see also Exercise 2(e) of Chapter IX].
(e) $S^{-1}(\text{Rad } A) = \text{Rad}(S^{-1}A)$.

5. Localization properties.
(a) Show that if R is an integral domain and if P runs through the maximal ideals of R, then $R = \bigcap R_P$.

(b) Show that if A is an ideal of any ring R and if P runs through the maximal ideals of R, then $A = \bigcap(AR_P \cap R)$.

(c) If A is an ideal of a ring R such that $\mathrm{Rad}(A)$ is a prime ideal Q of R, and if AR_P is QR_P-primary for each maximal ideal P containing A, prove that A is Q-primary.

(d) Show that the assertion of (c) is no longer true if we drop the assumption that $\mathrm{Rad}(A)$ is prime.

(e) Let x be an element of a ring R such that x is contained in only a finite number of maximal ideals $M_1, \ldots, M_n$ of R. Prove that if A is an ideal of R containing x and such that AR_{M_i} is finitely generated for $i = 1, \ldots, n$, then A is finitely generated.

(f) Use (e) to prove that if R is a ring such that R_M is Noetherian for each maximal ideal M of R and each nonzero element of R is contained in only a finite number of maximal ideals of R, then R is Noetherian.

(g) If, in a ring R, each ideal with prime radical is a power of its radical, prove that this property holds for R_P for each proper prime ideal P of R.

6. Saturation of a multiplicative system.
Let R be a ring and let S be a multiplicative system in R. Let $\bar{S}$ be its saturation.

(a) Prove that $R \backslash \bar{S}$ is the union of the prime ideals of R which do not meet S.

(b) Let $\{M_\alpha \mid \alpha \in I\}$ be the set of maximal ideals of $S^{-1}R$ and let $P_\alpha = M_\alpha \cap R$. Prove that $S^{-1}R = T^{-1}R$ where $T = R \backslash \bigcup P_\alpha$.

(c) If R is an integral domain and P is a prime ideal of R not meeting S, show that $R_P = S^{-1}R_{S^{-1}P}$.

(d) If R is an integral domain, show that every ring of quotients $S^{-1}R$ of R is an intersection of localizations of R; in particular, $S^{-1}R = \bigcap R_{P_\alpha}$ where $\{P_\alpha\}$ is the set of ideals of R maximal with respect to not meeting S.

7. Transitivity of quotient ring formation.

(a) Let R be a ring and let P and Q be proper prime ideals of R such that $P \subseteq Q$. Prove that

$$R_P \cong (R_Q)_{PR_Q}.$$

(b) Let S be a multiplicative system in R. Let S' be a multiplicative system in $S^{-1}R$ and let S'' be the multiplicative system in R generated by

$$S \cup \{s'' \mid s'' \in R \text{ and } s''/s \in S' \text{ for some } s \in S\}.$$

Show that $S''^{-1}R \cong S'^{-1}(S^{-1}R)$.

(c) The formulation of S'' in (b) is important. Show that the following result is *false*: Let S be a multiplicative system in R, S' a multiplicatively closed subset in $S^{-1}R$. Let S'' be the multiplicative system in R generated by $S \cup \{s'' \in R \mid \eta(s'') \in S'\}$. Then $S''^{-1}R \cong S'^{-1}(S^{-1}R)$. [Hint. Take $R = Z \oplus Z$, $S = \{(2, 2)^n \mid n \geq 0\}$, and $S' = \left\{\frac{(1,0)}{(2,2)^n} \mid n \geq 0\right\}$.]

8. A tensor product.

Let R be a ring, A an ideal of R, and M an R-module. Make $R/A \otimes_R M$ into an R-module by setting $a((b + A) \otimes x) = (ab + A) \otimes x$ [see Exercise 7(b) of Chapter I].

(a) Show that there is a homomorphism $\phi: R/A \otimes_R M \to M/AM$ such that $\phi((a + A) \otimes x) = ax + AM$ for all $a \in R$ and $x \in M$.

(b) Show that there is a homomorphism $\psi: M/AM \to R/A \otimes_R M$ such that $\psi(x + AM) = (1 + A) \otimes x$ for all $x \in M$. Show that $\phi\psi$ and $\psi\phi$ are identity isomorphisms, so that ϕ and ψ are group isomorphisms. Verify that they are, in fact, isomorphisms of R-modules.

(c) Let S be a multiplicative system in R. Treating R/A as an R-module, show that there is an isomorphism of R-modules $S^{-1}(R/A) \cong S^{-1}R/S^{-1}A$. (Compare with Theorem 3.17. Are the isomorphisms the same?)

9. The prime divisors of an ideal.

(a) Let R be a ring and A an ideal of R. Let

$$U(A) = \{b \mid b \in R \text{ and } b + A \text{ is a regular element of } R/A\}.$$

Show that $U(A)$ is multiplicatively closed and that $U(A) \cap A$ is empty. An ideal P of R which is maximal with respect to the properties $A \subseteq P$ and $U(A) \cap P$ is empty is called a **maximal prime divisor** of A.

(b) A prime ideal P of R is called a **prime divisor** of A if there is a multiplicatively closed set S in R such that $A \cap S$ is empty and $S^{-1}P$ is a maximal prime divisor of $S^{-1}A$. Show that a maximal prime divisor of A is a maximal element in the set of all prime divisors of A.

(c) Let P be a minimal prime divisor of A. Show that $AR_P \cap R$ is P-primary; it is called the **isolated P-primary component** of A. Show that if A has a primary decomposition, then the isolated P-primary component of A is the P-primary ideal occurring in every reduced primary decomposition of A.

(d) Show that a minimal prime divisor of A is a minimal element in the set of prime divisors of A.

(e) Show that if A has a primary decomposition, then the prime divisors of A, as defined in the text after Proposition 2.25, are the same as the prime divisors of A, as defined in this exercise.

10. Families of quotient rings.

Let $\{S_\alpha \mid \alpha \in I\}$ be a family of multiplicative systems in a ring R such that for each maximal ideal P of R there is an $\alpha \in I$ such that $P \cap S_\alpha$ is empty.

(a) Show that if A and B are ideals of R then $A = B$ if and only if $S_\alpha^{-1}A = S_\alpha^{-1}B$ for all $\alpha \in I$.

(b) Show that if, for each $\alpha \in I$, S_α consists entirely of regular elements, then $R = \bigcap_{\alpha \in I} S_\alpha^{-1}R$.

(c) Let M and N be R-modules and $f: M \to N$ a homomorphism. Show that f is surjective (respectively, injective, bijective, the zero homomorphism) if and only if the same is true for $S_\alpha^{-1}f$ for all $\alpha \in I$.

(d) If P be a proper prime ideal of R, let $S(P)$ be the set of regular elements of $R \backslash P$. Then $S(P)$ is a multiplicative system in R, and we denote $S(P)^{-1}R$ by $R_{S(P)}$. Show that if A and B are ideals of R, then $A = B$ if and only if $AR_{S(P)} = BR_{S(P)}$ for every regular maximal ideal P of R. Show that if P runs over the set of regular maximal ideals of R, then $R = \bigcap R_{S(P)}$.

11. The support of a module.

Let R be a Noetherian ring and M a finitely generated R-module. The **support** of M is the set $\text{Supp}(M) = \{P \mid P$ is a proper prime ideal of R and $S^{-1}M \neq 0$ where $S = R \backslash P\}$.

(a) Show that if $M \neq 0$, then $\text{Supp}(M)$ is not empty.

(b) Show that a proper prime ideal of R is in $\text{Supp}(M)$ if and only if it contains an element of $\text{Ass}(M)$. Conclude that $\text{Ass}(M) \subseteq \text{Supp}(M)$, and show that these two sets have the same minimal elements.

(c) Show that there is a chain of submodules of M, $M = M_0 \supset M_1 \supset \cdots \supset M_n = 0$, such that for $i = 0, \ldots, n-1$, $M_i/M_{i+1} \cong R/P_i$ for some prime ideal P_i of R. Show that $\text{Ass}(M) \subseteq \{P_0, \ldots, P_{n-1}\} \subseteq \text{Supp}(M)$, and that the minimal elements of these sets coincide. Show that these minimal elements are precisely the minimal ones among the proper prime ideals of R which contain $\text{Ann}(M) = \bigcap_{x \in M} \text{Ann}(x)$.

12. $\text{Ass}(M)$ and primary submodules.

Let R be a Noetherian ring and M a finitely generated R-module. Let A be an ideal of R.

(a) Show that $A^n M = 0$ for some positive integer n if and only if $A \subseteq P$ for every $P \in \text{Ass}(M)$.

(b) Let N be a submodule of M. Show that N is primary submodule if and only if $\text{Ass}(M/N)$ consists of a single prime ideal P, and in this case $P = \text{Rad}(N)$.

(c) Show that if N is a submodule of M, then $\text{Ass}(M/N)$ is the set of prime divisors of N.

13. Further properties of $\text{Ass}(M)$.

(a) Let R be a Noetherian ring and M a finitely generated R-module. Let S be a multiplicatively closed set in R and let $\mathcal{T}$ be the prime ideals of R which do not intersect S. Show that the mapping $P \mapsto S^{-1}P$ is a one-to-one mapping from $\text{Ass}(M) \cap \mathcal{T}$ into $\text{Ass}(S^{-1}M)$ ($S^{-1}M$ being considered as an $S^{-1}R$-module).

(b) Show that if $P \in \mathcal{T}$ and $S^{-1}P \in \text{Ass}(S^{-1}M)$, then $P \in \text{Ass}(M)$. Show that the mapping in part (a) is onto.

(c) Show that the mapping $x \mapsto sx/s$ from M into $S^{-1}M$, where

$s \in S$, is independent of s, and is a homomorphism; let N be its kernel. Let $\mathscr{T}' = \text{Ass}(M) \cap \mathscr{T}$. Show that N is the unique submodule of M such that $\text{Ass}(M/N) = \mathscr{T}'$ and $\text{Ass}(N) = \text{Ass}(M) \backslash \mathscr{T}'$.

14. The ring $R(X)$.

Let R be a ring and X an indeterminate. Let S be the set of all $a_0 + a_1 X + \cdots + a_n X^n \in R[X]$ such that $R = (a_0, a_1, \ldots, a_n)$.

(a) Show that S is a multiplicative system in $R[X]$, consisting entirely of regular elements. Denote the ring $S^{-1}R[X]$ by $R(X)$.

(b) Show that if A is an ideal of R, then $R(X)/AR(X) \cong (R/A)(X)$.

(c) Let Q be a P-primary ideal of R. Show that $QR(X)$ is a $PR(X)$-primary ideal of $R(X)$, and that $QR(X) \cap R = Q$.

(d) Let $A_1, \ldots, A_k$ be ideals of R. Show that

$$(A_1 \cap \cdots \cap A_k)R(X) = A_1 R(X) \cap \cdots \cap A_k R(X).$$

(e) Show that an ideal M' of $R(X)$ is maximal if and only if $M' = MR(X)$ for some maximal ideal M of R.

(f) Let A be a proper ideal of R. Show that $AR(X)$ is a proper ideal of $R(X)$.

(g) Show that $R(X)$ is a flat R-module.

15. Totally ordered sets of ideals.

Let T be an overring of a ring R and let M be a proper prime ideal of T. Let $P = M \cap R$ and assume that the set of ideals of R_P is totally ordered. Prove that the set of ideals of T_M is totally ordered. [Hint. T_M is an overring of $h(R_P)$, where $h: R_P \to T_M$ is given by $h(x/y) = x/y$.]

16. Symbolic powers of primary ideals.

Let P be a proper prime ideal of a ring R and let Q be a P-primary ideal of R. If n is a positive integer then $Q^{(n)} = Q^n R_P \cap R$ is called the ***n*th symbolic power** of Q.

(a) Show that $Q^{(n)}$ is P-primary and that if Q^n is primary then $Q^{(n)} = Q^n$.

(b) Show that P is the unique minimal prime divisor of Q^n.

(c) Let m and n be positive integers. Show that P is the unique minimal prime divisor of $Q^{(m)}Q^{(n)}$, and that the isolated P-primary component of $Q^{(m)}Q^{(n)}$ is $Q^{(m+n)}$.

17. Rings which are almost Noetherian.
A ring R is said to be **almost Noetherian** if the localization R_M is Noetherian for each maximal ideal M of R.

(a) Show that if R is almost Noetherian, then (ACC) holds in the set of prime ideals of R. [Hint. Let $P_1 \subseteq P_2 \subseteq \cdots$ be an ascending chain of prime ideals of R and set $P = \bigcup P_i$.]

(b) Suppose that R is almost Noetherian and that each finitely generated ideal of R has only finitely many minimal prime divisors. Show that if P is a proper prime ideal of R, then P is the unique minimal prime divisor of some finitely generated ideal of R.

(c) Suppose that R is almost Noetherian. Show that R is Noetherian if and only if each finitely generated ideal A of R has only a finite number of minimal prime divisors and Rad(A) is finitely generated.

(d) Prove that for an almost Noetherian ring R the following statements are equivalent:

(1) R is Noetherian;

(2) Each finitely generated ideal of R is an intersection of a finite number of primary ideals of R;

(3) Each finitely generated ideal of R has only a finite number of prime divisors.

[Hint. To prove that (3) implies (1), use the fact that if P is the unique minimal prime divisor of a finitely generated ideal A and if $PR_P = (x_1, \ldots, x_n)R_P$, then P is the unique minimal prime divisor of $B = A + (x_1, \ldots, x_n)$ and $BR_P = PR_P$.]

(e) Show that conditions (2) and (3) in (d) and the condition of (c) can hold in a ring R without R being Noetherian. [Hint. Consider the ring of polynomials in a countable number of indeterminates over a field.]

CHAPTER

IV

Integral Dependence

1 DEFINITION OF INTEGRAL DEPENDENCE

Let R' be a ring and let R be a subring of R'. If $a_1, \ldots, a_n \in R'$ we denote by $R[a_1, \ldots, a_n]$ the set of polynomial expressions in $a_1, \ldots, a_n$ with coefficients in R. Thus, if $X_1, \ldots, X_n$ are indeterminates, then

$$R[a_1, \ldots, a_n] = \{f(a_1, \ldots, a_n) \mid f(X_1, \ldots, X_n) \in R[X_1, \ldots, X_n]\}.$$

The mapping $f(X_1, \ldots, X_n) \mapsto f(a_1, \ldots, a_n)$ is a homomorphism from $R[X_1, \ldots, X_n]$ into R'; consequently its image $R[a_1, \ldots, a_n]$ is a subring of R'. It is clear that $R \subseteq R[a_1, \ldots, a_n]$.

4.1. Proposition. *Let R be a subring of a ring R' and let $a \in R'$. Then the following statements are equivalent:*

(1) *There are elements $b_0, b_1, \ldots, b_{n-1} \in R$ $(n \geq 1)$ such that $b_0 + b_1 a + \cdots + b_{n-1} a^{n-1} + a^n = 0$.*
(2) *$R[a]$ is a finitely generated R-module.*
(3) *There is a subring R'' of R' such that $a \in R''$ and R'' is a finitely generated R-module.*

Proof. Suppose (1) holds and let $f(X) \in R[X]$. Suppose $\deg f(X) = d > n$. If $f(X) = c_0 + c_1 X + \cdots + c_d X^d$, then

$$\begin{aligned} f(a) &= c_0 + c_1 a + \cdots + c_{d-1} a^{d-1} + c_d a^{d-n}(-b_0 - b_1 a - \cdots - b_{n-1} a^{n-1}) \\ &= c_0' + c_1' a + \cdots + c_{d-1}' a^{d-1}. \end{aligned}$$

If necessary we repeat this argument until we see finally that $f(a)$ is in the R-module generated by $1, a, \ldots, a^n$. Thus, as an R-module, $R[a] = R1 + Ra + \cdots + Ra^n$; hence (2) holds. Clearly, (2) implies (3), taking $R'' = R[a]$. Now suppose (3) holds and let $a_1, \ldots, a_n$ generate R'' as an R-module. For $i = 1, \ldots, n$,

$$aa_i = \sum_{j=1}^{n} b_{ij} a_j, \qquad b_{ij} \in R,$$

or

$$\sum_{j=1}^{n} (b_{ij} - \delta_{ij} a) a_j = 0.$$

If $d = \det[b_{ij} - \delta_{ij} a]$ then $da_j = 0$ for $j = 1, \ldots, n$ by Lemma 2.22. Hence $dc = 0$ for all $c \in R''$. With $c = 1$ we get $d = 0$. Since d is a polynomial in a with coefficients in R such that the coefficient of a^n is ± 1, we see that (1) holds.

4.2. Definition. *Let R be a subring of a ring R' and let $a \in R'$. If the equivalent statements of Proposition* 4.1 *hold for a, we say that a is* **integral** *over R. If every element of R' is integral over R we say that R' is* **integral** *over R. If the elements of R are the only elements of R' which are integral over R, we say that R is* **integrally closed in** *R'. If R is integrally closed in its total quotient ring, we say simply that R is* **integrally closed**.

4.3. Proposition. *Let R be a subring of a ring R' and let*

$$R_0 = \{a \mid a \in R' \text{ and } a \text{ is integral over } R\}.$$

Then R_0 is a subring of R' and $R \subseteq R_0$.

Proof. Clearly $R \subseteq R_0$. Let $a, b \in R_0$. Then $R[a]$ is a finitely generated R-module and $R[a, b] = R[a][b]$ is a finitely generated $R[a]$-module. Hence $R[a, b]$ is a finitely generated R-module. Since $a - b, ab \in R[a, b]$ they are integral over R; that is, $a - b$, $ab \in R_0$. Therefore, R_0 is a subring of R'.

In the notation of this proposition, R_0 is called the **integral closure** of R in R'. It follows from the next proposition that R_0 is integrally closed in R'.

4.4. Proposition. *Let R be a subring of a ring R', and R' a subring of a ring R''. If R' is integral over R and if $a \in R''$ is integral over R', then a is integral over R.*

Proof. Suppose $b_0 + b_1 a + \cdots + b_{n-1}a^{n-1} + a^n = 0$ where $b_0, b_1, \ldots, b_{n-1} \in R'$. Then a is integral over $R[b_0, \ldots, b_{n-1}]$. Therefore $R[b_0, \ldots, b_{n-1}, a]$ is a finitely generated R-module. It follows that a is integral over R.

4.5. Proposition. *Let R be a subring of R' and let S be a multiplicative system in R. Then $S^{-1}R$ may be considered as a subring of $S^{-1}R'$ and if R' is integral over R, then $S^{-1}R'$ in integral over $S^{-1}R$.*

Proof. Let 0_S and $0_S'$ be the S-components of 0 in R and R', respectively. We certainly have $0_S \subseteq 0_S' \cap R$. If $a \in 0_S' \cap R$, then $sa = 0$ for some $s \in S$, and since $S \subseteq R$ this implies that $a \in 0_S$. Thus $0_S = 0_S' \cap R$. Therefore the mapping taking $a/s \in S^{-1}R$ onto $a/s \in S^{-1}R'$ is an injective homomorphism; if we identify $a/s \in S^{-1}R$ with its image $a/s \in S^{-1}R'$, then $S^{-1}R$ may be considered as a subring of $S^{-1}R'$. Now suppose that R' is integral over R and let $a/s \in S^{-1}R'$, $a \in R'$, $s \in S$. There are elements $b_0, b_1, \ldots, b_{n-1} \in R$ such that $b_0 + b_1 a + \cdots + b_{n-1}a^{n-1} + a^n = 0$. Then

$$\begin{aligned} b_0/s^n + (b_1/s^{n-1})(a/s) + \cdots + (b_{n-1}/s)(a/s)^{n-1} + (a/s)^n \\ = (b_0 + b_1 a + \cdots + b_{n-1}a^{n-1} + a^n)/s^n = 0. \end{aligned}$$

Therefore a/s is integral over $S^{-1}R$.

2 INTEGRAL DEPENDENCE AND PRIME IDEALS

Now we are ready to prove some of the principal theorems concerning integral dependence. These theorems deal with the relations between the ideals of a ring R and the ideals of a ring R' which has R as a subring and is integral over R. If A is an ideal of R and A' is an ideal of R' such that $A = A' \cap R$, then A' is said to **lie over** A.

4.6. Theorem (The Lying-Over Theorem). *Let R be a subring of a ring R' which is integral over R. If P is a prime ideal of R, then there* *is a prime ideal P' of R' that lies over P. Moreover, if P' and P'' are prime ideals of R' that lie over P and if $P' \subseteq P''$, then $P' = P''$.*

Proof. The set of ideals A' of R' such that $A' \cap R \subseteq P$ is not empty, and it follows from Zorn's lemma that this set has a maximal element P'. Then $P' \cap R \subseteq P$. Suppose $P' \cap R \subset P$ and let $a \in P$, $a \notin P'$. Then $P' \subset P' + R'a$ and consequently, by our choice of P', $(P' + R'a) \cap R \nsubseteq P$. Hence there is an element $c \in P'$, and an $r \in R'$, such that $c + ra = b \notin P$ but $b \in R$. Let $d_0, \ldots, d_{n-1} \in R$ be such that $d_0 + d_1 r + \cdots + d_{n-1} r^{n-1} + r^n = 0$. Then

$$\begin{aligned} b^n + d_{n-1}ab^{n-1} + \cdots + d_1 a^{n-1} b + d_0 a^n \\ &= (c + ra)^n + d_{n-1}a(c + ra)^{n-1} + \cdots + d_1 a^{n-1}(c + ra) + d_0 a^n \\ &= f(c) + a^n(r^n + d_{n-1}r^{n-1} + \cdots + d_1 r + d_0) \\ &= f(c) \in P' \cap R \subseteq P; \end{aligned}$$

here $f(c)$ is a polynomial in c with coefficients in R'. Hence, since $a \in P$, we have $b^n \in P$, so $b \in P$, a contradiction. Therefore we must have $P' \cap R = P$.

Next, we show that P' is a prime ideal. Let $S = R \backslash P$; then S is a multiplicative system in R'. (If $P = R$, then $P' = R'$ which is prime, so we may assume $P \neq R$.) Let A' be an ideal of R' with $P' \subset A'$. Then $A' \cap R \nsubseteq P$ so $A' \cap R$ meets S; hence $A' \cap S$ is not empty. Thus P' is maximal in the set of ideals of R' whose intersection with S is empty. We know that such an ideal is prime (see Proposition 2.12).

Now suppose that P' and P'' are prime ideals of R' that lie over P and that $P' \subset P''$. Choose $a \in P''$ with $a \notin P'$. Since a is integral over R there is a least positive integer n such that there are elements $b_0, \ldots, b_{n-1} \in R$ for which $a^n + b_{n-1}a^{n-1} + \cdots + b_1 a + b_0 \in P'$. Then $b_0 \in P'' \cap R = P = P' \cap R$. Hence

$$a(a^{n-1} + b_{n-1}a^{n-2} + \cdots + b_1) \in P',$$

but $a \notin P'$, so $a^{n-1} + b_{n-1}a^{n-2} + \cdots + b_1 \in P'$. This contradicts our choice of n. Thus, if $P' \subseteq P''$, we must have $P' = P''$.

4.7. Corollary (The Going-Up Theorem). *Let R and R' be as in Theorem 4.6. Let $P_0 \subset P_1 \subset \cdots \subset P_r$ be a chain of prime ideals of R. If the prime ideal P_0' of R' lies over P_0, then there is a chain*

$P_0' \subset P_1' \subset \cdots \subset P_r'$ of prime ideals of R' such that P_i' lies over P_i for $i = 0, \ldots, r$. If, for a given i, there is no prime ideal of R strictly between P_i and P_{i+1}, then there is no prime ideal of R' strictly between P_i' and P_{i+1}'.

Proof. Suppose that $0 \leq k < r$ and that we have shown that there is a chain $P_0' \subset \cdots \subset P_k'$ of prime ideals of R' such that P_i' lies over P_i for $i = 0, \ldots, k$. Then R/P_k may be considered as a subring of R'/P_k' which, in turn, is integral over R/P_k [Exercise 2(a)]. Hence, by Theorem 4.6, there is a prime ideal P_{k+1}' of R' such that $P_k' \subset P_{k+1}'$ and P_{k+1}'/P_k' lies over P_{k+1}/P_k. Then P_{k+1}' lies over P_{k+1}. This proves the first assertion of the corollary. Suppose that P' is a prime ideal of R' and that $P_i' \subset P' \subset P_{i+1}'$. By Theorem 4.6, P' cannot lie over either P_i or P_{i+1}. Hence the prime ideal $P' \cap R$ of R is strictly between P_i and P_{i+1}.

It is natural to ask whether, in the notation of the corollary, we can start with a prime ideal P_r' of R' lying over P_r and show the existence of a chain $P_0' \subset \cdots \subset P_r'$ of prime ideals of R', such that P_i' lies over P_i for each i. The answer is that we can do this, under somewhat more restrictive hypotheses than those of the going-up theorem.

Let R be an integral domain and let K be the quotient field of R. Let K' be a normal algebraic extension of K and let R' be the integral closure of R in K'. By Proposition 4.3, R' is a subring of K' and $R \subseteq R'$. Let $G(K'/K)$ be the group of K-automorphisms of K'. If $a \in R'$ and $\sigma \in G(K'/K)$ then $\sigma(a) \in R'$. For, if $b_0 + b_1 a + \cdots + b_{n-1}a^{n-1} + a^n = 0$, where $b_0, \ldots, b_{n-1} \in R$, then

$$\begin{aligned} 0 &= \sigma(b_0 + b_1 a + \cdots + b_{n-1}a^{n-1} + a^n) \\ &= b_0 + b_1\sigma(a) + \cdots + b_{n-1}\sigma(a)^{n-1} + \sigma(a)^n. \end{aligned}$$

Furthermore $a = \sigma(\sigma^{-1}(a))$. Thus $\sigma(R') = R'$ for all $\sigma \in G(K'/K)$. Let P be a prime ideal of R and let P' be a prime ideal of R' lying over P. Then $\sigma(P')$ is a prime ideal of R' and $\sigma(P')$ also lies over P. By Theorem 4.6 either $\sigma(P') = P'$ or $P' \nsubseteq \sigma(P')$ and $\sigma(P') \nsubseteq P'$.

4.8. Theorem. *Let R be an integrally closed integral domain and let K, K', and R' be as above. If the prime ideals P' and P'' of R' lie over the same prime ideal of R then $P'' = \sigma(P')$ for some $\sigma \in G(K'/K)$.*

Proof. Let $P' \cap R = P'' \cap R = P$. First, we assume that K' is a finite extension of K and we let $G(K'/K) = \{\sigma_1, \ldots, \sigma_n\}$. Further, we assume that $P'' \neq \sigma_i(P')$ for $i = 1, \ldots, n$. Then $P'' \nsubseteq \sigma_i(P')$ for $i = 1, \ldots, n$ and so, by Exercise 5(c) of Chapter II, there is an $a \in P''$ such that $a \notin \sigma_i(P')$ for $i = 1, \ldots, n$. Then $\sigma_i(a) \notin P'$ for $i = 1, \ldots, n$. If d is the degree of inseparability of K'/K, then

$$N_{K'/K}(a) = \left(\prod_{i=1}^{n} \sigma_i(a)\right)^d \in R' \cap K.$$

Since R is integrally closed, $R' \cap K = R$. Therefore, $N_{K'/K}(a) \in P'' \cap R = P \subseteq P'$, a contradiction.

Now let K' be of arbitrary degree over K. Let $\mathcal{S}$ be the set of all pairs (T, τ), where T is a subring of R' with $R \subseteq T$, T is integrally closed, the quotient field L of T in K' is normal over K, and τ is a K-automorphism of L such that $\tau(P' \cap T) = P'' \cap T$ (note that $P' \cap T$ and $P'' \cap T$ are prime ideals of T lying over P). The set $\mathcal{S}$ is not empty since (R, i) belongs to $\mathcal{S}$, where i is the identity isomorphism of K. Partially order $\mathcal{S}$ by writing $(T, \tau) \leq (T', \tau')$ if $T \subseteq T'$ and $\tau'(a) = \tau(a)$ for all $a \in T$. By Zorn's lemma $\mathcal{S}$ has a maximal element which we denote henceforth by (T, τ). If $T = R'$, we are finished. If $T \neq R'$, let $a \in R'$, $a \notin T$. Let L be the field of quotients of T in K'. Since K'/K is normal, there is a normal extension L' of K such that $L \subseteq L'$, $a \in L'$, and L'/L is finite. Let T' be the integral closure of T in L'. Then T' is integrally closed, and $T \subset T'$ since $a \in T'$ but $a \notin T$. The K-automorphism τ can be extended to a K-automorphism of L' which we also denote by τ. Then $P' \cap T'$ and $\tau^{-1}(P'' \cap T')$ are prime ideals of T' both lying over $P' \cap T$, and so by the first part of the proof there is an L-automorphism ρ of L' such that $\rho(P' \cap T') = \tau^{-1}(P'' \cap T')$; then $\tau\rho\,(P' \cap T') = P'' \cap T'$. Hence $(T', \tau\rho)$ is in $\mathcal{S}$, and $(T, \tau) < (T', \tau\rho)$, which contradicts our choice of (T, τ). Therefore we must have $T = R'$.

4.9. Theorem (The Going-Down Theorem). *Let R be an integrally closed integral domain and let R' be a ring such that:*

(i) *R is a subring of R';*
(ii) *R' is a integral over R;* and
(iii) *no nonzero element of R is a zero-divisor in R'.*

Let $P_0 \subset P_1 \subset \cdots \subset P_r$ be a chain of prime ideals of R and let P_r' be a prime ideal of R' lying over P_r. Then there is a chain $P_0' \subset P_1' \subset \cdots \subset$

P_r' of prime ideals of R' such that P_i' lies over P_i for $i = 0, \ldots, r$. If $P_0 = 0$ and if P' is a prime ideal of R' with $P' \cap R = 0$ and $P' \subseteq P_r'$, then there is such a chain with $P_0' = P'$.

Proof. We may assume that P_r is a proper prime ideal of R. Let $S = \{ab \mid a \in R, a \neq 0, \text{ and } b \in R' \backslash P_r'\}$. Then S is a multiplicative system in R' because of (iii). Let P' be a maximal element in the set of ideals of R' that do not intersect S. Then P' is a prime ideal and $P' \subseteq P_r'$. Furthermore, if $a \in R$, $a \neq 0$, then $a \in S$ and so $a \notin P'$; thus $P' \cap R = 0$. Therefore we may replace R' by R'/P' and assume from the start that R' is an integral domain. Then, to prove the theorem, it is sufficient to verify the first assertion under the assumption that $P_0 = 0$.

Let K and K' be quotient fields of R and R', respectively, with $K \subseteq K'$. We leave it to the reader to show that K' is an algebraic extension of K. Let K'' be a normal algebraic extension of K such that $K' \subseteq K''$, and let R'' be the integral closure of R in K''; then $R \subseteq R' \subseteq R''$. By the lying-over theorem, there is a prime ideal P_r'' of R'' which lies over P_r' and a prime ideal Q_0 of R'' which lies over $P_0 = 0$. By the going-up theorem there is a chain $Q_0 \subset Q_1 \subset \cdots \subset Q_r$ of prime ideals of R'' such that Q_i lies over P_i for $i = 0, \ldots, r$. Since P_r'' and Q_r both lie over P_r there is a K-automorphism σ of K'' such that $P_r'' = \sigma(Q_r)$. Let $P_i' = \sigma(Q_i) \cap R'$ for $i = 0, \ldots, r$. Then, for each i,

$$P_i' \cap R = \sigma(Q_i) \cap R = Q_i \cap R = P_i.$$

3 INTEGRAL DEPENDENCE AND FLAT MODULES

If R is a ring and K is its total quotient ring, then any ring T such that $R \subseteq T \subseteq K$ is called an **overring** of R. If R is an integral domain and T is an overring of R, then T is an integral domain and K is the quotient field of T as well as of R.

Let T be an overring of R. If A is an ideal of R then

$$AT = \{\text{finite sums } \textstyle\sum a_i t_i \mid a_i \in A, t_i \in T\}$$

is an ideal of T. In fact, it is the ideal of T generated by A.

4.10. Proposition. *If T is an overring of an integral domain R, then the following statements are equivalent:*

(a) *For every prime ideal P of R, either $PT = T$ or $T \subseteq R_P$.*
(b) *$((y):(x))T = T$ for all $x/y \in T$.*

Proof. First, we shall assume that (a) holds. Let $x/y \in T$ and suppose that $((y):(x))T \neq T$. Then there is a prime ideal P of R such that $(y):(x) \subseteq P$ and $PT \neq T$. By (a), $T \subseteq R_P$ and consequently $x/y \in R_P$. Thus $x/y = a/s$ where $s \notin P$; hence $xsu = yau$ for some $u \in R\backslash P$. This implies that $su \in (y):(x) \subseteq P$, a contradiction. Thus we do have $((y):(x))T = T$.

Conversely, assume that (b) holds. Let P be a prime ideal of R and suppose that $PT \neq T$. Let $x/y \in T$. Then $((y):(x))T = T$ and so $(y):(x) \nsubseteq P$. Let $s \in (y):(x)$, $s \notin P$. Then $sx = ay$ for some $a \in R$ and consequently $x/y = a/s \in R_P$. Thus $T \subseteq R_P$.

Note that if $A_1, \ldots, A_k$ are ideals of R such that $A_i T = T$ for $i = 1, \ldots, k$, then $(A_1 \cap \cdots \cap A_k)T = T$. For if $1 \leq i \leq k$, there are elements $a_{ij} \in A_i$, $t_{ij} \in T$, for $j = 1, \ldots, n_i$, such that $1 = a_{i1}t_{i1} + \cdots + a_{in_i}t_{in_i}$. Then

$$1 = \prod_{i=1}^{k} \sum_{j=1}^{n_i} a_{ij} t_{ij},$$

and each term on the right-hand side is the product of one element from each of the A_i and k elements of the form t_{ij}. Hence

$$1 \in (A_1 \cap \cdots \cap A_k)T,$$

which gives the desired equality.

4.11. Definition. *An overring T of a ring R is a* **flat overring** *if T is a flat R-module.*

4.12. Proposition. *An overring T of an integral domain R is a flat overring if and only if it satisfies the equivalent conditions of Proposition* 4.10.

Proof. Suppose that T satisfies (b) of Proposition 4.10. By Exercise 11(b) of Chapter I and Proposition 1.18 in order to show that T is a flat R-module it is sufficient to show that for an arbitrary ideal A of R the homomorphism $A \otimes_R T \to T$ given by $a \otimes b \mapsto ab$ is injective.

Let $c \in A \otimes_R T$; then $c = \sum_{i=1}^{s} a_i \otimes b_i$, $a_i \in A$, $b_i \in T$. There are elements $b, c_1, \ldots, c_s \in R$ such that $b_i = c_i/b$ for $i = 1, \ldots, s$; hence $c = \sum_{i=1}^{s} a_i \otimes c_i/b$. By assumption, $((b):(c_i))T = T$ for $i = 1, \ldots, s$, and so if $C = \bigcap_{i=1}^{s} (b):(c_i)$, then $CT = T$. Now suppose that the image of c is 0, that is, $\sum_{i=1}^{s} a_i c_i/b = 0$. If $d \in C$ we have $dc_i/b \in R$ for $i = 1, \ldots, s$; hence

$$dc = \sum_{i=1}^{s} a_i \otimes dc_i/b = \sum_{i=1}^{s} (da_i c_i/b \otimes 1)$$

$$= \left(d \sum_{i=1}^{s} a_i c_i/b\right) \otimes 1 = 0.$$

Therefore $cC = 0$ and consequently $0 = 0T = cCT = cT$. But $c \in cT$, so $c = 0$. Thus T is a flat R-module.

Conversely, suppose that T is a flat R-module. We shall show that (a) of Proposition 4.10 holds. If $x/y \in T$ then $y(x/y) - x1 = 0$. Hence by Exercise 13(c) of Chapter I, there are elements $z_{jk} \in R$, $j = 1, \ldots, r$, $k = 1, 2$, and elements $b_1, \ldots, b_r \in T$, such that

$$x/y = \sum_{j=1}^{r} b_j z_{j1},$$

$$1 = \sum_{j=1}^{r} b_j z_{j2},$$

$$z_{j1} y - z_{j2} x = 0, \qquad j = 1, \ldots, r.$$

Let P be a prime ideal of R. If $z_{j2} \in P$ for $j = 1, \ldots, r$, then $PT = T$. Suppose $z_{j2} \notin P$ for some j; then $(y):(x) \nsubseteq P$. Thus, we conclude that either $PT = T$ or $(y):(x) \nsubseteq P$ for all $x/y \in T$. Suppose the latter. Let $x/y \in T$; then there are elements $a \in R$ and $s \in R \backslash P$ such that $ay = sx$, or $x/y = a/s \in R_P$. Therefore $T \subseteq R_P$.

4.13. Proposition. *Let T and T' be overrings of an integral domain R with $T \subseteq T'$.*

(1) *If T' is a flat overring of R, then T' is a flat overring of T.*
(2) *If T' is a flat overring of T and T is a flat overring of R, then T' is a flat overring of R.*

Proof. Suppose that T' is a flat overring of R. Let $a, b \in T$ be such that $a/b \in T'$. Write $a = c/s$ and $b = d/s$, where $c, d, s \in R$. Then $c/d \in T'$ and so $((d):(c))T' = T'$ by Proposition 4.12. Hence $1 = t_1 u_1 + \cdots + t_k u_k$, where $t_1, \ldots, t_k \in T'$, $u_1, \ldots, u_k \in R$, and $u_i c \in (d)$ for $i = 1, \ldots, k$. Then $u_i a \in Tb$ for $i = 1, \ldots, k$, so $(Tb : Ta)T' = T'$. Therefore T' is a flat overring of T, again by Proposition 4.12. The second assertion of the proposition is a consequence of Exercise 9(d) of Chapter I.

4.14. Proposition. *If T is an overring of an integral domain R, then the following statements are equivalent:*

(1) *T is a flat overring of R.*
(2) *$T_P = R_{P \cap R}$ for all maximal ideals P of T.*
(3) *$T = \bigcap R_{P \cap R}$, where P runs over all maximal ideals of T.*

Proof. (1) $\Rightarrow$ (2). Suppose that T is a flat overring of R and let P be a maximal ideal of T. Certainly we have $R_{P \cap R} \subseteq T_P$. Let $x/y \in T_P$, where $x, y \in T$ and $y \notin P$; then there are elements $u, v, s \in R$ such that $x = u/s$ and $y = v/s$. Let $C = (s):(u) \cap (s):(v)$. By Proposition 4.12, $CT = T$; hence $C \nsubseteq P \cap R$. Let $z \in C$, $z \notin P \cap R$. Then zx, $zy \in R$ and $zy \notin P$, so $zy \notin P \cap R$. Hence $x/y = zx/zy \in R_{P \cap R}$. Thus $T_P \subseteq R_{P \cap R}$.

(2) $\Rightarrow$ (3). By Exercise 5(a) of Chapter III, $T = \bigcap T_P$, where P runs over all maximal ideals of T; hence, if (2) holds, so does (3).

(3) $\Rightarrow$ (1). Suppose that (3) holds and let Q be a prime ideal of R such that $QT \neq T$. Then $QT \subseteq P$ for some maximal ideal P of T and $Q \subseteq P \cap R$. Hence $R_{P \cap R} \subseteq R_Q$. But $T \subseteq R_{P \cap R}$ so that $T \subseteq R_Q$. Thus T is a flat overring of R.

4.15. Theorem. *Let T be an overring of an integral domain R. If T is both integral over R and a flat overring of R, then $T = R$.*

Proof. Let $x/y \in T$; then $((y):(x))T = T$ by Proposition 4.12. If P is a proper prime ideal of R, then by the lying-over theorem there is a proper prime ideal P' of T lying over P. Since $PT \subseteq P'$, we must have $PT \neq T$. It follows that $(y):(x)$ is not contained in any proper prime ideal of R; hence $(y):(x) = R$. Thus $x = ay$ for some $a \in R$. Then $x/y = ay/y = a \in R$. Thus $T = R$.

4 ALMOST INTEGRAL DEPENDENCE

Let R be a subring of a ring R' and let $a \in R'$. By the third of the equivalent conditions of Proposition 4.1, a is integral over R if there is a subring R'' of R' such that $a \in R''$ and R'' is a finitely generated R-module. An example referred to below shows that it is necessary to insist that R'' be a subring of R' and not just a submodule of the R-module R'. However, this weaker condition turns out to be an important one, as we shall see in Chapter VIII, and is worthy of investigation.

4.16. Definition. *Let R be a subring of a ring R'. An element $a \in R'$ is* **almost integral** *over R if there exists a finitely generated submodule of the R-module R' which contains all powers of a.*

Clearly, an element of R' which is integral over R is also almost integral over R. The converse is not true; an example is given in Exercise 11. However, the converse is true if R is Noetherian [see Exercise 1(c)].

4.17. Definition. *Let R be a subring of a ring R'. The set R_0 of elements of R' which are almost integral over R is the* **complete integral closure** *of R in R'. If $R_0 = R$, the R is* **completely integrally closed in** *R. If R is completely integrally closed in its total quotient ring, then we say simply that R is* **completely integrally closed**.

It is immediately clear that the complete integral closure R_0 of R in R' is a subring of R'. However, R_0 is not necessarily itself completely integrally closed; an example is given in Exercise 14. Furthermore, if R, T, and T' are rings with $R \subseteq T \subseteq T'$, then an element $a \in T$ may be almost integral over R as an element of T', but not almost integral over R as an element of T; an example is given in Exercise 13.

Even though the complete integral closure of one ring in another may not be completely integrally closed, we do have the following:

4.18. Proposition. *Let R be a subring of a ring R' and let R_0 be the complete integral closure of R in R'. Then R_0 is integrally closed in R'.*

Proof. Let $x \in R'$ be integral over R_0; say $x^m + a_{m-1}x^{m-1} + \cdots + a_0 = 0$, where $a_0, \ldots, a_{m-1} \in R_0$. Then x is integral over the ring

$$R[a_0, \ldots, a_{m-1}].$$

For $i = 0, \ldots, m-1$, a_i is contained in some finitely generated submodule of the R-module R'; call it $M_i = Rx_{i1} + \cdots + Rx_{ik_i}$, where each $x_{ij} \in R'$. Then $R[a_0, \ldots, a_{m-1}] \subseteq M_0 \cdots M_{m-1}$, which is the submodule of the R-module R' generated by all products

$$x_{0j_0}x_{1j_1} \cdots x_{m-1, j_{m-1}}$$

where each j_i runs between 1 and k_i. Hence x is contained in

$$\begin{aligned} R[x] &\subseteq R[a_0, \ldots, a_{m-1}, x] \\ &= \sum_{h=0}^{m-1} R[a_0, \ldots, a_{m-1}]x^h \\ &\subseteq \sum_{h=0}^{m-1} \sum_{i=0}^{m-1} \sum_{j_i=1}^{k_i} Rx_{0j_0}x_{1j_1} \cdots x_{m-1, j_{m-1}}x^h, \end{aligned}$$

which is a finitely generated submodule of the R-module R'. Thus x is almost integral over R, and consequently $x \in R_0$. Therefore R_0 is integrally closed.

4.19. Corollary. *Let R, R_1, and R_2 be rings such that $R \subseteq R_1 \subseteq R_2$. If every element of R_1 is almost integral over R, and if R_2 is integral over R_1, then every element of R_2 is almost integral over R.*

We have the following useful characterization of the complete integral closure of a ring.

4.20. Theorem. *The complete integral closure of a ring R with total quotient ring K is*

$$R_0 = \{x \mid x \in K \text{ and there exists a regular element } r \in R \text{ such that } rx^n \in R \text{ for all positive integers } n\}.$$

Proof. Let x be an element of K which is almost integral over R. Then there exist elements $k_1, \ldots, k_s \in K$ such that $R[x] \subseteq \sum_{i=1}^{s} Rk_i$. If r is the product of the denominators of the k_i's, then $rx^n \in R$ for all positive integers n. Conversely, if $x \in R_0$, say $rx^n \in R$ for all positive integers n and some regular element $r \in R$, then $R[x] \subseteq Rr^{-1}$. Therefore x is almost integral over R.

4.21. Corollary. *If R is completely integrally closed, then for all nonunits $x \in R$, $\bigcap_{n=1}^{\infty} (x^n)$ consists entirely of zero-divisors.*

Proof. Let K be the total quotient ring of R. If x is a regular element of R then x has an inverse in K. If $r \in \bigcap_{n=1}^{\infty} (x^n)$ then $r(x^{-1})^n \in R$ for all n. If r is a regular element of R then x^{-1} is almost integral over R and so belongs to R, contrary to assumption. Hence r is a zero-divisor. If x itself is a zero-divisor the assertion is clearly true.

EXERCISES

1. Integral dependence.

(a) Let R be a subring of a ring R' and let $a \in R'$. Let A be an ideal of R and suppose that there is a finitely generated submodule M of the R-module R' such that $aM \subseteq AM$, and in addition, if $bM = 0$ for some $b \in R[a]$ then $b = 0$. Show that there exist a positive integer n and elements c_0, c_1, $\ldots, c_{n-1} \in A$ such that $c_0 + c_1 a + \cdots + c_{n-1}a^{n-1} + a^n = 0$.

(b) Show that a is integral over R if and only if there is a finitely generated submodule M of the R-module R' such that $aM \subseteq M$ and $bM = 0$ for some $b \in R[a]$ implies $b = 0$.

(c) Show that if R is Noetherian then a is integral over R if and only if $R[a]$ is contained in some finitely generated submodule of the R-module R'.

2. On integral extensions.

(a) Let R be a subring of a ring R'. If A' is an ideal of R' and $A = A' \cap R$, we may consider R/A as a subring of R'/A'. Show that if R' is integral over R, then R'/A' is integral over R/A.

(b) Show that if R and R' are integral domains and if R' is integral over R, then R' is a field if and only if R is a field.

(c) Let R be a subring of R' and assume R' is integral over R. Let M' be a maximal ideal of R' and let $M = R \cap M'$. Prove that $R'_{M'}$ is not necessarily integral over R_M. [Hint. Consider $R = K[X^2 - 1]$ and $R' = K[X]$ for a field K. Let $M' = (X - 1)R'$.]

3. Integral extensions and ideals.

(a) Suppose R' is integral over R and that P' is a prime ideal of R' lying over a prime ideal P of R. Without reference to the lying-over theorem, show that P' is maximal if and only if P is maximal. [Hint. Use Exercise 2(b).]

(b) Let R, R', P, and P' be as in (a). Show that there is no ideal A' of R', prime or otherwise, such that $A' \cap R = P$ and $P' \subset A'$.

(c) Show that if R' is integral over R, and if A is a proper ideal of R, then $AR' \neq R'$.

(d) Let R' be integral over R and let A be an ideal of R. Show that $\text{Rad}(AR') = \{a \mid a \in R'$ and there is a positive integer n and elements $b_0, \ldots, b_{n-1} \in A$ such that $b_0 + b_1 a + \cdots + b_{n-1}a^{n-1} + a^n = 0\}$.

(e) Suppose that R' is an integral domain and is integral over R. Show that if A' is a nonzero ideal of R' then $A' \cap R \neq 0$.

4. Local integral closure.

Let R be an integral domain. Prove the equivalence of the following assertions; that is, prove that integral closure of an integral domain is a property of localizations.

(1) R is integrally closed.

(2) R_P is integrally closed for each proper prime ideal P of R.

(3) R_M is integrally closed for each maximal ideal M of R.

5. Simple ring extensions.

(a) Let the ring R be a subring of the ring R'. Let P be a prime ideal of R. Show that there is a prime ideal of R' lying over P if and only if $PR' \cap R = P$.

(b) Let R be an integral domain and F any field having R as a subring. Let P be a prime ideal of R. Show that if $a \in F$,

$a \neq 0$, then either $R[a]$ or $R[1/a]$ contains a prime ideal lying over P.

6. The pseudo-radical of a ring.
Let R be an integral domain and let Rad*(R) be the intersection of all of the nonzero prime ideals of R; Rad*(R) is called the **pseudo-radical** of R.
(a) Let R be a subring of an integral domain R'. Show that if the conclusion of the lying-over theorem holds for R and R', then Rad*$(R) \neq 0$ implies that Rad*$(R') \neq 0$. Give an example to show that the converse is not true.
(b) Let K be the quotient field of an integral domain R. Show that the following statements are equivalent:
(1) Rad*$(R) \neq 0$.
(2) $K = R[a]$ for some $a \in K$.
(3) $K = R[a_1, \ldots, a_n]$ for some $a_1, \ldots, a_n \in K$.

7. Integrally closed domains.
Let R be an integral domain.
(a) Prove that if R is a unique factorization domain, then R is integrally closed.
(b) Let S be a multiplicative system in R. Show that if R is integrally closed, then $S^{-1}R$ is integrally closed.
(c) Let P be a prime ideal of R. Show by an example that we may have R integrally closed but R/P not integrally closed.

8. Integral closure and polynomial rings.
(a) Suppose that R is an integrally closed domain. Let K be the quotient field of R and let $f(X)$ be a monic polynomial in $R[X]$. Show that if $f(X) = g(X)h(X)$, where $g(X)$ and $h(X)$ are monic polynomials in $K[X]$, then $g(X), h(X) \in R[X]$. [Hint. Take a field L containing K such that in $L[X]$, $f(X)$ and $g(X)$ can be factored into linear factors: $f(X) = \prod (X - a_i)$, $g(X) = \prod (X - b_i)$.]
(b) Prove the result of (a) without assuming R is an integral domain.
(c) Let R be a subring of a ring R' and let R_0 be the integral closure of R in R'. Prove that $R_0[X]$ is the integral closure of $R[X]$ in $R'[X]$. [Hint. Suppose $f^n + r_{n-1}f^{n-1} + \cdots + r_0 = 0$ with $r_i \in R[X]$, and $f \in R'[X]$. Pick an integer

$m \geq \{n, \deg r_0, \ldots, \deg r_{n-1}\}$. Consider $g(X) = f(X) - X^m$. Put this in the integral expression for f and simplify. Apply Exercise 8(b).]

9. Integral closure and field extensions.
Let R be an integral domain with quotient field K and let L be an algebraic extension of K. Let T be the integral closure of R in L. Let $T_0 = T \cap K$. Prove the following:
(a) T_0 is the integral closure of R.
(b) L is the field of quotients of T.
(c) If $x \in L$ is integral over R and $f(X)$ is the minimal polynomial over K which has x as a root, then $f(X) \in T_0[X]$. Furthermore, the ideal of $T_0[X]$ consisting of those polynomials which have x as a root is principal. (Note that this says that if R is integrally closed, $x \in L$ is integral over R if and only if $f(X) \in R[X]$.)

10. A proof of the going-down theorem.
(a) Let R and R' be as in Theorem 4.9. Let P and Q be proper prime ideals of R with $Q \subset P$ and let P' be a prime ideal of R' lying over P. Let $S = \{rs \mid r \in R \backslash Q \text{ and } s \in R' \backslash P'\}$. Show that S is multiplicatively closed and that $S \cap QR'$ is empty.
(b) Let $\mathscr{W} = \{A' \mid A' \text{ is an ideal of } R', QR' \subseteq A', \text{ and } S \cap A' \text{ is empty}\}$. Let Q' be a maximal element of $\mathscr{W}$ (Q' exists by Zorn's lemma). Show that Q' is a prime ideal of R', that $Q' \subset P'$, and that $Q' \cap R = Q$.

11. Almost integral closure.
Let R be an integrally closed integral domain with quotient field K. Let $T = R + XK[X]$.
(a) Prove that T is integrally closed.
(b) Prove that $K[X]$ is the complete integral closure of R in $K[X]$.

12. Completely integrally closed rings.
(a) Prove that each unique factorization domain is completely integrally closed.
(b) If R is completely integrally closed and S is a multiplicative system in R, prove that $S^{-1}R$ is completely integrally closed.

13. On almost integrity.
Let R, T_1, and T_2 be rings such that $R \subseteq T_1 \subseteq T_2$. For $i = 1, 2$, let R_i be the complete integral closure of R in T_i. Clearly, $R_1 \subseteq R_2 \cap T_1$.

(a) Show that if T_2 is a submodule of some T_1-module M such that T_1 is a direct summand of M, then $R_1 = R_2 \cap T_1$.

(b) Show that the same conclusion holds if every finitely generated T_1-module M with $T_1 \subseteq M \subseteq T_2$ is a submodule of a T_1-module of which T_1 is a direct summand.

(c) Show that $R_1 = R_2 \cap T_1$ if T_1 is a principal ideal domain.

(d) Let K be a field and let X and Y be indeterminates over K. Let $R = K[\{XY^n \mid n \geq 1\}]$, $T_1 = R[Y]$, and $T_2 = T_1[1/X]$. With R_1 and R_2 as above, show that $R_1 \subset R_2 \cap T_1$.

14. The complete integral closure need not be completely integrally closed.
Let K be a field and let X and Y be indeterminates. Let

$$R = K[\{X^{2n+1}Y^{n(2n+1)} \mid n \geq 0\}].$$

(a) Show that the quotient field of R is $K(X, Y)$.

(b) Show that if $R' = K[\{XY^n \mid n \geq 0\}]$, then $R \subset R' \subseteq R^* \subseteq K[X, Y]$, where R^* is the complete integral closure of R.

(c) Show that Y is almost integral over R', and hence is almost integral over R^*, but that $Y \notin R^*$. [Hint. Show that for any element of R, the exponent of Y in any of the monomials of that element is less than or equal to the square of the exponent of X in the same monomial.]

CHAPTER

V

Valuation Rings

1 THE DEFINITION OF A VALUATION RING

5.1. Definition. *A* **valuation ring** *is an integral domain V with the property that if A and B are ideals of V then either $A \subseteq B$ or $B \subseteq A$.*

5.2. Proposition. *For an integral domain V the following statements are equivalent:*

(1) *V is a valuation ring.*
(2) *If $a, b \in V$ then either $(a) \subseteq (b)$ or $(b) \subseteq (a)$.*
(3) *If x belongs to the quotient field K of V, then either $x \in V$ or $x^{-1} \in V$.*

Proof. That (1) implies (2) is clear. Assume that (2) holds and let x belong to K. Then $x = a/b$ where $a, b \in V$. If $(a) \subseteq (b)$ then $a = br$ where $r \in V$; hence $x = br/b = r \in V$. If $(b) \subseteq (a)$ then $x \neq 0$ and $x^{-1} = b/a \in V$ by the same argument. Finally, suppose (3) holds and let A and B be ideals of V. Suppose $A \nsubseteq B$ and choose $a \in A$ with $a \notin B$. If $b \in B$ and $b \neq 0$ then $a/b \notin V$, for if $a/b \in V$ then $a \in (b) \subseteq B$. Hence $b/a \in V$ and $b \in (a) \subseteq A$. Thus $B \subseteq A$.

5.3. Corollary. *Each overring of a valuation ring is a valuation ring.*

5.4. Proposition. *If V is a valuation ring then the nonunits of V form an ideal of V, which is the unique maximal ideal of V.*

Proof. Let P be the set of nonunits of V. Let $a, b \in P$ and $c \in V$. Then ac is a nonunit of V, that is, $ac \in P$. We may assume that $a/b \in V$. Then $a - b = (a/b - 1)b \in P$. Thus P is an ideal of V. If A is a proper ideal of V then every element of A is a nonunit, so $A \subseteq P$. Thus P is the unique maximal ideal of V.

5.5. Proposition. *Valuation rings are integrally closed.*

Proof. Let V be a valuation ring and let K be its quotient field. Let $x \in K$ be integral over V, say

$$x^n + a_{n-1}x^{n-1} + \cdots + a_0 = 0,$$

for $a_{n-1}, \ldots, a_0 \in V$. If $x \notin V$, then $x^{-1} \in V$. Hence

$$x = -(a_{n-1} + a_{n-2}x^{-1} + \cdots + a_0 x^{-n+1}),$$

and thus $x \in V$, a contradiction.

Let K be a field. A **partial homomorphism** of K is a homomorphism ϕ from a subring K_ϕ of K into an algebraically closed field. Let $\mathscr{S}$ be the set of all such pairs (ϕ, K_ϕ). Define a partial ordering $\leq$ on $\mathscr{S}$ by writing

$$(\phi, K_\phi) \leq (\psi, K_\psi)$$

if $K_\phi \subseteq K_\psi$ and ψ is an extension of ϕ to K_ψ, that is, $\psi(a) = \phi(a)$ for all $a \in K_\phi$. Zorn's lemma guarantees that if $(\phi, K_\phi) \in \mathscr{S}$, then there is a maximal element of $\mathscr{S}$ which is greater than or equal to (ϕ, K_ϕ). Maximal elements of $\mathscr{S}$ are called **maximal partial homomorphisms**. We shall prove below that (ϕ, K_ϕ) is a maximal partial homomorphism if and only if K_ϕ is a valuation ring with quotient field K.

Let $\mathscr{T}$ be the set of all pairs (V, P), where V is a subring of the field K and P is a proper prime ideal of V. Define a partial ordering $\ll$ on $\mathscr{T}$ by writing

$$(V_1, P_1) \ll (V_2, P_2)$$

if $V_1 \subseteq V_2$ and $P_2 \cap V_1 = P_1$. If this is the case, we say that (V_2, P_2) **dominates** (V_1, P_1). Zorn's lemma guarantees that if $(V, P) \in \mathscr{T}$, then there is a maximal element of $\mathscr{T}$ which dominates (V, P). Maximal elements of $\mathscr{T}$ are called **maximal pairs**. We shall prove below that (V, P) is a maximal element of $\mathscr{T}$ if and only if V is a valuation ring and P is the maximal ideal of V.

5.6. Lemma. *Let x be a nonzero element of a field K. Let V be a subring of K and let P be the unique maximal ideal of V. Then either $PV[x] \neq V[x]$ or $PV[x^{-1}] \neq V[x^{-1}]$.*

Proof. Suppose that $PV[x] = V[x]$ and $PV[x^{-1}] = V[x^{-1}]$. Then there exist $a_0, \ldots, a_m, b_0, \ldots, b_n \in P$ such that

$$a_0 + a_1 x + \cdots + a_m x^m = 1 \tag{1}$$

and

$$b_0 + b_1 x^{-1} + \cdots + b_n x^{-n} = 1. \tag{2}$$

Choose m and n as small as possible, and suppose $m \geq n$. Multiply (2) by x^n. This gives

$$(1 - b_0)x^n = b_1 x^{n-1} + \cdots + b_n.$$

Now since $b_0 \in P$ and P is the unique maximal ideal of V, $1 - b_0$ is a unit in V. Thus

$$\begin{aligned} x^n &= (1 - b_0)^{-1} b_1 x^{n-1} + \cdots + (1 - b_0)^{-1} b_n \\ &= c_1 x^{n-1} + \cdots + c_n, \end{aligned}$$

where $c_1, \ldots, c_n \in P$. Use this in (1):

$$\begin{aligned} 1 &= a_0 + a_1 x + \cdots + a_{m-1} x^{m-1} + a_m x^{m-n}(c_1 x^{n-1} + \cdots + c_n) \\ &= d_0 + d_1 x + \cdots + d_{m-1} x^{m-1} \end{aligned}$$

and $d_0, \ldots, d_{m-1} \in P$. This contradicts our choice of m.

5.7. Theorem. *Let K be a field and let V be a subring of K. Then the following statements are equivalent:*

(1) *V is a valuation ring with quotient field K.*
(2) *V has a maximal ideal P such that (V, P) is a maximal pair.*

(3) *There exists a homomorphism ϕ from V into an algebraically closed field such that (ϕ, V) is a maximal partial homomorphism.*

Proof. (1) $\Rightarrow$ (2). Assume that V is a valuation ring with quotient field K. Let P be the unique maximal ideal of V. Suppose that $(V, P) \ll (V', P')$. Let x be a nonzero element of V'. If $x \notin V$, then $x^{-1} \in P \subseteq P'$, which is absurd since P' is a proper ideal of V'. Thus $x \in V$; so $V' = V$ and $P' = P$. Therefore, (V, P) is a maximal pair.

(2) $\Rightarrow$ (3). Assume that P is a maximal ideal of V and that (V, P) is a maximal pair. Let L be the algebraic closure of the field V/P, and let ϕ be the canonical homomorphism from V into L. If $(\phi, V) \leq (\phi', V')$, and if P' is the kernel of ϕ', then $P' \cap V = P$, so that $(V, P) \ll (V', P')$. Hence $V = V'$. Thus, (ϕ, V) is a maximal partial homomorphism.

(3) $\Rightarrow$ (1). Assume that there exists a homomorphism ϕ from V into an algebraically closed field L such that (ϕ, V) is a maximal partial homomorphism. Let P be the kernel of ϕ; since $\phi(1) = 1$, P is a proper prime ideal of V. If $s \in V \backslash P$ then $\phi(s)$ is a unit in L; hence ϕ can be extended to a homomorphism ϕ' from V_P into L by setting $\phi'(a/s) = \phi(a)\phi(s)^{-1}$ for all $a \in V$, $s \in V \backslash P$. Then $(\phi, V) \leq (\phi', V_P)$, and so we have $V_P = V$. It follows that P is the unique maximal ideal of V.

Let x be a nonzero element of K. We shall show that either x or x^{-1} is an element of V. By Lemma 5.6, we can assume that $PV[x] \neq V[x]$. Then there exists a maximal ideal M of $V[x]$ such that $PV[x] \subseteq M$; then $M \cap V = P$ since P is a maximal ideal of V. Hence the mapping $\sigma: V/P \rightarrow V[x]/M$ given by $\sigma(a + P) = a + M$ is an injective homomorphism. Clearly, we have $V[x]/M = \sigma(V/P)[x + M]$. Since $V[x]/M$ is a field, this implies that $x + M$ is algebraic over $\sigma(V/P)$. Thus, if $\bar{\phi}: V/P \rightarrow L$ is given by $\bar{\phi}(a + P) = \phi(a)$, then $\bar{\phi}\sigma^{-1}: \sigma(V/P) \rightarrow L$ can be extended to an injective homomorphism $\psi: V[x]/M \rightarrow L$. Let $\pi: V[x] \rightarrow V[x]/M$ be the canonical homomorphism. Then $(\phi, V) \leq (\psi\pi, V[x])$, since for all $a \in V$ we have

$$\psi\pi(a) = \psi(a + M) = \bar{\phi}(a + P) = \phi(a).$$

By the maximality of (ϕ, V) we have $V[x] = V$. Thus, $x \in V$, which is what we wanted to show. This implies that K is the quotient field of V.

This theorem has an interesting and useful corollary.

5.8. Corollary. *Let R be an integral domain with quotient field K. The integral closure of R is the intersection of all of the valuation rings of K containing R.*

Proof. Since valuation rings are integrally closed by Proposition 5.5, the integral closure of R is certainly contained in the intersection of all valuation rings of K containing R. Conversely, let x be an element of K which is not integral over R. Then $x \notin R[x^{-1}]$. For, suppose that $x \in R[x^{-1}]$. Then there is a polynomial $f(X) \in R[X]$, of degree n, such that $x = f(x^{-1})$; then $x^{n+1} - x^n f(x^{-1}) = 0$, which means that x is integral over R, contrary to assumption. Thus, x^{-1} is not a unit in $R[x^{-1}]$, so there is a maximal ideal P of $R[x^{-1}]$ such that $x^{-1} \in P$. Let L be the algebraic closure of $R[x^{-1}]/P$. The canonical homomorphism $R[x^{-1}] \to R[x^{-1}]/P$ furnishes us with a homomorphism $\pi\colon R[x^{-1}] \to L$. Let (ϕ, V) be a maximal partial homomorphism of K into L such that $(\pi, R[x^{-1}]) \leq (\phi, V)$. By Theorem 5.7, V is a valuation ring of K, and $R \subseteq V$. Since $\phi(x^{-1}) = 0$ we must have $x \notin V$. Thus x is not in the intersection of all valuation rings of K containing R. This establishes the desired equality.

For Noetherian rings, valuation rings can be characterized by much weaker conditions than for arbitrary rings.

5.9. Theorem. *Let V be a Noetherian integral domain which is not a field. Then the following statements are equivalent:*

(1) *V is a valuation ring.*
(2) *The nonunits of V form a nonzero principal ideal.*
(3) *V is integrally closed and has exactly one nonzero proper prime ideal.*

Proof. (1) $\Rightarrow$ (2). Let V be a valuation ring and let P be its ideal of nonunits. Then $P \neq 0$ since V is not a field. Since V is Noetherian, P is finitely generated, say $P = (a_1, \ldots, a_k)$. We may assume $(a_1) \subseteq (a_2) \subseteq \cdots \subseteq (a_k)$. Then $P = (a_k)$.

(2) $\Rightarrow$ (3). Assume that (2) holds and let P be the ideal of nonunits of V. Then P is the only maximal ideal of V and $P = (a)$ where $a \neq 0$. By Corollary 2.24, $\bigcap_{n=1}^{\infty} P^n = 0$. If A is any nonzero proper ideal of V, then $A \subseteq P$, so there is a positive integer n such that

$A \subseteq P^n$ and $A \nsubseteq P^{n+1}$. Let $b \in A$, $b \notin P^{n+1}$; then $b = a^n u$ where u is a unit in V. Now let $c \in P^n$. Then for some $d \in V$, $c = a^n d = bu^{-1}d \in A$. Therefore $A = P^n = (a^n)$. It follows immediately from this that P is the only nonzero proper prime ideal of V.

Now let c be an element of the quotient field of V and write $c = r/s$, where $r, s \in V$; assume $c \neq 0$. Since every nonzero element of V is a unit times some power of a, we may assume that either r is a unit or s is a unit. Suppose that c is integral over V. Then there are elements $b_0, \ldots, b_{n-1} \in V$ such that $b_0 + b_1 c + \cdots + b_{n-1}c^{n-1} + c^n = 0$. Multiplying by s^n we obtain

$$s^n b_0 + s^{n-1} b_1 r + \cdots + s b_{n-1} r^{n-1} + r^n = 0,$$

or

$$r^n = -s(s^{n-1}b_0 + s^{n-2}b_1 r + \cdots + b_{n-1}r^{n-1}).$$

If s is a unit in V, then $c \in V$. If s is not a unit in V, then $r^n \in P$, so $r \in P$; that is, r is not a unit. This is contrary to our assumption. Therefore, $c \in V$ and we conclude that V is integrally closed.

(3) $\Rightarrow$ (1). Assume that (3) holds. It is sufficient to show that the unique nonzero proper prime ideal P of V is principal. For once this is done, a repetition of the argument given in the proof that (2) implies (3) will show that every nonzero ideal of V is a power of P. Thus the ideals of V are totally ordered under inclusion, and consequently V is a valuation ring.

To prove that P is principal, we consider $P^* = \{x \mid x \in K$ and $xP \subseteq V\}$ where K is the quotient field of V. Then P^*P is an ideal of V such that $P \subseteq P^*P \subseteq V$. Hence $P^*P = P$ or $P^*P = V$. We will show below that $P^*P \neq P$; hence we have $P^*P = V$. Then there are elements $a_1, \ldots, a_k \in P$ and $b_1, \ldots, b_k \in P^*$ such that $a_1 b_1 + \cdots + a_k b_k = 1$. Hence, for some i, $a_i b_i \notin P$; that is, there are elements $a \in P$, $b \in P^*$ such that $ab = u$, where u is a unit in V. Then $abu^{-1} = 1$. Let $c \in P$; then $c = abcu^{-1}$. Since $bc \in V$, $c \in (a)$. Hence $P = (a)$.

Now we show that $P^*P \neq P$. Suppose $P^*P = P$. Let $P = (a_1, \ldots, a_k)$. If $a \in P^*$ then $aP \subseteq P$, so

$$aa_i = \sum_{j=1}^{k} r_{ij} a_j, \qquad i = 1, \ldots, k, \quad r_{ij} \in V.$$

Then $\sum_{j=1}^{k} (\delta_{ij} a - r_{ij}) a_j = 0$, $i = 1, \ldots, k$. Hence, since we are working in the quotient field of V and at least one $a_j \neq 0$, we have

$$\det[\delta_{ij}a - r_{ij}] = 0.$$

Thus, a is integral over V and so belongs to V. Therefore $P^* = V$.

To complete the proof we shall show that we cannot have $P^* = V$. Let $a \in P$, $a \neq 0$, and let $S = \{a^n | n$ is a positive integer$\}$. Then $S^{-1}V$ is the field of quotients of V. For, suppose it is not. Then $S^{-1}V$ has a nonzero maximal ideal P'. Since a is a unit in $S^{-1}V$, we have $a \notin P'$; hence $P' \cap V \neq P$ and consequently $P' \cap V = 0$. However, this is not true, for if $c/a^n \in P'$ and $c/a^n \neq 0$, then $c \neq 0$ and $c \in P' \cap V$. Thus every element in the field of quotients of V can be written in the form b/a^n for some $b \in V$.

Now let $c \in V$, $c \neq 0$. Then $1/c = b/a^n$ and so $a^n = cb \in (c)$. Thus, for each $a \in P$, some power of a is in (c). Since P is finitely generated, it follows that $P^n \subseteq (c)$ for some smallest positive integer n. Let $d \in P^{n-1}$, $d \notin (c)$. Then $dP \subseteq (c)$; that is, $(d/c)P \subseteq V$. Thus, $d/c \in P^*$, but $d/c \notin V$, so that $P^* \neq V$.

In the course of this rather long proof we have shown that if V is a Noetherian valuation ring and if P is its ideal of nonunits, then P is principal and every nonzero proper ideal of V is a power of P. Since P is maximal, each such ideal is P-primary [Exercise 6(e) of Chapter II].

2 IDEAL THEORY IN VALUATION RINGS

In this section we obtain several results concerning the ideals of valuation rings.

5.10. Theorem. *Let V be a valuation ring and let A be an ideal of V.*

(1) $\operatorname{Rad}(A)$ *is a prime ideal of* V.

(2) *If* $B = \bigcap_{n=1}^{\infty} A^n$, *then* B *is a prime ideal of* V *which contains every prime ideal of* V *which is properly contained in* A.

Proof. (1) $\operatorname{Rad}(A)$ is the intersection of the minimal prime divisors of A by Proposition 2.14. But since the set of ideals of V is totally ordered, A has only one minimal prime divisor, which must coincide with $\operatorname{Rad}(A)$.

(2) Let a and b be elements of V such that $a \notin B$, $b \notin B$. Then $a \notin A^n$, $b \notin A^m$ for some positive integers n and m. Thus $A^n \subset (a)$ and $A^m \subset (b)$, so $A^n(b) \subseteq (a)(b)$. Actually, we have $A^n(b) \neq (a)(b)$. For, since $A^n \subset (a)$, there exists $x \in V$ such that $xa \notin A^n$. If $A^n(b) = (a)(b)$, then there exists $y \in A^n$ such that $yb = xab$. But this means that $xa = y \in A^n$, contrary to our choice of x; thus $A^n(b) \subset (a)(b)$. Hence $A^{n+m} \subseteq A^n(b) \subset (ab)$, and so $ab \notin A^{n+m}$. Therefore, $ab \notin B$, and we conclude that B is a prime ideal.

If P is a prime ideal of V such that $P \subset A$, then P contains no power of A. Hence $P \subset A^n$ for each positive integer n. Thus $P \subseteq B$.

We now turn to the primary ideals of a valuation ring.

5.11. Theorem. *Let P be a proper prime ideal of a valuation ring V.*

(1) *If Q is P-primary and $x \in V \backslash P$, then $Q = Q(x)$.*

(2) *The product of P-primary ideals of V is a P-primary ideal. if $P \neq P^2$, then the only P-primary ideals are powers of P.*

(3) *The intersection of all P-primary ideals of V is a prime ideal of V and there are no prime ideals of V properly between it and P.*

Proof. (1) Since $x \notin P$, $Q \subset (x)$. Let K be the quotient field of V and let $A = \{y \mid y \in K \text{ and } yx \in Q\}$. Since $Q \subset (x)$, A is a subset of V. Furthermore, it is easy to check that A is an ideal of V and $Q = A(x)$. Moreover, since Q is P-primary and $(x) \nsubseteq P$, we have $A \subseteq Q$. Thus $Q = A$ and $Q = Q(x)$, as claimed.

(2) Let Q_1, Q_2 be P-primary ideals of V. Clearly $\operatorname{Rad}(Q_1Q_2) = P$. Let x, y be elements of V with $xy \in Q_1Q_2$ and $x \notin P$. By (1), $Q_1 = Q_1(x)$. Hence $xy \in (x)Q_1Q_2$. Since V is an integral domain, this implies that $y \in Q_1Q_2$. Thus Q_1Q_2 is P-primary.

Now suppose that $P \neq P^2$ and let Q be a P-primary ideal of V. By Exercise 5(c), Q contains a power of P^2, and so contains a power of P. Thus, there is a positive integer m such that $P^m \subseteq Q$ but $P^{m-1} \nsubseteq Q$. Let $x \in P^{m-1}$, $x \notin Q$; then $Q \subset (x)$. If we define A as in the proof of (1), then $Q = A(x)$. Since Q is P-primary and $x \notin Q$, $A \subseteq P$. Therefore, $Q = A(x) \subseteq P(x) \subseteq P^m$, and so we conclude that $Q = P^m$.

(3) If P is the only P-primary ideal of V, there is nothing to

prove. Suppose there is a P-primary ideal Q of V with $Q \neq P$, and let $\{Q_\alpha\}$ be the set of all P-primary ideals of V. By (2), Q^n is P-primary for each $n \geq 1$; hence $\bigcap_\alpha Q_\alpha \subseteq \bigcap_{n=1}^\infty Q^n$. However, by Exercise 5(c), each Q_α contains a power of Q; thus $\bigcap_\alpha Q_\alpha = \bigcap_{n=1}^\infty Q^n$. Hence, by (2) of Theorem 5.10, $\bigcap_\alpha Q_\alpha$ is a prime ideal of V which contains every prime ideal of V which is properly contained in Q. If P' is a prime ideal of V properly contained in P, then $Q \nsubseteq P'$, so $P' \subset Q$; hence $P' \subseteq \bigcap_\alpha Q_\alpha$.

3 VALUATIONS

By an **ordered Abelian group** we mean an Abelian group G on which there is defined a total ordering $\leq$ such that if $\alpha, \beta, \gamma \in G$ and $\alpha \leq \beta$, then $\alpha + \gamma \leq \beta + \gamma$. For example, the additive group of real numbers with the natural ordering of the real numbers is an ordered Abelian group. Each subgroup of this group (or, indeed, of any ordered Abelian group), with the induced ordering, is an ordered Abelian group.

Let $G_1, \ldots, G_n$ be ordered Abelian groups and let $G = G_1 \oplus \cdots \oplus G_n$. The elements of G may be denoted by n-tuples $(\alpha_1, \ldots, \alpha_n)$, where $\alpha_i \in G_i$ for $i = 1, \ldots, n$. If $(\alpha_1, \ldots, \alpha_n)$ and $(\beta_1, \ldots, \beta_n)$ are distinct elements of G we write

$$(\alpha_1, \ldots, \alpha_n) < (\beta_1, \ldots, \beta_n)$$

if $\alpha_1 < \beta_1$ or if, for some $k > 1$, $\alpha_i = \beta_i$ for $i = 1, \ldots, k-1$ and $\alpha_k < \beta_k$. We leave it to the reader to show that this is a total ordering on G and that G, together with this ordering, is an ordered Abelian group. We refer to this ordering as the **lexicographic ordering** of G.

Let G be an ordered Abelian group and let $\{\infty\}$ be a set whose single element is not an element of G. Let $G^* = G \cup \{\infty\}$ and make G^* into a commutative semigroup by defining, for α and β in G^*,

$$\alpha + \beta = \begin{cases} \text{their sum in } G & \text{if } \alpha, \beta \in G, \\ \infty & \text{if } \alpha = \infty \text{ or } \beta = \infty. \end{cases}$$

We extend the ordering of G to an ordering of G^* by defining $\alpha \leq \infty$ for all $\alpha \in G^*$. Then G^* is an **ordered semigroup** in the sense that if $\alpha, \beta, \gamma \in G^*$ and $\alpha \leq \beta$ then $\alpha + \gamma \leq \beta + \gamma$.

5.12. Definition. *Let K be a field. A* **valuation** *on K is a mapping v from K onto G^*, where G is an ordered Abelian group, such that*

(i) $v(a) = \infty$ *if and only if* $a = 0$,
(ii) $v(ab) = v(a) + v(b)$ *for all* $a, b \in K$,
(iii) $v(a + b) \geq \min\{v(a), v(b)\}$ *for all* $a, b \in K$.

The group G is called the **value group** of the valuation v.

The mapping v from a field K into G^* given by $v(a) = 0$ if $a \neq 0$ and $v(0) = \infty$ is clearly a valuation on K, called a **trivial valuation.**

Let v be a valuation on a field K and set

$$V = \{a \mid a \in K \quad \text{and} \quad v(a) \geq 0\}.$$

If $a, b \in V$, then $v(ab) = v(a) + v(b) \geq 0$ and

$$v(a + b) \geq \min\{v(a), v(b)\} \geq 0,$$

so that $ab, a + b \in V$. Since $v(-1) = v(1) = 0$, and hence $-1 \in V$, we see that V is a subring of K. Let $a \in K$, $a \neq 0$; if $a \notin V$ then $v(a) < 0$, so $v(1/a) = -v(a) > 0$. Thus $1/a \in V$. Therefore V is a valuation ring. Note that K is the quotient field of V. The maximal ideal of V is $\{a \mid a \in K$ and $v(a) > 0\}$. We shall now show that all valuation rings are determined by valuations in this way.

5.13. Proposition. *Let V be a valuation ring with quotient field K. Then there is a valuation v on K such that*

$$V = \{a \mid a \in K \quad \text{and} \quad v(a) \geq 0\}.$$

Proof. Let U be the (multiplicative) group of units of V. Then U is a subgroup of K^*, the multiplicative group of nonzero elements of K. Let $G = K^*/U$; we write G additively, so that $aU + bU = abU$. Define a relation on G by $bU \leq aU$ if $a/b \in V$. This is a well-defined relation. If $bU = b'U$ and $aU = a'U$, then $b'/b \in U$ and $a'/a \in U$; hence $a/b \in V$ if and only if $a'/b' \in V$. It is easy to see that we have defined a partial ordering on G; it is a total ordering since if $a, b \in K^*$, then either a/b or b/a is in V. Finally, G together with this ordering is an ordered Abelian group. For, if $bU \leq aU$ and if cU is any element of G, then $ac/bc \in V$ and so $bU + cU = bcU \leq acU = aU + cU$.

Now, define $v: K \to G^*$ by $v(0) = \infty$ and $v(a) = aU$ if $a \neq 0$. Then

(i) and (ii) from the definition of valuation hold, and it follows that $v(1)=0$. Furthermore, $v(a+b) \geq \min\{v(a), v(b)\}$ if either $a=0$ or $b=0$. Suppose $a, b \in K^*$ and that $bU \leq aU$. Then $a/b \in V$ so $a/b+1 \in V$. Hence $v(a/b+1) \geq v(1)=0$. It follows that

$$v(a+b)=v((a/b+1)b)=v(a/b+1)+v(b) \geq v(b)=\min\{v(a), v(b)\}.$$

Therefore, v is a valuation on K, and

$$V=\{a \mid a \in K \quad \text{and} \quad a/1 \in V\}=\{a \mid a \in K \quad \text{and} \quad v(a) \geq 0\}.$$

If V and v are related as in this proposition, we say that v is the **valuation determined by** V. If v is an arbitrary valuation on K, then $\{a \mid a \in K \text{ and } v(a) \geq 0\}$ is called the **valuation ring** *of* v.

Let v and v' be valuations on a field K, with value groups G and G', respectively. We say that v and v' are **equivalent** if there is an order-preserving isomorphism ϕ from G onto G' such that $v'(a)=\phi(v(a))$ for all $a \in K^*$. This relation between valuations is reflexive, symmetric, and transitive. It is clear that equivalent valuations have the same valuation ring.

Conversely, two valuations on a field K having the same valuation ring are equivalent. To verify this, we shall show that if v is a valuation on K, if V is the valuation ring of v, and if v' is the valuation determined by V, then v and v' are equivalent. Let G be the value group of v and U be the group of units of V. Define $\phi: G \to K^*/U$ by $\phi(v(a))=aU$. If $v(a)=v(b)$, then $v(a/b)=1$, so that $a/b \in U$ and $aU=bU$; thus ϕ is well-defined. Since $\phi(v(a))=v'(a)$ for all $a \in K^*$, it remains to show that ϕ is an order-preserving isomorphism. Now, ϕ is a homomorphism, since

$$\phi(v(a)+v(b))=\phi(v(ab))=abU=aU+bU=\phi(v(a))+\phi(v(b));$$

it is injective since $\phi(v(a))=0$ implies that $a \in U$ and so $v(a)=0$, and it is clearly surjective. Finally, if $v(a) \leq v(b)$, then $b/a \in V$ and consequently $aU \leq bU$, that is, $\phi(v(a)) \leq \phi(v(b))$.

Let G be an ordered Abelian group. A subgroup H of G is called an **isolated subgroup** if for each nonnegative element α of H, $0 \leq \beta \leq \alpha$ implies that $\beta \in H$. If H is an isolated subgroup of G and $H \neq G$, then H is termed **proper**.

5.14. Definition. *If an ordered Abelian group G has only a finite number of isolated subgroups, then the number of proper isolated subgroups of G is the* **rank** *of G.*

Thus, G has **rank one** if and only if $G \neq 0$ and G and 0 are the only isolated subgroups of G.

5.15. Proposition. *A nonzero ordered Abelian group G has rank one if and only if there is an order-preserving isomorphism from G onto a subgroup of the additive group of real numbers.*

Proof. Let G' be a nonzero subgroup of the additive group of real numbers; we shall show that G' has rank one, thus showing the sufficiency of the condition. Let H be a nonzero isolated subgroup of G'. Then H contains a positive element α. If β is a positive element of G', then by the Archimedian property of the ordering of the real numbers there is an integer n such that $0 < \beta < n\alpha$. But $n\alpha \in H$, so $\beta \in H$. Thus, H contains all positive elements of G' and so must coincide with G'.

Conversely, suppose that G has rank one. We shall show first that if α and β are positive elements, then there is an integer n such that $\beta \leq n\alpha$. Suppose that this assertion is not true for some pair of positive elements α and β of G. Let

$$S = \{\gamma \mid \gamma \in G, \gamma \geq 0, \quad \text{and} \quad \gamma \leq n\alpha \quad \text{for some integer} \quad n\}.$$

Then $\beta \notin S$ and if $\gamma_1, \gamma_2 \in S$, then $\gamma_1 + \gamma_2 \in S$. Let H be the subgroup of G generated by S; then H consists of all elements of G of the form $\gamma_1 - \gamma_2$ where $\gamma_1, \gamma_2 \in S$. Consider a positive element of H, say $\gamma_1 - \gamma_2$ where $\gamma_1, \gamma_2 \in S$, and suppose $0 \leq \delta \leq \gamma_1 - \gamma_2$. If $\gamma_1 \leq n\alpha$, then since $\gamma_2 \geq 0$, we have $\delta \leq \delta + \gamma_2 \leq \gamma_1 \leq n\alpha$; hence $\delta \in H$. Thus H is isolated, and $H \neq 0$, so $H = G$. But $\beta \notin H$ so we have arrived at a contradiction. Therefore, the assertion is true.

Suppose that the set of positive elements of G has a least element α. If β is any positive element then there is a positive integer n such that $(n-1)\alpha < \beta \leq n\alpha$. Then $0 < \beta - (n-1)\alpha \leq \alpha$, and so $\beta - (n-1)\alpha = \alpha$, or $\beta = n\alpha$. If γ is a negative element of G and if $-\gamma = m\alpha$, then $\gamma = (-m)\alpha$. Thus G is an infinite cyclic group, generated by α, and the mapping ϕ such that $\phi(n\alpha) = n$ is the desired order-preserving isomorphism.

On the other hand, suppose that the set of positive elements of G does not have a least element. Choose, and fix, a positive element α and set $\phi(\alpha)=1$. Let β be any other positive element of G and set

$$l(\beta)=\{m/n \mid m\alpha \leq n\beta\}$$

and

$$u(\beta)=\{m/n \mid m\alpha > n\beta\},$$

where m and n represent positive integers. There exist positive integers m and n such that $\alpha \leq n\beta$ and $\beta < m\alpha$. Then $1/n \in l(\beta)$ and $m/1 \in u(\beta)$, so $l(\beta)$ and $u(\beta)$ are not empty. Suppose that $m/n \in l(\beta)$ and $h/k \in u(\beta)$. Then $m\alpha \leq n\beta$ and $k\beta < h\alpha$; thus $mk\alpha < nh\alpha$. Hence $mk < nh$ and $m/n < h/k$. It follows that $l(\beta)$, together with the set of nonpositive rational numbers, and $u(\beta)$ are the lower and upper classes, respectively, of a Dedekind cut of the rational numbers. Hence, they determine a (positive) real number $\phi(\beta)$. We set $\phi(0)=0$ and if γ is a negative element of G, $\phi(\gamma)=-\phi(-\gamma)$. We leave it to the reader to show that ϕ is the desired order-preserving isomorphism.

5.16. Definition. *Let v be a valuation on a field K and let G be its value group. Let V be the valuation ring of v.*

(1) *If G has rank n, then v and V have* **rank** n.
(2) *If G is cyclic, v is a* **discrete valuation** *and V is a* **discrete valuation ring**.

Note that if v is a discrete valuation, then it has either rank one or rank zero; in the latter case, v is trivial.

It would be surprising if there were not some relationship between the ideal structure of a valuation ring and the group structure of the value group of the valuation determined by the valuation ring. We now explore this relationship.

5.17. Theorem. *Let v be a valuation on a field K. Let G be its value group and let V be its valuation ring. Then there exists a one-to-one order-reversing correspondence between the isolated subgroups of G and the proper prime ideals of V.*

Proof. Let $\mathscr{I}$ denote the set of isolated subgroups of G and let $\mathscr{P}$ be the set of proper prime ideals of V. For $H \in \mathscr{I}$, set

$$\pi(H) = \{x \mid x \in V \quad \text{and} \quad v(x) \notin H\}$$

and for $P \in \mathscr{P}$, set

$$\kappa(P) = \{\alpha \mid \alpha \in G \quad \text{and} \quad \alpha \notin v(P) \quad \text{and} \quad -\alpha \notin v(P)\}.$$

Using the fact that a proper ideal of V is a prime ideal if and only if $V \backslash P$ is multiplicatively closed and the fact that

$$\kappa(P) = \{\alpha \mid \alpha \in G \quad \text{and} \quad \alpha \text{ and } -\alpha \quad \text{belong to} \quad v(V \backslash P)\},$$

one can easily verify that π is a mapping of $\mathscr{I}$ into $\mathscr{P}$ and κ is a mapping of $\mathscr{P}$ into $\mathscr{I}$. Clearly π and κ are both order-reversing. We now show that they are inverse mappings of one another. If $x \in P \in \mathscr{P}$, when $v(x) \in v(P)$, so $v(x) \notin \kappa(P)$. Hence $x \in \pi(\kappa(P))$, which proves that $P \subseteq \pi(\kappa(P))$. On the other hand, $x \in V \backslash P$ implies that $v(x) \notin v(P)$, and so $v(x) \in \kappa(P)$; that is, $x \notin \pi(\kappa(P))$. Consequently, $\pi(\kappa(P)) \subseteq P$. A similar argument can be used to show that for each $H \in \mathscr{I}$, $\kappa(\pi(H)) = H$.

It is worthwhile to note that this theorem says that if a valuation ring has rank n, then there exists a chain $0 \subset P_1 \subset \cdots \subset P_n$ of proper prime ideals of V, but no longer such chain exists.

We now show that the only Noetherian valuation rings are those which are fields or which have rank one and are discrete.

5.18. Theorem. *A valuation ring which is not a field has rank one and is discrete if and only if it is Noetherian.*

Proof. Let V be a Noetherian valuation ring which is not a field. If P is the unique maximal ideal of V, then by Theorem 5.9, P is a nonzero principal ideal and $\bigcap_{n=1}^{\infty} P^n = 0$. Let $P = (a)$. If $b \in V$, $b \neq 0$, we can write uniquely $b = ua^n$, where u is a unit in V and n is a nonnegative integer. In fact, every element of K, the quotient field of V, can be written uniquely in this way if we allow n to be a negative integer. If U is the group of units of V, then it is easily seen that the mapping $bU \mapsto n$ is an order-preserving isomorphism from K^*/U onto the additive group of integers. Therefore, V has rank one and is discrete.

Conversely, suppose that V is a valuation ring that has rank one and is discrete, and let v be a valuation on K having V as its valuation ring and the additive group of integers as its value group. Let A be a nonzero ideal of V. There is an element $a \in A$ such that $v(a) = \min\{v(b) \mid b \in A\}$. Let $c \in A$, $c \neq 0$; then $v(a) \leq v(c)$, so $v(c/a) \geq 0$. Thus $c/a \in V$ and $c \in Va$. Therefore, $A = Va$. It follows that V is Noetherian.

Now we show the role of complete integral closure in the theory of valuation rings. All valuation rings are integrally closed, but those that are completely integrally closed have rank one or are fields.

5.19. Theorem. *Let V be a valuation ring which is not equal to its quotient field K. Then V is completely integrally closed if and only if V has rank one.*

Proof. Suppose that V is completely integrally closed. Let P be the maximal ideal of V and let Q be another proper prime ideal of V. If $x \in P \backslash Q$, then $Q \subset (x^n)$ for each positive integer n. Hence $Q \subseteq \bigcap_{n=1}^{\infty} (x^n)$. By Corollary 4.21, $\bigcap_{n=1}^{\infty} (x^n) = 0$. Hence $Q = 0$, and P is the only nonzero proper prime ideal of V. Therefore V has rank one.

Conversely, suppose that V has rank one and let $x \in K \backslash V$. Then $x^{-1} \in V$ and by Theorem 5.10, $\bigcap_{n=1}^{\infty} (x^{-n})$ is a prime ideal of V. It is a proper ideal of V, so either it is the zero ideal or it is the maximal ideal of V. In the latter case, $(x^{-2}) = (x^{-1})$ and it follows easily that $x \in V$, contrary to assumption. Hence $\bigcap_{n=1}^{\infty} (x^{-n}) = 0$. Let r be any nonzero element of V; there exists an integer n such that $(r) \nsubseteq (x^{-n})$, which implies that $r \notin (x^{-n})$, that is, $rx^n \notin V$. Therefore x is not in the complete integral closure of V by Theorem 4.20. Consequently V is completely integrally closed.

We close this section with two examples of valuations that will prove useful later.

Let K be a field and let X be an indeterminate. Let $p(X)$ be a nonconstant monic irreducible polynomial in $K[X]$. If $f(X)/g(X)$ is a nonzero element of the field $K(X)$ of rational functions in X over K, we can write

$$f(X)/g(X) = p(X)^n(f_1(X)/g_1(X)),$$

where $f_1(X)$ and $g_1(X)$ are not divisible by $p(X)$ and n is an integer which may be positive, negative, or zero, and is uniquely determined by $f(X)/g(X)$. Set

$$v(f(X)/g(X)) = n.$$

If we also set $v(0) = \infty$, then v is a valuation on $K(X)$. Clearly v has rank one and is discrete. It is called the **$p(X)$-adic valuation** on $K(X)$.

Again let K be a field and X an indeterminate. Let w be a valuation on K. If $f(X) = a_0 + a_1X + \cdots + a_nX^n \in K[X]$, we set

$$w'(f(X)) = \min\{w(a_i) \mid i = 0, \ldots, n\}.$$

Then w' is a mapping from $K[X]$ into the value group of w, and w' satisfies (i), (ii), and (iii) in the definition of valuation. Hence w' determines a valuation on $K(X)$ [see Exercise 9(d)], which we also denote by w', and call the **extension** of w to $K(X)$. Note that the value groups of w and w' coincide; hence w and w' have the same rank. In particular, if w has rank one and is discrete, the same is true of w'.

4 PROLONGATION OF VALUATIONS

5.20. Definition. *Let v be a valuation on a field K and let K' be an extension of K. A valuation v' on K' is a* **prolongation** *of v if there is an order-preserving injective homomorphism ϕ from the value group of v into the value group of v' such that $v'(a) = \phi(v(a))$ for all $a \in K^*$, the multiplicative group of nonzero elements of K.*

It is clear that if v' is a valuation on K' which is a prolongation of v, and if v'' is a valuation on K' which is equivalent to v', then v'' is a prolongation of v.

5.21. Proposition. *Let v be a valuation on a field K, let K' be an extension of K, and let v' be a valuation on K'. Let V and V' be the valuation rings of v and v', respectively, and let P be the maximal ideal of V and P' that of V'. Then the following statements are equivalent:*

(1) v' *is a prolongation of* v.
(2) $V' \cap K = V$.
(3) $P' \cap V = P$.

Proof. (1) $\Rightarrow$ (2). Let v' be a prolongation of v; then there is an order-preserving injective homomorphism ϕ from the value group of v into that of v' such that $v'(a) = \phi(v(a))$ for all $a \in K^*$. Hence if $a \in K$, then $v(a) \geq 0$ if and only if $v'(a) \geq 0$. Therefore $V' \cap K = V$.

(2) $\Rightarrow$ (3). If $V' \cap K = V$, then $P' \cap V = P' \cap V' \cap K = P' \cap K$; we shall show that $P' \cap K = P$. If $a \in P$, $a \neq 0$, then $1/a \notin V$; hence $1/a \notin V'$ and it follows that $a \in P' \cap K$. Conversely, if $a \in P' \cap K$, then $1/a \notin V'$; hence $1/a \notin V$ and consequently $a \in P$. Therefore $P = P' \cap K$.

(3) $\Rightarrow$ (1). Assume that $P' \cap V = P$. First we shall show that $U' \cap K^* = U$, where U and U' are the groups of units of V and V', respectively. Let $a \in U' \cap K^*$. Then $a \notin P$, so $v(a) \leq 0$. Also $1/a \in U' \cap K^*$ and it follows that $-v(a) = v(1/a) \leq 0$. Hence $v(a) = 0$ and so $a \in U$. On the other hand, if $a \in U$, then $a \notin P'$ so $v'(a) \leq 0$. Likewise $-v'(a) = v'(1/a) \leq 0$. Consequently $v'(a) = 0$ and $a \in U' \cap K^*$. Note that we can now conclude that $V \subseteq V'$.

Since $U' \cap K^* = U$, the mapping $\phi\colon K^*/U \to K'^*/U'$ given by $\phi(aU) = aU'$ is a well-defined injective homomorphism. If $bU \leq aU$, then $a/b \in V$ and we conclude that $a/b \in V'$; hence $bU' \leq aU'$. Thus ϕ is order-preserving. There is an order-preserving isomorphism ψ from the value group of v onto K^*/U such that $\psi(v(a)) = aU$ for all $a \in K^*$, and an order-preserving isomorphism ψ' from the value group of v' onto K'^*/U' such that $\psi'(v'(a)) = aU'$ for all $a \in K'^*$. Then $\psi'^{-1}\phi\psi$ is an order-preserving injective homomorphism and

$$\psi'^{-1}\phi\psi(v(a)) = \psi'^{-1}\phi(aU) = \psi'^{-1}(aU') = v'(a)$$

for all $a \in K$. Thus v' is a prolongation of v.

5.22. Theorem. *Let* v *be a valuation on a field* K, *and let* K' *be an algebraic extension of* K. *If* v' *is a valuation on* K' *which is a prolongation of* v, *then* v' *has rank* r *if and only if* v *has rank* r.

Proof. Let G and G' be the value groups of v and v', respectively. For purposes of this proof we may assume that G is a subgroup of G' and that $v'(a) = v(a)$ for all $a \in K$. Let $\alpha \in G'$ and let $a \in K'$ be such that $v'(a) = \alpha$. There are elements $b_0, \ldots, b_n \in K$, with $b_n = 1$, such that $b_0 + b_1 a + \cdots + b_n a^n = 0$. If $v'(b_i a^i) \neq v'(b_j a^j)$ whenever $i \neq j$, then

$$\infty = v'(0) = \min\{v'(b_i a^i) \mid i = 1, \ldots, n\} \in G'.$$

Hence, for some i and j with $i > j$ we must have $v'(b_i a^i) = v'(b_j a^j)$, and $b_i \neq 0 \neq b_j$. Then $(i-j)\alpha = v'(a^{i-j}) = v(b_j/b_i) \in G$. Thus every element of G'/G is of finite order, and so G' has rank r if and only if G has rank r by Exercise 14(a).

5.23. Corollary. *Let V be a valuation ring with quotient field K. Let K' be an algebraic extension of K and let R' be the integral closure of V in K'. Let V' be a valuation ring in K', with maximal ideal P', such that $R' \subseteq V'$ and $P' \cap R'$ is a maximal ideal of R'. Then V' has rank r if and only if V has rank r.*

Proof. If $P'' = P' \cap R'$ then $P'' \cap V$ is the maximal ideal of V by Exercise 3(a) of Chapter IV. Hence $P' \cap V$ is the maximal ideal of V, and the assertion follows from Proposition 5.21 and Theorem 5.22.

If K' is an extension of K which is not algebraic over K, then a valuation on K' which is a prolongation of a valuation of rank r on K can have rank greater than r. The next theorem considers this situation. It is stated in terms of valuation rings because of later applications.

5.24. Theorem. *Let V be a valuation ring of rank r, let K be the quotient field of V, and let K' be the quotient field of the polynomial ring $V[X_1, \ldots, X_n]$. Then there is a valuation ring V' contained in K' such that $V = V' \cap K$ and $X_1, \ldots, X_n \in V'$. Every such V' has rank which is $\leq r + n$, and at least one such V' has rank $r + n$.*

Proof. An induction argument shows that it is sufficient to prove the assertions when $n = 1$. Let v be the valuation determined by V and let v' be a valuation on $K(X)$ which is a prolongation. [Note that $K(X)$ is contained in the quotient field of $V[X]$, and so it must be the quotient field of $V[X]$.] Let G and G' be the value groups of v and v', respectively; we may assume that G is a subgroup of G'

and that $v'(a) = v(a)$ for all $a \in K$. If every element of G'/G has finite order then v' has rank r by Exercise 14(a). Suppose this is not the case; then there is a positive element $\alpha \in G'$ such that $k\alpha \notin G$ for all positive integers k. Let $a \in K(X)$ be such that $v'(a) = \alpha$. By the same argument as that used in the proof of Theorem 5.22, it follows that a is transcendental over K. Let v'' be the restriction of v' to $K(a)$. If we show that v'' has rank $\leq r+1$, the same will be true of v', since $K(X)$ is algebraic over $K(a)$ and v' is a prolongation of v''.

Therefore, to prove v' has rank $\leq r+1$ it is sufficient to do so under the assumption that $v'(X) = \alpha$. First, we shall show that $G' = G + \langle\alpha\rangle$, where $\langle\alpha\rangle$ is the subgroup of G' generated by α. It is clear that if $G' = G + \langle\alpha\rangle$, then the sum is direct. Let $f(X) = b_0 + b_1X + \cdots + b_nX^n \in K[X]$. Then $v'(b_iX^i) \neq v'(b_jX^j)$ if $i \neq j$. Hence

$$v'(f(X)) = \min\{(v(b_i) + i\alpha \mid i = 0, \ldots, n\} \in G + \langle\alpha\rangle.$$

Thus, $G' = G + \langle\alpha\rangle$.

Let $0 \subset H_1 \subset \cdots \subset H_s$ be a chain of distinct proper isolated subgroups of G', and consider the chain $0 \subseteq H_1 \cap G \subseteq \cdots \subseteq H_s \cap G$. If $H_i \nsubseteq G$ then $H_i \cap G \subset H_{i+1} \cap G$. For, let $\gamma = \beta + p\alpha \in H_i \backslash G$ and $\gamma' = \beta' + q\alpha \in H_{i+1} \backslash H_i$, where $\beta, \beta' \in G$ and p, q are integers. Since $\gamma' \notin H_i$ we have $p\gamma' \notin H_i$, and so $p\gamma' - q\gamma \in H_{i+1} \backslash H_i$. Hence, since $p\gamma' - q\gamma = p\beta' - q\beta \in G$, we have

$$p\gamma' - q\gamma \in (H_{i+1} \cap G) \backslash (H_i \cap G).$$

If $H_i \nsubseteq G$ for $i = 1, \ldots, s$, then

$$0 \subseteq H_1 \cap G \subset \cdots \subset H_s \cap G \subseteq G.$$

Otherwise, let j be the largest integer such that $H_j \subseteq G$; then

$$0 \subset H_1 \subset \cdots \subset H_j \subseteq H_{j+1} \cap G \subset H_{j+2} \cap G \subset \cdots \subset H_s \cap G \subseteq G.$$

If $H_s \cap G \neq G$, then in either of these two cases we have $s \leq r$; hence $s + 1 \leq r + 1$. Suppose, on the other hand, that $H_s \cap G = G$. We claim that $H_s = G$. If $H_s \neq G$ then H_s/G is isomorphic to a subgroup $\langle h\alpha\rangle$ of $\langle\alpha\rangle$. Then $G'/H_s \cong (G'/G)/(H_s/G) \cong \langle\alpha\rangle/\langle h\alpha\rangle$, which is a finite nontrivial group. Since H_s is a proper isolated subgroup of G' this is impossible. Since $H_s = G$, $j = s$, and

$$0 \subset H_1 \subset \cdots \subset H_s = G.$$

Thus $s \leq r$, and $s + 1 \leq r + 1$. It follows that G' has rank $\leq r + 1$.

To complete the proof of the theorem we shall show that there is a valuation v' on $K(X)$ such that v' has rank $r+1$, v' is a prolongation of v, and $v'(X) > 0$. Let Z be the ordered additive group of integers and let $G' = G \oplus Z$. Order G' lexicographically; by Exercise 14(b), G' has rank $r+1$. Let w be the X-adic valuation on the field of quotients of the polynomial ring $(V/P)[X]$, where P is the maximal ideal of V. Let $f(X) \in V[X], f(X) \neq 0$. Write $f(X) = df_1(X)$, where $d \in V$, $f_1(X) \in V[X]$, and at least one coefficient of $f_1(X)$ is a unit in V. Then d is uniquely determined to within a unit factor. Let $\bar{f}_1(X)$ be obtained by replacing each coefficient of $f_1(X)$ by its canonical image in V/P. Set $v'(f(X)) = v(d) + w(\bar{f}_1(X))$. We shall leave it to the reader to show that v' is a well-defined valuation on $K(X)$ having value group G'. Clearly it is a prolongation of v, and so the theorem is proved.

EXERCISES

1. Overrings of valuation rings.
Let V be a valuation ring.
(a) Show that if V' is an overring of V, then there is a prime ideal P of V such that $V' = V_P$.
(b) Prove that the set of overrings of V is totally ordered.

2. Valuations and homomorphisms.
Let v be a valuation on a field K and let V be its valuation ring. Let M be the maximal ideal of V.
(a) Prove that if ϕ is an isomorphism of a field L onto K, then $v\phi$ is a valuation on L with valuation ring $\phi^{-1}(V)$.
(b) Let π be the canonical homomorphism from V onto V/M. Let V' be a valuation subring of V/M. Prove that $W = \pi^{-1}(V')$ is a valuation ring and that $W_M = V$.
(c) Show that if V has rank r and V' has rank r', then W has rank $r + r'$.

3. Existence theorem for valuation rings.
(a) Let R be a subring of a field K and let P be a nonzero prime ideal of R. Prove that there exists a valuation ring V of K which contains R and which has a prime ideal M lying over P, that is, with $M \cap R = P$. [Hint. Consider the

set of all subrings T of K which contain R and are such that $PT \neq T$, and pick a maximal element of this set.]

(b) Extend (a) as follows: let R be a subring of a field K and let $0 \subset P_1 \subset \cdots \subset P_m$ be a chain of prime ideals of R. Prove that there is a valuation ring V of K containing R such that V has prime ideals $M_1, \ldots, M_m$ which lie over $P_1, \ldots, P_m$, respectively.

4. Places and valuation rings.
Let K be a field and consider the set $K \cup \{\infty\}$, where ∞ represents an element not in K. Define $a + \infty = \infty + a = \infty$ for $a \in K \cup \{\infty\}$ and $a\infty = \infty a = \infty$ for nonzero $a \in K \cup \{\infty\}$. Now let K and $\bar{K}$ be fields. A mapping $\phi: K \cup \{\infty\} \to \bar{K} \cup \{\infty\}$ which preserves addition and multiplication and is such that $\phi(1) = 1$ is called a **place of K having values in $\bar{K}$**.

(a) Let ϕ be a place of K. If we set $1/0 = \infty$ and $1/\infty = 0$, show that $\phi(0) = 0$, $\phi(\infty) = \infty$, and $\phi(1/a) = 1/\phi(a)$ for all $a \in K \cup \{\infty\}$.

(b) Let V be a valuation ring with quotient field K. Let P be the maximal ideal of V and set $\bar{K} = V/P$. Define $\phi: K \cup \{\infty\} \to \bar{K} \cup \{\infty\}$ by setting $\phi(a) = a + P$ if $a \in V$ and $\phi(a) = \infty$ if $a \notin V$. Show that ϕ is a place of K having values in $\bar{K}$.

(c) Let ϕ be a place of K having values in $\bar{K}$. Let $V = \{a \mid a \in K \text{ and } \phi(a) \in \bar{K}\}$. Show that V is a valuation ring and that the maximal ideal of V is $\{a \mid a \in K \text{ and } \phi(a) = 0\}$.

5. Ideals of valuation rings.
Let V be a valuation ring with quotient field K. Let A be an ideal of V.

(a) Show that if A is finitely generated, then A is principal.

(b) Show that every prime ideal of V which is properly contained in A is contained in every power of A.

(c) Show that if B is an ideal of V with $A \subset \text{Rad}(B)$, then B contains a power of A.

6. Primary ideals of valuation rings.
Let V be a valuation ring and let P be a nonzero proper prime ideal of V.

(a) Show that if there is a finitely generated P-primary ideal of V, then P is the maximal ideal of V.

(b) Show that the following statements are equivalent:
  (1) There is an ideal A of V such that $A \neq P$ and $\text{Rad}(A) = P$.
  (2) P is the radical of a principal ideal.
  (3) P is not the union of the chain of prime ideals properly contained in P.
  (4) There is a prime ideal M of V such that $M \subset P$ and such that there are no prime ideals M' of V with $M \subset M' \subset P$.
  (5) There exists a P-primary ideal of V distinct from P.

A prime ideal which satisfies these conditions is said to be **branched**. If P is the only P-primary ideal of V, then P is said to be **unbranched**.

7. Noetherian valuation rings.

(a) Let R be a Noetherian integral domain which is not a field and which has a unique maximal ideal M. Prove that the following statements are equivalent:
  (1) R is a valuation ring.
  (2) As a vector space over R/M, M/M^2 has dimension one.
  (3) Every nonzero ideal of R is a power of M.

(b) Let R be an integral domain and let $P = (a)$ be a nonzero principal prime ideal of R such that $\bigcap_{n=1}^{\infty} P^n = 0$. Show that R_P is a Noetherian valuation ring.

8. Primary ideals of a rank one valuation ring.

Let V be a rank one valuation ring and let P be its maximal ideal. Let Q and Q_1 be P-primary ideals of V.

(a) Show that $\bigcap_{n=1}^{\infty} Q^n = 0$.

(b) If $Q^n = Q^{n+1}$ for some positive integer n, show that $Q = Q^2 = P$.

(c) If $Q \subseteq Q_1 \subset P$, show that $Q_1{}^n \subseteq Q$ for some positive integer n.

(d) Show that if $Q \subset P$, then $Q^2 \subset QP$.

(e) If $Q \subset Q_1$, show that $Q : Q_1 = Q$ implies that $Q_1 = P = P^2$.

9. Valuations.

Let v be a valuation on a field K.

(a) Show that if $v(a) \neq v(b)$, then $v(a+b) = \min\{v(a), v(b)\}$.
(b) Show that if K is a finite field, then $v(a) = 0$ for all nonzero $a \in K$.
(c) Let G be an ordered Abelian group, and let w be a mapping from a field K onto G^* such that $w(a) = \infty$ if and only if $a = 0$, $w(ab) = w(a) + w(b)$ for all $a, b \in K$, and $w(a) \geq 0$ implies that $w(1 + a) \geq 0$. Show that w is a valuation on K.
(d) Let R be an integral domain and K its field of quotients. Let G be an ordered Abelian group. Let w' be a mapping from R into G^* which satisfies the conditions in the definition of valuation. Show that there is a unique valuation w on K such that $w(a) = w'(a)$ for all $a \in R$.

10. Valuations on the field of rational numbers.
(a) Let p be a prime integer. For each nonzero integer a we can write uniquely $a = p^n a'$, where a' is not divisible by p. Set $v_p(a) = n$, and set $v_p(0) = \infty$. Show that there is a valuation on the field of rational numbers which coincides with v_p when restricted to Z. We denote this valuation by v_p; it is called the **p-adic valuation**.
(b) Show that each nontrivial valuation on the field of rational numbers is equivalent to v_p for some prime p.

11. Valuations and places on $K(X)$.
Let K be a field and X an indeterminate.
(a) Let $p(X)$ be a nonconstant monic irreducible polynomial in $K[X]$. Prove that the $p(X)$-adic valuation on $K(X)$ is actually a valuation. (See p. 114 for definition.)
(b) Define v on $K[X]$ by $v(0) = \infty$ and $v(f(X)) = -\deg f(X)$ if $f(X) \neq 0$. Show that there is a unique valuation on $K(X)$ whose restriction to $K[X]$ coincides with v.
(c) Show that each nontrivial valuation v on K(X) such that $v(a) = 0$ for every nonzero $a \in K$ is equivalent to some $p(X)$-adic valuation or to the valuation of part (b).
(d) Let w be a valuation on K. Show that the extension of w to $K(X)$ is actually a valuation. (See p. 114, for definition.)
(e) Let K be a field and let $a \in K$. If X is an indeterminate define $\phi: K(X) \cup \{\infty\} \to K \cup \{\infty\}$ as follows: $\phi(\infty) = \infty$ and if $f(X) \in K(X)$ then $\phi(f(X)) = f(a)$. Show that ϕ is a place on $K(X)$ having values in K, and that V, as defined

in Exercise 4(c), is given by $V = K[X]_P$, where $P = (X - a)$.

12. Valuations with prescribed value groups.
Let G be an ordered Abelian group. In this exercise a valuation will be constructed having G as its value group. Let K be a field. Let $\{X_g | g \in G\}$ be a set of indeterminates. Define a mapping v from $K[\{X_g | g \in G\}]$ into G^* as follows: $v(0) = \infty$; if $m = aX_{g1}^{n_1} \cdots X_{gk}^{nk}$ is a nonzero monomial, then $v(m) = \sum_{i=1}^{k} n_i g_i$; and if $f = m_1 + \cdots + m_h$, where $m_1, \ldots, m_h$ are distinct nonzero monomials, then $v(f) = \min\{v(m_i) | i = 1, \ldots, h\}$. It is now necessary to verify that v satisfies (ii) and (iii) of Definition 5.12; it will follow from Exercise 9(d) that there is a valuation having G as its value group. Verify the following assertions.

(a) If $m_1, \ldots, m_h$ are nonzero monomials, then $v(m_1 + \cdots + m_h) \geq \min\{v(m_i) | i = 1, \ldots, h\}$.

(b) If $f_1, f_2 \in K[\{X_g | g \in G\}]$, then $v(f_1 + f_2) \geq \min\{v(f_1), v(f_2)\}$. This verifies (iii).

(c) If $f_1, f_2 \in K[\{X_g | g \in G\}]$, then $v(f_1 f_2) = v(f_1) + v(f_2)$. [Hint. Write $f_1 = m_1 + \cdots + m_r$ and $f_2 = m_1' + \cdots + m_s'$ as sums of distinct monomials, where

$$v(f_1) = v(m_1) = \cdots = v(m_h) < v(m_{h+1}) \leq \cdots \leq v(m_r)$$

and

$$v(f_2) = v(m_1') = \cdots = v(m_k') < v(m_{k+1}') \leq \cdots \leq v(m_s').$$

Set $f_1^* = m_1 + \cdots + m_h$ and $f_2^* = m_1' + \cdots + m_k'$, and consider $f_1 f_2 = f_1^* f_2^* + (f_1 f_2 - f_1^* f_2^*)$.]

13. Valuation ideals.
An ideal A of an integral domain R with quotient field K is a **valuation ideal** if there is a valuation ring V of K containing R such that $A = B \cap R$ for some ideal B of V. If v is the valuation determined by V, then A is a **v-ideal.**

(a) If v is a valuation on K which is nonnegative on R, and if A is an ideal of R, prove that the following statements are equivalent:

(1) A is a v-ideal.
(2) If $a, b \in R$, $a \in A$, and $v(b) \geq v(a)$, then $b \in A$.
(3) If V is the valuation ring of v, then $AV \cap R = A$.

(b) Prove that every prime ideal of an integral domain is a valuation ideal.
(c) Show that a primary ideal need not be a valuation ideal. {Hint. Consider the ideal (X^2, Y^2) of $K[X, Y]$.}
(d) If A and B are v-ideals of an integral domain R and C is any ideal of R, prove that Rad(A), $A \cap B$, and $A : C$ are v-ideals.
(e) If A is a v-ideal of an integral domain R, prove that Rad(A) is a prime ideal of R.
(f) Show that a valuation ideal need not be primary. {Hint. Consider the polynomial ring $K[X, Y]$. Let $G = Z \oplus Z$, ordered lexicographically, and let v be the valuation on the quotient field of $K[X, Y]$, having G as its value group, constructed as in Exercise 12. Show that (X^2, XY) is a v-ideal; it is not primary [see Exercise 11(b) of Chapter II].}

14. Ordered Abelian groups.
(a) Let G be an ordered Abelian group and let H be a subgroup of G (with the induced ordering). Show that if every element of G/H has finite order, then G has rank r if and only if H has rank r. [Hint. Show that $H' \mapsto H \cap H'$ is a one-to-one mapping from the set of isolated subgroups of G onto the set of isolated subgroups of H.]
(b) Let G_1 and G_2 be ordered Abelian groups and let $G_1 \oplus G_2$ be ordered lexicographically. Show that if G_1 has rank r_1 and G_2 has rank r_2, then $G_1 \oplus G_2$ has rank $r_1 + r_2$.
(c) Show that the direct sum of r copies of the additive group of integers, ordered lexicographically, has rank r.

CHAPTER

VI

Prüfer and Dedekind Domains

The various aspects of ring theory developed in the first five chapters will now be brought together to study an important class of integral domains, the Prüfer domains. The impetus for the study of these domains comes from algebraic number theory. Since the pioneer work of Dedekind and others, including Prüfer, an immense literature on Prüfer domains and especially on the Noetherian domains in this class, the Dedekind domains, has been developed. In the text of this chapter, we will present a number of fundamental results concerning these domains, with further properties being touched upon in the exercises.

1 FRACTIONAL IDEALS

In this section, we will discuss the notion of fractional ideal relative to an arbitrary ring, although in the remaining sections of the chapter we will be concerned exclusively with integral domains.

6.1. Definition. *A* **fractional ideal** *of a ring R is a subset A of the total quotient ring K of R such that:*

(i) *A is an R-module, that is, if $a, b \in A$ and $r \in R$, then $a - b$, $ra \in A$; and*
(ii) *there exists a regular element d of R such that $dA \subseteq R$.*

Note that for (ii) to hold it is enough for there to exist a regular element x of K such that $xA \subseteq R$. For, if such an x exists, then $x = d/s$, where d and s are regular elements of R, and $dA = sxA \subseteq R$.

Each ideal of R is a fractional ideal of R. Some authors call such fractional ideals **integral**. If $x \in K$, the total quotient ring of R, then Rx is a fractional ideal of R and will be denoted by (x); fractional ideals of this type are called **principal**.

Let A and B be fractional ideals of R. Their **sum**

$$A + B = \{a + b \mid a \in A, b \in B\}$$

and their **product**

$$AB = \{\text{finite sums} \quad \sum a_i b_i \mid a_i \in A, b_i \in B\},$$

as well as their intersection $A \cap B$, are fractional ideals of R. Furthermore, if B contains a regular element of R, then

$$[A : B] = \{x \mid x \in K \quad \text{and} \quad Bx \subseteq A\}$$

is a fractional ideal of R. It is clear that $[A : B]$ is an R-module. Let b be a regular element of R contained in B and let d be a regular element of R such that $dA \subseteq R$. Then $bd[A : B] \subseteq dA \subseteq R$. Note that B contains a regular element of R if and only if it contains a regular element of K. Note also that if A and B are ideals of R, then $[A : B]$ is not necessarily the same as $A : B$. In fact, $A : B = [A : B] \cap R$.

Denote by $\mathscr{F}(R)$ the set of all nonzero fractional ideals of R.

6.2. Definition. *A fractional ideal A of a ring R is* **invertible** *if there exists a fractional ideal B of R such that $AB = R$.*

6.3. Proposition. *Let R be a ring with total quotient ring K.*

(1) *If $A \in \mathscr{F}(R)$ is invertible, then A is finitely generated (as an R-module) and A contains a regular element of R.*

(2) *If $A, B \in \mathscr{F}(R)$ and $A \subseteq B$ and B is invertible, then there is an ideal C of R such that $A = BC$.*

(3) *If $A \in \mathscr{F}(R)$, then A is invertible if and only if there is a fractional ideal B of R such that AB is principal and generated by a regular element of K.*

Proof. (1) Let $B \in \mathscr{F}(R)$ be such that $AB = R$. Then there exist $a_1, \ldots, a_n \in A$ and $b_1, \ldots, b_n \in B$ such that $1 = \sum_{i=1}^{n} a_i b_i$. For each $x \in A$, $xb_i \in R$, for $i = 1, \ldots, n$, and $x = \sum_{i=1}^{n} a_i(xb_i)$. Thus, $a_1, \ldots, a_n$ generate A as an R-module. Let d be a regular element of R such that $dB \subseteq R$. Then $d \in dR = dAB \subseteq AR = A$.

(2) If $B' \in \mathscr{F}(R)$ is such that $BB' = R$, and if $C = AB'$, then $C \subseteq R$ and $BC = BB'A = A$.

(3) If x is a regular element of K and $B \in \mathscr{F}(R)$ is such that $AB = (x)$, then $A(Bx^{-1}) = R$. The necessity of the condition is obvious.

If A is an invertible fractional ideal of R, it follows from (1) of Proposition 6.3 that $[R: A]$ is a fractional ideal of R.

6.4. Proposition. *If $A \in \mathscr{F}(R)$ is invertible and if $B \in \mathscr{F}(R)$ is such that $AB = R$, then $B = [R: A]$.*

Proof. Since $AB = R$ we have $B \subseteq [R: A]$. Also, $A[R: A] \subseteq R$ so that $[R: A] = AB[R: A] \subseteq BR = B$.

If A is an invertible fractional ideal of R, we shall denote $[R: A]$ by A^{-1}. It is clear that if B is another invertible fractional ideal of R, then AB is invertible and $(AB)^{-1} = A^{-1}B^{-1}$. If x is an element of the total quotient ring of R, then (x) is invertible if and only if x is regular, in which case $(x)^{-1} = (x^{-1})$. Finally observe that a product, $A = A_1 \cdots A_n$ of fractional ideals of R is invertible if and only if A_i is invertible for $i = 1, \ldots, n$. If the product A is invertible, then $A_1^{-1} = A^{-1}A_2 \cdots A_n$.

2 PRÜFER DOMAINS

Let R be an integral domain. If $A, B \in \mathscr{F}(R)$, then each of $A + B$, AB, $A \cap B$, and $[A: B]$ is in $\mathscr{F}(R)$. We have seen that invertible fractional ideals of R are finitely generated. In this section we investigate integral domains for which the converse is true.

6.5. Definition. *An integral domain R is a* **Prüfer domain** *if each nonzero finitely generated ideal of R is invertible.*

We leave it to the reader to show that if R is a Prüfer domain, then each finitely generated fractional ideal of R is invertible. We shall now obtain a number of equivalent conditions for an integral domain to be a Prüfer domain.

6.6. Theorem. *If R is an integral domain, then the following statements are equivalent:*

(1) *R is a Prüfer domain.*
(2) *Every nonzero ideal of R generated by two elements is invertible.*
(3) *If $AB = AC$, where A, B, C are ideals of R and A is finitely generated and nonzero, then $B = C$.*
(4) *For every proper prime ideal P of R the ring of quotients R_P is a valuation ring.*
(5) *$A(B \cap C) = AB \cap AC$ for all ideals A, B, C of R.*
(6) *$(A + B)(A \cap B) = AB$ for all ideals A, B of R.*
(7) *If A and C are ideals of R, with C finitely generated, and if $A \subseteq C$, then there is an ideal B of R such that $A = BC$.*
(8) *$(A + B) : C = A : C + B : C$ for all ideals A, B, C of R with C finitely generated.*
(9) *$C : (A \cap B) = C : A + C : B$ for all ideals A, B, C of R with A and B finitely generated.*
(10) *$A \cap (B + C) = A \cap B + A \cap C$ for all ideals A, B, C of R.*

Proof. We begin by showing that (1) and (2) are equivalent. Clearly (1) $\Rightarrow$ (2).

(2) $\Rightarrow$ (1). Let $C = (c_1, \ldots, c_n)$ be a nonzero ideal of R; we shall show that C is invertible by induction on n, assuming that (2) holds. This is true for $n = 1$, and also for $n = 2$ by assumption. Suppose $n > 2$ and that every nonzero ideal generated by $n - 1$ elements is invertible. We may assume that $c_1, \ldots, c_n$ are all nonzero. Let

$$A = (c_1, \ldots, c_{n-1}), \qquad B = (c_2, \ldots, c_n),$$
$$D = (c_1, c_n), \qquad E = c_1 A^{-1} D^{-1} + c_n B^{-1} D^{-1}.$$

Then

$$\begin{aligned} CE &= (A + (c_n))c_1 A^{-1} D^{-1} + ((c_1) + B)c_n B^{-1} D^{-1} \\ &= c_1 D^{-1} + c_n c_1 A^{-1} D^{-1} + c_1 c_n B^{-1} D^{-1} + c_n D^{-1} \\ &= c_1 D^{-1}(R + c_n B^{-1}) + c_n D^{-1}(R + c_1 A^{-1}). \end{aligned}$$

Since $c_n B^{-1} \subseteq R$ and $c_1 A^{-1} \subseteq R$, this gives

$$CE = c_1 D^{-1} + c_n D^{-1} = (c_1, c_n)D^{-1} = R.$$

Thus C is invertible.

(1) $\Rightarrow$ (3). Suppose $AB = AC$ where A is finitely generated and nonzero. If A is invertible, then

$$B = A^{-1}AB = A^{-1}AC = C.$$

(3) $\Rightarrow$ (4). Suppose that (3) holds for R. If A, B, C are ideals of R with A finitely generated and nonzero, and if $AB \subseteq AC$, then $B \subseteq C$. For we have $AC = AB + AC = A(B + C)$; hence $C = B + C$ and consequently $B \subseteq C$.

Let P be a proper prime ideal of R. We must show that if a/s, $b/t \in R_P$, then $(a/s) \subseteq (b/t)$ or $(b/t) \subseteq (a/s)$. However, since we may assume that $s, t \notin P$, and therefore that $1/s$ and $1/t$ are units in R_P, it is sufficient to show that either $aR_P \subseteq bR_P$ or $bR_P \subseteq aR_P$. This is certainly true if either $a = 0$ or $b = 0$, so we may assume $a \neq 0$ and $b \neq 0$. We have $(ab)(a, b) \subseteq (a^2, b^2)(a, b)$, and it follows that $(ab) \subseteq (a^2, b^2)$. Then $ab = xa^2 + yb^2$ for some $x, y \in R$. Thus, $(yb)(a, b) \subseteq (a)(a, b)$, and so $(yb) \subseteq (a)$. Let $yb = au$. Then $ab = xa^2 + uab$ or $xa^2 = ab(1 - u)$. If $u \notin P$, then $a = b(y/u) \in bR_P$. If $u \in P$, then $1 - u \notin P$ and $b = a(x/(1 - u)) \in aR_P$.

(4) $\Rightarrow$ (5). Suppose that (4) holds for R and let P be a maximal ideal of R. Then R_P is a valuation ring and it follows easily that (5) holds for ideals of R_P. Let A, B, C be ideals of R. Then, using Exercise 4(a) and (b) of Chapter III,

$$\begin{aligned} A(B \cap C)R_P &= (AR_P)(BR_P \cap CR_P) \\ &= (AR_P)(BR_P) \cap (AR_P)(CR_P) \\ &= (AB)R_P \cap (AC)R_P \\ &= (AB \cap AC)R_P. \end{aligned}$$

This equality holds for every maximal ideal P of R. Therefore by Proposition 3.13, $A(B \cap C) = AB \cap AC$.

(5) $\Rightarrow$ (6). If (5) holds we have for all ideals A and B of R,

$$(A + B)(A \cap B) = (A + B)A \cap (A + B)B \supseteq AB,$$

and the reverse inclusion always holds.

(6) $\Rightarrow$ (2). Let $C = (c_1, c_2)$ be a nonzero ideal of R. If $c_1 = 0$ or

$c_2 = 0$, then C is invertible; hence we shall assume $c_1 \neq 0$ and $c_2 \neq 0$. Then $A = (c_1)$ and $B = (c_2)$ are invertible, and

$$C(A \cap B)B^{-1}A^{-1} = (A + B)(A \cap B)B^{-1}A^{-1} = ABB^{-1}A^{-1} = R$$

if (6) holds. Thus, C is invertible.

Up to this point in the proof we have shown the equivalence of (1) through (6).

(1) $\Rightarrow$ (7). Assume that R is a Prüfer domain. Let A and C be ideals of R with C finitely generated and with $A \subseteq C$. If $C = 0$, then $A = BC$ for every ideal B of R. If $C \neq 0$, then C is invertible and $A = BC$ for some ideal B of R by Proposition 6.3.

(7) $\Rightarrow$ (4). Let P be a proper prime ideal of R. We must show that, under the assumption of (7), if $a, b \in R$, then either $aR_P \subseteq bR_P$ or $bR_P \subseteq aR_P$. We have $(a) \subseteq (a, b)$, so $(a) = (a, b)B$ for some ideal B of R. Let $a = ax + by$ where $x, y \in B$. If $x \in P$, then $1 - x \notin P$, and so $a = b(y/(1 - x)) \in bR_P$. Since $bB \subseteq (a)$, we have $bx \in (a)$ so that if $x \notin P$, then $b \in aR_P$.

(4) $\Rightarrow$ (8). Let A, B, C be ideals of R, with C finitely generated, and let P be a maximal ideal of R. If (4) holds, then the equality in question holds for ideals of R_P. Hence, by Exercise 4(c) and (d) of Chapter III,

$$\begin{aligned}((A + B)\colon C)R_P &= (A + B)R_P\colon CR_P \\ &= (AR_P + BR_P)\colon CR_P \\ &= AR_P\colon CR_P + BR_P\colon CR_P \\ &= (A\colon C)R_P + (B\colon C)R_P \\ &= (A\colon C + B\colon C)R_P.\end{aligned}$$

Therefore, $(A + B)\colon C = A\colon C + B\colon C$ by Proposition 3.13.

(4) $\Rightarrow$ (9). If we assume (4), then for every maximal ideal P of R, and for ideals A, B, C of R with A and B finitely generated,

$$\begin{aligned}(C\colon (A \cap B))R_P &\subseteq CR_P\colon (A \cap B)R_P \\ &= CR_P\colon AR_P + CR_P\colon BR_P \\ &= (C\colon A)R_P + (C\colon B)R_P \\ &= (C\colon A + C\colon B)R_P \\ &\subseteq (C\colon (A \cap B))R_P.\end{aligned}$$

Thus $(C:(A \cap B))R_P = (C:A + C:B)R_P$ for every maximal ideal P of R. Therefore, the desired equality holds.

$(4) \Rightarrow (10)$. This is proved by a similar argument.

$(8) \Rightarrow (2)$. Let $a, b \in R$. If (8) holds, then

$$\begin{aligned} R = (a, b):(a, b) &= ((a) + (b)):(a, b) \\ &= (a):(a, b) + (b):(a, b) \\ &= (a):(b) + (b):(a). \end{aligned}$$

Let $1 = x + y$ where $xb \in (a)$ and $ya \in (b)$. Then $(xb)b \subseteq (ab)$ and $(ya)a \subseteq (ab)$. Hence $(a, b)(bx, ay) \subseteq (ab)$. But $ab = abx + aby$, so $(ab) = (a, b)(bx, ay)$. We may assume $a \neq 0$ and $b \neq 0$. Then (ab) is invertible. Therefore, (a, b) is invertible.

$(9) \Rightarrow (2)$. Let $a, b \in R$. If (9) holds then

$$\begin{aligned} R &= ((a) \cap (b)):((a) \cap (b)) \\ &= ((a) \cap (b)):(a) + ((a) \cap (b)):(b) \\ &= (b):(a) + (a):(b). \end{aligned}$$

Proceed as above.

$(10) \Rightarrow (4)$. Assume (10) holds. Let P be a proper prime ideal of R and let $a, b \in R$. Since $a \in (b) + (a - b)$ we have

$$(a) = (a) \cap ((b) + (a - b)) = ((a) \cap (b)) + ((a) \cap (a - b)).$$

Let $a = t + c(a - b)$ where $t \in (a) \cap (b)$, $c \in R$, and $c(a - b) \in (a)$. Then $cb \in (a)$ and $(1 - c)a = t - cb \in (b)$. If $c \notin P$, then $b \in aR_P$. If $c \in P$, then $1 - c \notin P$ and $a \in bR_P$. Thus, R_P is a valuation ring.

This completes the proof of Theorem 6.6.

6.7. Corollary. *An integral domain R is a Prüfer domain if and only if for every maximal ideal P of R the ring of quotients R_P is a valuation ring.*

Proof. We need to prove only the sufficiency of the condition. Let P be a proper prime ideal of R and let P' be a maximal ideal of R such that $P \subseteq P'$. Then $R \backslash P' \subseteq R \backslash P$ and so $R_{P'} \subseteq R_P$. If $R_{P'}$ is a valuation ring, it follows from Corollary 5.3 that R_P is also a valuation ring.

To conclude this section we prove some elementary results on the ideal theory in Prüfer domains. One result which is immediate, but quite important, is that given two primary ideals A and B of a Prüfer domain R, either A and B are comaximal, or they are contained in a maximal ideal of R and hence $A \subseteq B$ or $B \subseteq A$; this assertion follows from Corollary 6.7 and Proposition 3.9.

6.8. Theorem. *Let R be a Prüfer domain and let P be a prime ideal of R.*

(1) *If Q is P-primary and $x \in R \backslash P$, then $Q = Q[Q + (x)]$.*
(2) *The product of P-primary ideals of R is P-primary.*

Proof. (1) We will show that $QR_M = [Q^2 + Q(x)]R_M$ for each maximal ideal M of R. If $Q \nsubseteq M$, then $QR_M = Q^2R_M = R_M$. Let M be a maximal ideal with $Q \subseteq M$. Then QR_M is a PR_M-primary ideal of the valuation ring R_M. Since $x \notin PR_M$, $QR_M = Q(x)R_M$ by Theorem 5.11. Since $Q^2 \subseteq Q$, $Q = Q^2 + Q(x)$.

(2) Let Q_1, Q_2 be P-primary ideals of R. Then for each maximal ideal M of R such that $P \subseteq M$, Q_1R_M and Q_2R_M are PR_M-primary ideals of the valuation ring R_M. By Theorem 5.11, $Q_1R_MQ_2R_M = Q_1Q_2R_M$ is PR_M-primary. This proves that Q_1Q_2 is P-primary by Exercise 5(c) of Chapter III.

Now let R be a Prüfer domain and let P be a prime ideal of R such that P is not the only P-primary ideal of R. We can use the relation between the P-primary ideals of R and the PR_P-primary ideals of R_P to obtain information about the P-primary ideals of R. For example, the set of P-primary ideals of R is totally ordered, and it follows from Theorem 5.11 that if P_1 is the intersection of the ideals in this set, then P_1 is prime and there are no prime ideals of R properly between P and P_1. Thus, the valuation ring R_P/P_1R_P has rank one.

If A is an ideal of R let A^* be its image under the composition of the canonical homomorphisms

$$R \to R_P \to R_P/P_1R_P.$$

There is a one-to-one order-preserving correspondence between the P-primary ideals Q of R and the P^*-primary ideals of R^*, the correspondence being $Q \leftrightarrow Q^*$. By Theorem 6.8, this correspondence preserves products. It also preserves residuals; this follows from

Exercise 4(d) of Chapter III, using the fact the P-primary ideals of R are contractions of ideals of R_P (see Proposition 3.9). Hence, if Q_1 and Q_2 are P-primary ideals of R, then $(Q_1Q_2)^* = Q_1^*Q_2^*$ and $(Q_1 : Q_2)^* = Q_1^* : Q_2^*$.

The correspondence of the preceding paragraph reduces the study of the P-primary ideals of R to the study of the primary ideals of a rank one valuation ring. We can state the following theorem, which follows from Exercise 8 of Chapter V. Note that the assertions hold also when P is the only P-primary ideal of R.

6.9. Proposition. *Let Q and Q_1 be P-primary ideals of a Prüfer domain R. Then:*

(1) $\bigcap_{n=1}^{\infty} Q^n$ *is a prime ideal of R;*
(2) *If $Q^n = Q^{n+1}$ for some positive integer n, then $Q = Q^2 = P$;*
(3) *If $Q \subseteq Q_1 \subset P$, then $Q_1{}^n \subseteq Q$ for some positive integer n;*
(4) *If $P \neq P^2$, then $Q = P^n$ for some positive integer n;*
(5) *If $Q \subset P$, then $Q^2 \subset QP$;*
(6) *If $Q \subset Q_1$, then $Q : Q_1 = Q$ implies that $Q_1 = P = P^2$.*

3 OVERRINGS OF PRÜFER DOMAINS

In this section we give two characterizations of Prüfer domains in terms of their overrings.

6.10. Theorem. *An integral domain R is a Prüfer domain if and only if every overring of R is a flat R-module.*

Proof. Suppose that every overring of R is a flat overring and let P be a maximal ideal of R. By Proposition 4.13, every overring of R_P is a flat overring. Let a and b be elements of R_P and suppose that $aR_P \nsubseteq bR_P$. Then $bR_P : aR_P \neq R_P$, so $bR_P : aR_P \subseteq PR_P$, the unique maximal ideal of R_P. Now consider the ring

$$R_P[a/b] = \{f(a/b) \mid f(X) \in R_P[X]\}.$$

(We may assume $b \neq 0$, for if $b = 0$, then $bR_P \subseteq aR_P$, which is what we are trying to show.) This ring is an overring of R_P and so is a flat overring of R_P. Since $a/b \in R_P[a/b]$, we have

$$(bR_P : aR_P)R_P[a/b] = R_P[a/b]$$

by Proposition 4.12. Thus there are elements $x_1, \ldots, x_n \in bR_P : aR_P$ and $b_1, \ldots, b_n \in R_P[a/b]$ such that $x_1 b_1 + \cdots + x_n b_n = 1$. There is an integer s and elements $a_{ij} \in R_P$, $1 \leq i \leq n$, $0 \leq j \leq s$, such that

$$1 = \sum_{i=1}^{n} x_i \sum_{j=0}^{s} a_{ij}(a/b)^j = \sum_{j=0}^{s} d_j(a/b)^j,$$

where $d_j = x_1 a_{1j} + \cdots + x_n a_{nj} \in bR_P : aR_P$. Note that d_0 is not a unit in R_P, since $bR_P : aR_P \neq R_P$; hence $1 - d_0$ is a unit in R_P. If we multiply by $(1 - d_0)^{s-1}(b/a)^s$, we get

$$((1 - d_0)(b/a))^s - d_1((1 - d_0)(b/a))^{s-1} - \cdots - d_s(1 - d_0)^{s-1} - 0.$$

Thus $(1 - d_0)(b/a)$ is integral over R_P. But $R_P[(1 - d_0)(b/a)]$ is a flat overring of R_P, so it equals R_P by Theorem 4.15. Hence $(1 - d_0)b \in aR_P$, and since $1 - d_0$ is a unit in R_P, this implies that $b \in aR_P$; that is, $bR_P \subseteq aR_P$. Therefore R_P is a valuation ring. Since P is an arbitrary maximal ideal of R, we conclude that R is a Prüfer domain.

Conversely, assume that R is a Prüfer domain, Let T be an overring of R and let P be a maximal ideal of T. Then T_P is an overring of the valuation ring $R_{P \cap R}$ and so T_P is a valuation ring by Corollary 5.3. Let $x \in T_P$. If $x \notin R_{P \cap R}$, then $1/x \in R_{P \cap R}$; and since it is not a unit in $R_{P \cap R}$, we have $1/x \in (P \cap R)R_{P \cap R} \subseteq PT_P$, which is impossible since $x \in T_P$. Therefore $T_P = R_{P \cap R}$, and so T is a flat overring of R by Proposition 4.14.

We can immediately draw two corollaries from this theorem, its proof, and Proposition 4.14.

6.11. Corollary. *Every overring of a Prüfer domain is a Prüfer domain.*

6.12. Corollary. *Let T be an overring of a Prüfer domain R. Then there is a set Δ of prime ideals of R such that*

$$T = \bigcap_{P \in \Delta} R_P.$$

In fact, Δ is the set of all prime ideals P of R such that $PT \neq T$.

We shall now give a second characterization of Prüfer domains in terms of their overrings.

6.13. Theorem. *An integral domain R is a Prüfer domain if and only if every overring of R is integrally closed.*

Proof. If R is a Prüfer domain, then every overring of R is an intersection of valuation rings (Corollary 6.12). Each of these valuation rings is integrally closed, so the same is true of their intersection.

Conversely, suppose that every overring of R is integrally closed. Let P be a maximal ideal of R; we shall show that R_P is a valuation ring. Let $a \neq 0$ belong to the quotient field of R. By hypothesis, $R_P[a^2]$ is integrally closed, and since a is integral over $R_P[a^2]$, we conclude that $a \in R_P[a^2]$. Then there are elements $b_0, \ldots, b_n \in R_P$ such that $a = b_0 + b_1 a^2 + \cdots + b_n a^{2n}$. If we multiply by b_0^{2n-1}/a^{2n}, we obtain

$$(b_0/a)^{2n} - (b_0/a)^{2n-1} + b_1 b_0 (b_0/a)^{2n-2} + \cdots + b_n b_0^{2n-1} = 0.$$

Thus b_0/a is integral over R_P; hence $b_0/a \in R_P$. If b_0/a is a unit in R_P, then $a \in R_P$. If b_0/a is not a unit in R_P, then $1 - (b_0/a)$ is a unit in R_P. If we multiply the equation expressing a in terms of powers of a^2 by $1/a^{2n}$, we obtain

$$(1 - (b_0/a))(1/a)^{2n-1} - b_1(1/a)^{2n-2} - \cdots - b_n = 0.$$

Since $1 - (b_0/a)$ is a unit in R_P, it follows that $1/a$ is integral over R_P. Hence $1/a \in R_P$. Therefore R_P is a valuation ring. Since P is an arbitrary maximal ideal of R, it follows that R is a Prüfer domain.

4 DEDEKIND DOMAINS

This section will be devoted to the study of an important class of integral domains called Dedekind domains. The importance of this class lies in the fact that the ring of integers of a finite algebraic number field is a Dedekind domain. We shall show that the class of Dedekind domains is precisely the class of Noetherian Prüfer domains. Then we shall obtain a large number of equivalent conditions for a Noetherian integral domain to be a Dedekind domain; the equivalence of some of these conditions will follow from Theorem 6.6. We shall also obtain some results concerning overrings and extensions of Dedekind domains.

6.14. Definition. *An integral domain R is a* **Dedekind domain** *if every ideal of R is a product of prime ideals.*

6.15. Proposition. *For $i=1, \ldots, k$, let P_i be an invertible proper prime ideal of an integral domain R. Let $A=P_1 \cdots P_k$. Then this is the only way of writing A as a product of proper prime ideals of R, except for the order of the factors.*

Proof. Let $A=P_1' \cdots P_h'$, where P_i' is a proper prime ideal of R for $i=1, \ldots, h$. Assume P_1 is minimal among $P_1, \ldots, P_k$. Since $P_1' \cdots P_h' \subseteq P_1$, some P_i' is contained in P_1, say $P_1' \subseteq P_1$. Since $P_1 \cdots P_h \subseteq P_1'$, some P_i is contained in P_1'. But then we must have $i=1$, and so $P_1'=P_1$. Since P_1 is invertible, $P_2 \cdots P_k=P_2' \cdots P_h'$, and we can repeat the argument. Since the P_i and P_i' are proper ideals of R we must have $h=k$.

6.16. Theorem. *Let R be a Dedekind domain. Then every nonzero proper ideal of R can be written as a product of proper prime ideals of R in one and only one way, expect for the order of the factors.*

Proof. Suppose that we know that every invertible proper prime ideal of R is a maximal ideal. We shall show that every nonzero prime ideal of R is invertible. Then the theorem will follow from Proposition 6.15. Let P be a nonzero prime ideal of R. If $P=R$, then P is invertible, so we can assume that $P \neq R$. Let $a \in P$, $a \neq 0$, and write $(a)=P_1 \cdots P_k$, where each P_i is a proper prime ideal of R. Since (a) is invertible, each P_i is invertible and so maximal. Since $P_1 \cdots P_k \subseteq P$, $P_i \subseteq P$ for some i. Hence $P_i=P$, and P is invertible.

Now let P be an invertible proper prime ideal of R; we shall show that P is maximal. To do this we shall show that if $a \in R \backslash P$, then $P+(a)=R$. Suppose that for some $a \in R \backslash P$ we have $P+(a) \neq R$. Then $P+(a)=P_1 \cdots P_k$ and $P+(a^2)=Q_1 \cdots Q_n$, where the P_i and Q_i are proper prime ideals of R. Let $R'=R/P$, $P_i'=P_i/P$, $Q_i'=Q_i/P$, and $a'=a+P$. Then

$$a'R'=P_1' \cdots P_k', \qquad a'^2R'=Q_1' \cdots Q_n'.$$

Since $a' \neq 0$, the ideals $a'R'$ and a'^2R' of R' are invertible, so that the same is true of each P_i' and each Q_i'. We have

$$P_1'^2 \cdots P_k'^2 = Q_1' \cdots Q_n'.$$

Hence by Proposition 6.15, $n = 2k$, and we may so number the Q_i that for $i = 1, \ldots, k$, $Q_{2i-1} = Q_{2i} = P_i$. Thus, $(P + (a))^2 = P + (a^2)$. Then $P \subseteq (P + (a))^2$, and so if $b \in P$, we have $b = c + da$, where $c \in P^2$. Then $da \in P$, but $a \notin P$, so $d \in P$. Hence $P \subseteq P^2 + Pa$. Since P is invertible, there is a fractional ideal A such that $PA = R$. Then $R \subseteq P^2A + PAa = P + (a)$. Thus, our assumption that $P + (a) \neq R$ is false.

Since an invertible ideal is finitely generated (Proposition 6.3), and the product of invertible ideals is invertible, we have the following corollary.

6.17. Corollary. *If R is a Dedekind domain, then R is Noetherian and every nonzero proper prime ideal of R is a maximal ideal.*

Let R be a Dedekind domain. Recall that $\mathscr{F}(R)$ is the set of nonzero fractional ideals of R. As pointed out in Section 6.1, $\mathscr{F}(R)$ is closed under multiplication of fractional ideals. In fact, it is clear that with respect to this multiplication, $\mathscr{F}(R)$ is a commutative semigroup with identity element R. By Theorem 6.16 and its proof, every nonzero ideal of R has an inverse in $\mathscr{F}(R)$. Let A be an arbitrary element of $\mathscr{F}(R)$; then dA is an ideal of R for some nonzero $d \in R$. Then, if $(dA)B = R$, we have $A(dB) = R$, so that A has an inverse in $\mathscr{F}(R)$. Therefore $\mathscr{F}(R)$ is a group. We shall now prove that this is also a sufficient condition for R to be a Dedekind domain.

6.18. Proposition. *Let R be an integral domain. Then R is a Dedekind domain if and only if $\mathscr{F}(R)$ is a group with respect to multiplication.*

Proof. Let $\mathscr{S}$ be the set of all nonzero proper ideals of R which are not products of prime ideals. Under the assumption that $\mathscr{F}(R)$ is a group, we shall show that $\mathscr{S}$ is empty. Suppose that $\mathscr{S}$ is not empty. Since every nonzero ideal of R is invertible, R is Noetherian. Hence $\mathscr{S}$ has a maximal element A. Let $A \subseteq P$ where P is a maximal ideal of R; then $A \neq P$. Let $PB = R$ where $B \in \mathscr{F}(R)$. Then $AB \subseteq R$ and $A \subseteq AB$ since $R \subseteq B$. If $A \subset AB$, then AB is a product of prime ideals of R

and the same is true of $(AB)P = A$. Since this is not true, we have $A = AB$ and so $AP = A$. However, A is invertible in $\mathscr{F}(R)$, and consequently $P = R$, contrary to our choice of P. Therefore, $\mathscr{S}$ must be empty.

We can now state:

6.19. Theorem. *An integral domain R is a Dedekind domain if and only if every nonzero ideal of R is invertible.*

In the next theorem we shall give a number of equivalent conditions for a Noetherian integral domain to be a Dedekind domain.

6.20. Theorem. *If R is a Noetherian integral domain, then the following statements are equivalent:*

(1) *R is a Dedekind domain.*
(2) *R is integrally closed and every nonzero proper prime ideal of R is maximal.*
(3) *Every nonzero ideal of R generated by two elements is invertible.*
(4) *If $AB = AC$, where A, B, C are ideals of R and $A \neq 0$, then $B = C$.*
(5) *For every maximal ideal P of R, the ring of quotients R_P is a valuation ring.*
(6) *$A(B \cap C) = AB \cap AC$ for all ideals A, B, C of R.*
(7) *$(A + B)(A \cap B) = AB$ for all ideals A, B of R.*
(8) *If A and C are ideals of R and if $A \subseteq C$, then there is an ideal B of R such that $A = BC$.*
(9) *$(A + B) : C = A : C + B : C$ for all ideals A, B, C of R.*
(10) *$C : (A \cap B) = C : A + C : B$ for all ideals A, B, C of R.*
(11) *$A \cap (B + C) = A \cap B + A \cap C$ for all ideals A, B, C or R.*
(12) *If P is a maximal ideal of R, then there are no ideals of R strictly between P and P^2.*
(13) *If P is a maximal ideal of R, then every P-primary ideal of R is a power of P.*
(14) *If P is a maximal ideal of R, then the set of P-primary ideals of R is totally ordered by inclusion.*
(15) *Every overring of R is a flat overring.*
(16) *Every overring of R is integrally closed.*

Proof. Let R be a Noetherian integral domain. By Theorem 6.19, R is a Dedekind domain if and only if it is a Prüfer domain. Hence, by Theorems 6.6, 6.10, and 6.13, it follows that (1), (3)–(11), (15), and (16) are equivalent.

(1) $\Rightarrow$ (2). Assume R is a Dedekind domain. Then R is integrally closed since it is a Prüfer domain. By Corollary 6.17, every nonzero proper prime ideal of R is maximal.

(2) $\Rightarrow$ (5). Suppose that (2) holds and let P be a nonzero maximal ideal of R. Then R_P is Noetherian, and it is integrally closed by Exercise 7(b) of Chapter IV. Furthermore, R_P has exactly one nonzero proper prime ideal, namely PR_P. Therefore R_P is a valuation ring by Theorem 5.9.

(5) $\Rightarrow$ (12). Let P be a maximal ideal of R. If $P = 0$, then the assertion of (12) holds. Suppose $P \neq 0$ and that R_P is a valuation ring. Let A be an ideal of R with $P^2 \subseteq A \subseteq P$. Then A is P-primary, so $AR_P \cap R = A$ (see Proposition 3.9 and Corollary 3.10). But, either $AR_P = P^2R_P$ or $AR_P = PR_P$. Hence either $A = P^2$ or $A = P$. Note that the same argument shows that for all integers $n \geq 1$, there are no ideals of R strictly between P^n and P^{n+1}.

(12) $\Rightarrow$ (5). Assume that (12) holds and let P be a nonzero maximal ideal of R. Then PR_P is a nonzero ideal of R_P and $\bigcap_{n=1}^{\infty} P^nR_P = 0$ by Corollary 2.24. Hence $P^2R_P \neq PR_P$. Also, by (12), there are no ideals of R_P strictly between P^2R_P and PR_P. Let $P' = PR_P$ and let $a \in P'$, $a \notin P'^2$. Then $P' = aR_P + P'^2$ and by induction we show that $P' = aR_P + P'^n$ for all positive integers n. Hence

$$P' = \bigcap_{n=1}^{\infty} (aR_P + P'^n).$$

But then

$$\begin{aligned} P'/aR_P &= \left(\bigcap_{n=1}^{\infty} (aR_P + P'^n)\right) \Big/ aR_P \\ &= \bigcap_{n=1}^{\infty} (aR_P + P'^n)/aR_P \\ &= \bigcap_{n=1}^{\infty} (P'/aR_P)^n = 0, \end{aligned}$$

by Corollary 2.24, since P'/aR_P is the unique maximal ideal of the

Noetherian ring R_P/aR_P. Therefore $P' = aR_P$, and since P' is precisely the set of nonunits of R_P, it follows from Theorem 5.9 that R_P is a valuation ring.

(1) $\Rightarrow$ (13). Assume that R is a Dedekind domain and let P be a maximal ideal of R. If $P = 0$, then P is the only P-primary ideal of R. Suppose $P \neq 0$ and let Q be a P-primary ideal of R. Then Q is a product of prime ideals, and since P is maximal, each of these factors must equal P. Hence Q is a power of P.

(13) $\Rightarrow$ (12). This is clear since every ideal between a maximal ideal P and P^2 is P-primary.

(5) $\Rightarrow$ (14). This follows from the fact that the set of ideals of a valuation ring is totally ordered by inclusion.

(14) $\Rightarrow$ (12). Assume that (14) holds and let P be a maximal ideal of R. Then P/P^2 is a vector space over the field R/P. If $P = 0$, then the desired conclusion certainly holds. Suppose $P \neq 0$. The set of subspaces of P/P^2, each being of the form A/P where A is an ideal of R with $P^2 \subseteq A \subseteq P$, is totally ordered by inclusion by (14). Hence P/P^2 is one-dimensional. Therefore if A is an ideal of R with $P^2 \subseteq A \subseteq P$, then either $A/P^2 = P^2/P^2$ or $A/P^2 = P/P^2$; that is, either $A = P^2$ or $A = P$.

This completes the proof of Theorem 6.20.

It is important to note that there are Prüfer domains which are not Dedekind domains. In fact, there are integral domains R which are not Noetherian and which have the property that for each nonzero maximal ideal P of R the ring of quotients R_P is a Noetherian valuation ring (see Chapter IX).

We now show that overrings of Dedekind domains are Dedekind domains.

6.21. Theorem. *Every overring of a Dedekind domain is a Dedekind domain.*

Proof. Let R be a Dedekind domain which is not a field and let T be an overring of R other than the quotient field of R. Let M be a maximal ideal of T. Since R is a Prüfer domain, $T_M = R_{M \cap R}$ by Proposition 4.14. Thus $R_{M \cap R}$ is not a field, and so $M \cap R \neq 0$. Hence $M \cap R$ is a maximal ideal of R and therefore $R_{M \cap R} = T_M$ is a Noetherian valuation ring.

Now let A be a nonzero ideal of T. Then, for each maximal ideal M of T, $AT_M = M^{s(M)}T_M$ for some nonnegative integer $s(M)$. There are only a finite number of maximal ideals of T which contain A; this follows from Exercise 5(c). Since $s(M) > 0$ if and only if $A \subseteq M$, we have $s(M) > 0$ for only a finite number of maximal ideals M. By Exercise 5(b) of Chapter III,

$$A = \bigcap (AT_M \cap T) = \bigcap M^{s(M)}.$$

The maximal ideals of T are pairwise comaximal, so by Exercise 2(b) of Chapter II,

$$A = \prod M^{s(M)}.$$

This proves that T is a Dedekind domain.

5 EXTENSION OF DEDEKIND DOMAINS

Let R be a Dedekind domain and let K be its quotient field. Let K' be an extension of K and let R' be the integral closure of R in K'. Then R' is called an **extension** of R. We shall now set about to show that if K'/K is finite, then R' is a Dedekind domain. Since K' can be obtained as a purely inseparable extension of a separable extension of K, we can prove this assertion in two steps, first under the assumption that K'/K is separable, and then under the assumption that K'/K is purely inseparable.

6.22. Proposition. *If K'/K is finite and separable, then R' is a Dedekind domain.*

Proof. We shall show that R' is Noetherian and integrally closed and that every nonzero proper prime ideal of R' is maximal. By definition, R' is integrally closed, and every nonzero proper prime ideal of R' is maximal by the lying-over theorem. It remains to show that R' is Noetherian.

First, K' is the quotient field of R'. For, let $a \in K'$; then there are elements $b_0, \ldots, b_{k-1} \in K$ such that

$$a^k + b_{k-1}a^{k-1} + \cdots + b_1 a + b_0 = 0.$$

There are elements $c_0, \ldots, c_{k-1}, s \in R$ such that $b_i = c_i/s$, $i = 0, \ldots, k-1$; then

$$(sa)^k + c_{k-1}(sa)^{k-1} + \cdots + (c_1 s^{k-2})(sa) + c_0 s^{k-1} = 0.$$

Hence $sa \in R'$, and $a = sa/s$.

Now let $u_1, \ldots, u_n$ be a basis of K'/K. Then there are elements $v_1, \ldots, v_n, s \in R'$ such that $u_i = v_i/s$, $i = 1, \ldots, n$, and we see from the preceding paragraph that we may choose $s \in R$. Then $v_1, \ldots, v_n$ are linearly independent over K and so form a basis of K'/K. Thus without loss of generality, we may assume that $u_1, \ldots, u_n \in R'$.

Let $M = \{a_1 u_1 + \cdots + a_n u_n \mid a_1, \ldots, a_n \in R\}$. Then M is an R-module and $M \subseteq R'$. Let

$$M^* = \{b \mid b \in K' \text{ and } T_{K'/K}(ab) \in R \text{ for all } a \in M\},$$

where $T_{K'/K}$ is the trace mapping of K'/K.

Define R'^* in like manner. It follows from properties of the trace mapping that R'^* and M^* are R-modules, and we have $M \subseteq R' \subseteq R'^* \subseteq M^*$. If we can show that M^* is a finitely generated R-module, then R' and all of its ideals will be finitely generated R-modules by Theorem 1.12. Hence we will be able to conclude that R' is Noetherian.

Let $w_1, \ldots, w_n \in K$ and consider the following n equations in n unknowns:

$$\sum_{j=1}^{n} T_{K'/K}(u_i u_j) x_j = w_i, \qquad i = 1, \ldots, n.$$

Since K'/K is separable, $\det[T_{K'/K}(u_i u_j)] \neq 0$. Hence this system of equations has a unique solution $a_1, \ldots, a_n$ in K. Then $a = a_1 u_1 + \cdots + a_n u_n$ is the unique common solution of the equations

$$T_{K'/K}(u_i x) = w_i, \qquad i = 1, \ldots, n.$$

Thus, for fixed j, the n equations

$$T_{K'/K}(u_i x) = \delta_{ij}, \qquad i = 1, \ldots, n,$$

have a unique common solution u_j'. Suppose $c_1 u_1' + \cdots + c_n u_n' = 0$ where $c_1, \ldots, c_n \in K$. Then, for $i = 1, \ldots, n$,

$$\begin{aligned} 0 &= T_{K'/K}(u_i(c_1 u_1' + \cdots + c_n u_n')) \\ &= \sum_{j=1}^{n} c_j T_{K'/K}(u_i u_j') = c_i . \end{aligned}$$

Therefore $u_1', \ldots, u_n'$ are linearly independent over K, and consequently form a basis of K'/K.

We shall complete the proof by showing that $u_1', \ldots, u_n' \in M^*$ and that they generate M^* as an R-module. Let $a \in M$ and write $a = a_1 u_1 + \cdots + a_n u_n$ where $a_1, \ldots, a_n \in R$. Then for $j = 1, \ldots, n$,

$$T_{K'/K}(au_j') = T_{K'/K}((a_1 u_1 + \cdots + a_n u_n)u_j')$$

$$= \sum_{i=1}^{n} a_i T_{K'/K}(u_i u_j') = a_j \in R;$$

thus $u_j' \in M^*$ for $j = 1, \ldots, n$. Finally let $b \in M^*$ and write $b = b_1 u_1' + \cdots + b_n u_n'$ where $b_1, \ldots, b_n \in K$. Then for $i = 1, \ldots, n$,

$$b_i = T_{K'/K}(u_i(b_1 u_1' + \cdots + b_n u_n')) \in R.$$

This completes the proof of Proposition 6.22.

Now suppose that K'/K is finite and purely inseparable. Then K has prime characteristic p, and there is a nonnegative integer e such that $a^{p^e} \in K$ for all $a \in K'$. If f is a positive integer, set $K_f = \{a \mid a \in K'$ and $a^{p^f} \in K\}$. Then K_f is a subfield of K', we have $K \subseteq K_1 \subseteq K_2 \subseteq \cdots \subseteq K_e = K'$, and $a^p \in K_{f-1}$ for all $a \in K_f$. Therefore it is sufficient to prove that R' is a Dedekind domain under the assumption that $a^p \in K$ for all $a \in K'$.

6.23. Proposition. *If $a^p \in K$ for all $a \in K'$, then R' is a Dedekind domain.*

Proof. First we note that $R' = \{a \mid a \in K'$ and $a^p \in R\}$. If $a \in K'$ and $a^p \in R$, then a is a root of $x^p - a^p \in R[X]$, so $a \in R'$. Conversely if $a \in R'$, then $a^p \in R' \cap K = R$.

Let C be an algebraic closure of K which contains K'. Let $K'' = \{c \mid c \in C$ and $c^p \in K\}$ and let R'' be the integral closure of R in K''. Then $R'' = \{c \mid c \in K''$ and $c^p \in R\}$. The mapping from K'' onto K given by $c \mapsto c^p$ is an isomorphism, and its restriction to R'' maps R'' isomorphically onto R. Therefore R'' is a Dedekind domain. Note that $K' \subseteq K''$ and $R' \subseteq R''$.

Let A be a nonzero ideal of R'. Then AR'' is invertible; hence by Proposition 6.4, $(AR'')[R'' : AR''] = R''$. Let $a_1, \ldots, a_k \in A$ and $b_1, \ldots, b_k \in [R'' : AR'']$ be such that $a_1 b_1 + \cdots + a_k b_k = 1$; then

$a_1{}^p b_1{}^p + \cdots + a_k{}^p b_k{}^p = 1$. For $i = 1, \ldots, k$, $b_i{}^p \in K$ and $b_i{}^p a \in R'' \cap K' = R'$ for all $a \in A$; hence $b_i{}^p \in [R' : A]$. Since $a_i{}^p \in A$ for $i = 1, \ldots, k$, we conclude that $A[R' : A] = R'$; that is, A is invertible. Therefore, by Theorem 6.19, R' is a Dedekind domain.

We summarize all of this as:

6.24. Theorem. *Let R be a Dedekind domain, K the quotient field of R, K' a finite extension of K, and R' the integral closure of R in K'. Then R' is a Dedekind domain.*

6.25. Theorem. *Let v be a discrete rank one valuation on a field K and let K' be a finite extension of K. Then there is a valuation on K' which is a prolongation of v. Furthermore, up to equivalence of valuations, there is only a finite number of valuations on K' which are prolongations of v. Finally, the valuations on K' which are prolongations of v have rank one and are discrete.*

Proof. Let V be the valuation ring of v. Then V is a Dedekind domain, so by Theorem 6.24 the integral closure V' of V in K' is a Dedekind domain. If P is the maximal ideal of V, let $P_1, \ldots, P_k$ be the maximal ideals of V' containing PV'. Let $V_i' = V_{P_i}$ and $P_i' = P_i V_i'$ for $i = 1, \ldots, k$. By condition (5) of Theorem 6.20 each V_i' is a Noetherian valuation ring, and P_i' is its maximal ideal. Now $P_i \cap V = P$ for $i = 1, \ldots, k$, and $P_i' \cap V' = P_i$ by Proposition 3.9. Hence for $i = 1, \ldots, k$, $P_i' \cap V = P_i' \cap V' \cap V = P_i \cap V = P$, and consequently, by Proposition 5.21, the valuation v_i' on K' determined by V_i' is a prolongation of v. Since V_i' is Noetherian, v_i' has rank one and is discrete.

Let v'' be a valuation on K' which is a prolongation of v and let V'' be its valuation ring. Then $v''(b) \geq 0$ for all $b \in V$. Let $a \in V'$, $a \neq 0$; then there are elements $b_0, b_1, \ldots, b_{n-1} \in V$ such that

$$b_0 + b_1 a + \cdots + b_{n-1} a^{n-1} + a^n = 0.$$

Suppose $v''(a) < 0$. Then for $i = 0, \ldots, n-1$, we have $v''(b_i a^i) > v''(a^n)$. Therefore we have

$$v''(b_0 + b_1 a + \cdots + b_{n-1} a^{n-1}) > v''(a^n),$$

which is not true. Thus $v''(a) \geq 0$; that is, $a \in V''$. Therefore $V' \subseteq V''$, and if P'' is the maximal ideal of V'', we must have $P'' \cap V' = P_i$ for some i, since $P = P'' \cap V$. By condition (15) of Theorem 6.20, V'' is a flat overring of V', and therefore $V'' = V''_{P''} = V'_{P_i} = V_i'$ by Proposition 4.14. Thus v'' and v_i' are equivalent for some i, and all is proved.

EXERCISES

1. Criterion for invertibility of ideals.
Let R be a ring, let A be an ideal of R, let $\{M_\alpha \mid \alpha \in I\}$ be the set of maximal ideals of R, and let S be a multiplicative system in R.
(a) If A is invertible, prove that $S^{-1}A$ is invertible.
(b) If A is regular and finitely generated and such that AR_{M_α} is invertible for each $\alpha \in I$, prove that A is invertible.
(c) Assume that R is integrally closed. If $a, b \in R$, with a regular, and if there exists an integer $n > 1$ such that $a^{n-1}b \in (a^n, b^n)$, prove that (a, b) is an invertible ideal.

2. Invertible prime ideals.
Let R be a ring and let P be an invertible prime ideal of R.
(a) Show that if $x \in P^s \backslash P^{s+1}$ and $y \in P^t \backslash P^{t+1}$, then $xy \in P^{s+t} \backslash P^{s+t+1}$; therefore, $\bigcap_{n=1}^{\infty} P^n$ is a prime ideal.
(b) Show that the powers of P are P-primary and every P-primary ideal of R is a power P.
(c) If Q is a primary ideal of R with $\operatorname{Rad}(Q) \subset P$, show that $Q \subseteq P^n$ for each positive integer n.
(d) If A is an invertible ideal of R such that $P \subset A$, show that $A = R$.

3. More on invertible ideals.
Let R be a ring with a finite number of maximal ideals $M_1, \ldots, M_n$ and let A be an invertible ideal of R.
(a) Prove that A is principal. [Hint. Find $a \in A \backslash \bigcup_{i=1}^{n} AM_i$ and show that $A = (a)$.]
(b) If R has a unique maximal ideal and $A = (a_1, \ldots, a_n)$, show that $A = (a_s)$ for some $s (1 \leq s \leq n)$.

4. Prüfer domains.
Let R be a Prüfer domain with quotient field K.
(a) Let L be an algebraic extension of field of K and let R' be

the integral closure of R in L. Show that R' is a Prüfer domain.

(b) If P is a prime ideal of R, show that R/P is a Prüfer domain.

(c) Let $\{R_\alpha\}$ be an ascending chain of Prüfer domains all with common quotient field K. Show that $\bigcup R_\alpha$ is a Prüfer domain.

5. Overrings of Prüfer domains.
Let R be a Prüfer domain and let R' be an overring of R. Let $\Delta = \{P \mid P \text{ is a prime ideal of } R \text{ and } PR' \neq R'\}$.

(a) Show that if P is a proper prime ideal of R, then $P \in \Delta$ if and only if $R' \subseteq R_P$.

(b) Let P' be a maximal ideal of R' and $P = P' \cap R$. Show that $R_P = R'_{P'}$ and $P' = PR_P \cap R'$.

(c) Show that if A is an ideal of R', then $A = (A \cap R)R'$.

(d) Show that $\{PR' \mid P \in \Delta\}$ is the set of proper prime ideals of R'.

6. Another characterization of Prüfer domains.
Let R be an integral domain. Prove that the following are equivalent. [Hint: use Exercise 1(c).]

(1) R is a Prüfer domain.

(2) R is integrally closed and there is a positive integer $n > 1$ such that $(a, b)^n = (a^n, b^n)$ for any $a, b \in R$.

(3) R is integrally closed and there exists a positive integer $n > 1$ such that $a^{n-1}b \in (a^n, b^n)$ for any $a, b \in R$.

7. Valuation ideals and Prüfer domains.
Let R be an integral domain. An ideal A of R is a **valuation ideal** if there is a valuation overring V of R and an ideal B of V such that $B \cap R = A$. Refer to Exercise 13 of Chapter V.

(a) Prove that if $A_1, \ldots, A_n$ are v-ideals of R and $x_1, \ldots, x_n$ are elements of R such that $x_i \notin A_i$ for $i = 1, \ldots, n$, then $x_1 \cdots x_n \notin A_1 \cdots A_n$. In particular, $x^s \in A_1{}^s$ implies $x \in A_1$.

(b) Use (a) to prove that if (a^n, b^n) is the intersection of valuation ideals, then $(a^n, b^n) = (a, b)^n$.

(c) Prove that R is a Prüfer domain if and only if every ideal of R is the intersection of valuation ideals.

8. Completion of ideals and Prüfer domains.

Let R be an integral domain with quotient field K. Let $\{V_\alpha\}$ be the set of all valuation overrings of R. If A is an ideal of R, $A' = \bigcap AV_\alpha$ is called the **completion** of A and if $A = A'$, A is called **complete**. Let $A^* = \bigcap (AV_\alpha \cap R)$; then A^* is the intersection of all valuation ideals of R containing A.

(a) Prove that the following are equivalent:
   (1) R is integrally closed.
   (2) Each principal ideal of R is complete.
   (3) There exists a nonzero principal ideal of R which is complete.
   (4) Each principal ideal of R is an intersection of valuation ideals of R.

(b) Prove the following conditions are equivalent:
   (1) R is a Prüfer domain.
   (2) Each ideal of R is complete.
   (3) Each finitely generated ideal of R is an intersection of valuation ideals.

9. Prüfer domains, projective modules, and torsion.

Let R be an integral domain.

(a) Show that a nonzero ideal A of R is invertible if and only if it is a projective R-module. Thus, R is a Prüfer domain if and only if every nonzero finitely generated ideal of R is a projective R-module.

(b) An R-module M is said to be **without torsion** if $ax = 0$, where $a \in R$ and $x \in M$, implies that either $a = 0$ or $x = 0$. Show that a flat R-module is without torsion.

(c) Show that R is a Prüfer domain if and only if every finitely generated R-module without torsion is projective.

(d) Let K be the quotient field of R and let M be an R-module. Regarding K as an R-module and R as one of its submodules, show that if M is without torsion, then there is an exact sequence

$$0 \to M \to K \otimes_R M \to K/R \otimes_R M \to 0.$$

(e) Let R be a Prüfer domain and let M and N be R-modules, each without torsion. Show that $M \otimes_R N$ is without torsion.

10. Quasi-principal ideals.

An ideal M of a ring R is **quasi-principal** if $(A \cap B: M)M = AM \cap B$ and $(A + BM): M = A: M + B$ for all ideals A and B of R.

(a) Show that every principal ideal of R is quasi-principal.

(b) Show that the product of quasi-principal ideals of R is quasi-principal.

(c) Show that if R is a Noetherian ring with unique maximal ideal, then every quasi-principal ideal of R is principal. [Hint. Use Exercise 7(c) of Chapter II.]

(d) Show that every quasi-principal ideal of R is finitely generated.

(e) Show that if R is an integral domain, then R is a Prüfer domain if and only if every finitely generated ideal of R is quasi-principal.

11. Dedekind domains.

Let R be a Dedekind domain.

(a) Show that each nonzero element of R is contained in only finitely many maximal ideals of R.

(b) Let A and B be ideals of R with $B \neq 0$. Show that there is an ideal C of R and an element $a \in R$ such that $B + C = R$, $AC = (a)$, and $A = AB + (a)$.

(c) Show that if A is an ideal of R, then there exist elements $a, b \in A$ such that $A = (a, b)$.

(d) Show that if R has only a finite number of maximal ideals, then R is a principal ideal ring.

(e) Show that if A is a nonzero ideal of R, then R/A is a principal ideal ring.

12. Integral domains with quotient overrings.

Let R be an integral domain. We say R has the QR-**property** if every overring of R is a ring of quotients of R.

(a) Show that if R has the QR-property, then R is a Prüfer domain.

(b) Show that if T is an overring of R such that for every $a \in T$, $R[a]$ is a ring of quotients of R, then T is a ring of quotients of R.

(c) Show that if $a, b \in R$, $b \neq 0$, and if $(a, b)^n$ is a principal ideal for some positive integer n, then $R[a/b]$ is a ring of quotients of R. Conclude that R has the QR-property if for all $a, b \in R$, some power of (a, b) is a principal ideal.

(d) Suppose that R has the QR-property and let A be a finitely generated ideal of R. Show that some power of A is contained in a principal ideal.

(e) Suppose that R has the QR-property and let P be a finitely generated prime ideal of R. Show that $P = \text{Rad}((a))$ for some $a \in R$.

(f) Let R be a Noetherian integral domain. Show that R has the QR-property if and only if R is a Dedekind domain such that for every ideal A of R, some power of A is a principal ideal.

13. Cancellation of ideals.
Let R be a ring, not necessarily an integral domain. We say a fractional ideal A of R **can be canceled** if $AB = AC$ for fractional ideals B, C of R implies that $B = C$.

(a) Show that if $r \in R$, then (r) can be canceled if and only if r is a regular element of the total quotient ring of R.

(b) Show that a fractional ideal A of R can be canceled if and only if for fractional ideals B, C of R, $AB \subseteq AC$ implies that $B \subseteq C$.

(c) Show that if $A_1, \ldots, A_k$ are fractional ideals of R such that $A_1 \cdots A_k$ can be canceled, then A_i can be canceled for each $i = 1, \ldots, k$.

(d) If A is a fractional ideal of R which can be canceled, if $A = A_1 + \cdots + A_n$ for fractional ideals $A_1, \ldots, A_n$ of R, and if k is a positive integer, prove that $A^k = A_1{}^k + \cdots + A_n{}^k$.

(e) Give an example of an ideal which contains a regular element, but which can not be canceled.

14. More on cancellation of ideals.
Let R be a ring, not necessarily an integral domain.

(a) Let A be an ideal of R which can be canceled. Show that if A can be written as a product of proper prime ideals of R, then it can be so written in only one way, except for the order of the factors.

(b) Suppose that every finitely generated regular ideal of R can be canceled. Show that R is integrally closed.

(c) Let R be an integral domain and assume that there is a set $\mathscr{S}$ of nonzero proper ideals of R such that every nonzero proper ideal of R can be written as a product of ideals in $\mathscr{S}$ in one and only one way, except for the order of the factors. Show that R is a Dedekind domain and that $\mathscr{S}$ is the set of nonzero proper prime ideals of R.

15. A construction of Prüfer domains.
Let K be a field with prime subring k. Let K_0 be the integral closure of k in K. Let

$$f(X) = X^n + a_{n-1}X^{n-1} + \cdots + a_1X + a_0$$

be a polynomial of positive degree over K which has no roots in K. Let $S = \{1/f(a) \mid a \in K\}$, let R be the subring of K generated by S, and let R' be the integral closure of R in K. Prove the following statements which constitute a proof that R' is a Prüfer domain.

(a) If $a \in K$ and $g(X) \in K_0[X]$ is such that $\deg g(X) \leq n$, then $g(a)/f(a) \in R'$. Furthermore K is the quotient field of R'.
(b) If $a \in K$, then $f(a)R' = \{1, a, \ldots, a^n\}R'$. If $r > 0$, then $a^{n+r} \in f(a)R'$ if and only if $a \in R'$.
(c) If $a_1, a_2, \ldots, a_s \in K$, where $s \leq 2^t$, and if F is a fractional ideal of R' which is generated by $a_1, a_2, \ldots, a_s$, then F^{nt} is principal. Hence R' is a Prüfer domain. [Hint. Use Exercise 13(d) and induct on s.]

16. The ideal transform.
Let R be a ring with total quotient ring K. If A is an ideal of R, the **transform** $T(A)$ of A is $\bigcup_{n=1}^{\infty}\{x \mid x \in K \text{ and } xA^n \subseteq R\}$.

(a) Prove that if A is invertible, then $AT(A) = T(A)$.
(b) If A, B are ideals such that $A^n \subseteq B$ for some positive integer n, prove that $T(B) \subseteq T(A)$.
(c) If A is invertible and P is a prime ideal of R, show that $PT(A) = T(A)$ if and only if $A \subseteq P$.
(d) Let R' be an overring of R such that $R \subseteq R' \subseteq T(A)$. Show that there is a one-to-one correspondence between the prime ideals P' of R' which do not contain AR' and the prime ideals P of R which do not contain A. Show that if P corresponds to P', then $P = P' \cap R$ and $R'_{P'} = R_P$.

(e) Under the hypothesis of (d), show that if $AR' = R'$, then $R' = \bigcap R_{P_\alpha}$ where $\{P_\alpha\}$ is the collection of all prime ideals of R which do not contain A.

(f) Let A be a finitely generated ideal of R, say

$$A = (a_1, a_2, \ldots, a_n).$$

Prove that $T(A) = \bigcap_{i=1}^{n} T((a_i))$.

(g) Show that if A is a finitely generated ideal of R and $\{P_\alpha\}$ is the collection of prime ideals of R which do not contain A, then $T(A) = \bigcap R_{P_\alpha}$.

(h) For $u \in K$, denote by B_u the ideal $\{r \mid r \in R \text{ and } ru \in R\}$. If R is a Prüfer domain, show that $T(B_u) = R[u]$ and prove that u is almost integral over R if and only if $\bigcap_{n=1}^{\infty} (B_u)^n \neq 0$. [Hint. Show that if R_1 and R_2 are integral domains such that $R_1 \subseteq R_2$ and such that $\{x \mid x \in R_1 \text{ and } xR_2 \subseteq R_1\} \neq 0$, then R_1 and R_2 have the same complete integral closure.]

(i) Let R be a Prüfer domain. Prove that R is completely integrally closed if and only if for each proper finitely generated ideal A of R, $\bigcap_{n=1}^{\infty} A^n = 0$.

17. The ideal transform and overrings.
Let R be an integral domain which does not have a unique maximal ideal.

(a) If $\{x_\alpha\}$ is the collection of nonunits of R, prove that $R = \bigcap T((x_\alpha))$.

(b) Let x be a nonzero element of R and let

$$S = \{x^n \mid n = 1, 2, \ldots\}.$$

Show that $T((x)) = S^{-1}R$.

(c) Show that R is integrally closed if and only if the transforms of the principal ideals generated by nonzero nonunits of R are integrally closed.

(d) Prove that R is a Prüfer domain if and only if $T((x))$ is a Prüfer domain for every nonzero nonunit x of R.

18. Arithmetical rings.
A ring R is called an **arithmetical ring** if for all ideals A, B, C of R we have $A \cap (B + C) = (A \cap B) + (A \cap C)$.

(a) Show that R is an arithmetical ring if and only if for all ideals A, B, C of R we have $A + (B \cap C) = (A + B) \cap (A + C)$.

(b) Show that R is an arithmetical ring if and only if for every maximal ideal P of R, the set of ideals of R_P is totally ordered by inclusion.

(c) Show that the following statements are equivalent:

(1) R is an arithmetical ring.

(2) $(A+B):C = A:C + B:C$ for all ideals A, B, C of R with C finitely generated.

(3) $C:(A \cap B) = C:A + C:B$ for all ideals A, B, C of R with A and B finitely generated.

(d) Show that R is an arithmetical ring if and only if $A \subseteq C$, where A and C are ideals of R with C finitely generated, implies that there is an ideal B of R such that $A = BC$.

19. More on arithmetical rings.

(a) Show that every overring of an arithmetical ring is an arithmetical ring.

(b) Let R be an arithmetical ring with only a finite number of maximal ideals. Show that every finitely generated ideal of R is a principal ideal.

(c) Show that in an arithmetical ring every primary ideal is irreducible.

(d) Let R be an arithmetical ring. Show that every finitely generated regular ideal of R can be canceled. Thus R is integrally closed [see Exercise 14(b)].

(e) Let R be an arithmetical ring and P a proper prime ideal of R. Show that R/P is a Prüfer domain.

20. The Grothendieck group.

Let R be a ring. Denote by $\mathscr{K}(R)$ the set of equivalence classes of finitely generated R-modules under the relation of isomorphism. Denote the equivalence class containing M by $[M]$. If $[M]$, $[N] \in \mathscr{K}(R)$, set $[M]+[N]=[L]$ if there is an exact sequence $0 \to M \to L \to N \to 0$.

(a) Show that this is a well-defined binary operation on $\mathscr{K}(R)$, and that with this operation $\mathscr{K}(R)$ is a group: it is called the **Grothendieck group** of R.

(b) Show that if R is Noetherian, then $\mathscr{K}(R)$ is generated by the set $\{[R/P]\}$, where P runs through the set of proper nonzero prime ideals of R.

(c) Show that if R is a Dedekind domain, then $\mathscr{K}(R) \cong \mathscr{F}(R)$.

(d) Show that if R is a field, then $\mathscr{K}(R) \cong Z$.

21. Noetherian domains.

Let R be a Noetherian integral domain.

(a) Assume that R has a unique maximal ideal P. Show that if P is the only nonzero proper prime ideal of R, then $R \subset [R: P]$, that $P[R: P] \neq R$ implies that $P[R: P]^n \neq R$ for each positive integer n, and that each element of $[R: P]$ is integral over R.

(b) With the same assumption on P, show that if P is not the only nonzero proper prime ideal of R, then $P[R: P] = P$ and that each element of $[R: P]$ is integral over R.

(c) Now assume that not every nonzero proper prime ideal of R is maximal. Let $a \in R$, $a \neq 0$, and suppose that (a) has a prime divisor P which is not minimal. Show that for some $b \in R$ we have $b/a \notin R$ and b/a integral over R. Show also that P is a prime divisor of (c) for every $c \in P$ (see Exercise 9 of Chapter III).

22. Rings having few zero-divisors.

In this exercise, and the four which follow, some of the results which have been obtained for integral domains will be extended to a certain class of rings with zero-divisors. Many of these results will be obtained under more general conditions in Chapter X. A ring R **has few zero-divisors** if the set of zero-divisors of R is a union of a finite number of prime ideals of R. Thus, Noetherian rings and integral domains have few zero-divisors.

(a) Show that a ring R has few zero-divisors if and only if the ideal 0 has only a finite number of maximal prime divisors (see Exercise 9 of Chapter III). Show that this is equivalent to the total quotient ring of R having only a finite number of maximal ideals.

(b) Suppose that every ideal of a ring R with prime radical is a power of a prime ideal. Show that R has few zero-divisors if and only if its zero ideal has only a finite number of minimal prime divisors.

For the rest of this exercise, assume the ring R has few zero-divisors.

(c) Show that if $x, y \in R$, and if x is regular, then there is an element $u \in R$ such that $y + ux$ is a regular element of R.

(d) Show that every regular ideal of R is generated by its regular elements. Show that if a regular ideal of R can be generated by n elements, then it can be generated by $n+1$ regular elements.

(e) Let T and T' be overrings of R. Show that $T \subseteq T'$ if and only if every regular element of T is contained in T'.

23. A ring having more than a few zero-divisors.

In this exercise, an example is given which shows that, in general, the properties enjoyed by rings which have few zero-divisors do not hold for arbitrary rings. The ring of this exercise will be used for counterexamples in Chapter X.

Let F be a field and let X and Y be indeterminates. Let G be the set of irreducible polynomials $f(X, Y)$ in $F[X, Y]$ such that $f(0, 0) = 0$ and $f(X, 0) \neq 0$. For each $g \in G$, let Z_g be an indeterminate. Let $T = F[X, Y, \{Z_g | g \in G\}]$, and let I be the ideal of T generated by

$$\{gZ_g | g \in G\} \cup \{Z_f Z_g | f, g \in G\}.$$

Let $R = T/I$, and denote the residue classes of X, Y, and Z_g by x, y, and z_g for all $g \in G$, and let N be the ideal of R generated by y and z_g for all $g \in G$.

(a) Prove that M is a maximal ideal of R properly containing the regular prime ideal N, and that every element of $M \backslash N$ is a zero-divisor. Thus, M and N have the same regular elements but are not equal.

(b) Let $f \in F[X, Y]$ be such that $f(X, Y) \notin F[Y]$ and let n be the least integer such that $f(X, Y)$ has a term of the form aY^nX^p for some nonzero $a \in F$ and some positive integer p. Prove that $f(x, y)$ is a regular element of R if and only if $f(X, Y)$ has a term of the form bY^m, where b is a nonzero element of F and $0 \leq m \leq n$.

(c) Let Q be the ideal of R generated by $\{z_g | g \in G\}$. If $f(X, Y) \in F[X, Y]$ and $w \in Q$, prove that $f(x, y) + w$ is regular if and only if $f(x, y)$ is regular.

24. Quasi-valuation rings.

A ring R is a **quasi-valuation ring** if it has few zero-divisors and if for every pair of regular elements $a, b \in R$, either $(a) \subseteq (b$ or $(b) \subseteq (a)$. Let R be a ring which has few zero-divisors.

(a) Show that R is a quasi-valuation ring if and only if for every regular element x of the total quotient ring of R, either $x \in R$ or $x^{-1} \in R$.

(b) Suppose that R is integrally closed and that R has a unique regular maximal ideal M. Show that if $xy \in (x^2, y^2)$ for every pair of regular elements $x, y \in M$, then R is a quasi-valuation ring.

For the rest of this exercise, assume that R is a quasi-valuation ring.

(c) Show that the set of regular ideals of R is totally ordered by inclusion; conclude that either R has a unique regular maximal ideal or R is its own total quotient ring.

(d) Show that every overring of R is a quasi-valuation ring.

(e) Suppose R is not its own total quotient ring K, and let T be an overring of R with $T \neq K$. Let M be the unique regular maximal ideal of T, and let $P = M \cap R$. Show that $T = R_{S(P)}$. (For the notation used here, see Exercise 10 of Chapter III.)

(f) Let A be a regular ideal of R. Show that $\bigcap_{n=1}^{\infty} A^n$ is a prime ideal of R.

(g) Give an example of a quasi-valuation ring which is not a valuation ring.

25. Quasi-valuation overrings of a ring.

Let R be a ring which has few zero-divisors.

(a) Suppose that $R_{S(M)}$ is a quasi-valuation ring for every regular maximal ideal M of R. Show that if T is an overring of R, then T has this same property, and that $T = \bigcap R_{S(M)}$, where the intersection is over the set of all regular maximal ideals M of R such that $T \subseteq R_{S(M)}$.

(b) Show that a quasi-valuation ring is integrally closed.

(c) Show that the integral closure of R is the intersection of all of the quasi-valuation overrings of R.

26. P-rings.

A ring R is a ***P*-ring** if every overring of R is integrally closed. Throughout this exercise, assume that R has few zero-divisors.

(a) Show that R is a P-ring if and only if $R_{S(M)}$ is a quasi-valuation ring for every regular maximal ideal M of R.

(b) Show that the following statements are equivalent:
  (1) R is a P-ring.
  (2) Every overring of R is a flat R-module.
  (3) $A(B \cap C) = AB \cap AC$ for all ideals A, B, C of R with B and C regular.
  (4) $(A + B)(A \cap B) = AB$ for all regular ideals A, B of R.
  (5) Every finitely generated regular ideal of R is invertible.
  (6) If $AB = AC$, where A, B, C are ideals of R and A is finitely generated and regular, then $B = C$.

(c) Show that R is a P-ring if and only if whenever A and B are ideals of R, with B finitely generated and regular, and $A \subseteq B$, then there is an ideal C of R such that $A = BC$.

(d) Formulate analogs of statements (8)–(10) of Theorem 6.6, which are equivalent to the statement that R is a P-ring, and prove this equivalence.

CHAPTER

VII

Dimension of Commutative Rings

1 THE KRULL DIMENSION

Let R be a ring and consider a chain

$$P_0 \subset P_1 \subset \cdots \subset P_r$$

of $r + 1$ proper prime ideals of R. The **length** of such a chain is the integer r. Its **first term** is P_0 and its **last term** is P_r.

7.1. Definition. *The* **Krull dimension** *of R is the supremum of the lengths of all chains of distinct proper prime ideals of R. The Krull dimension of R is denoted by* $\dim R$.

Either $\dim R = \infty$ or $\dim R = d$, where d is a nonnegative integer such that R has a chain of $d + 1$ distinct proper prime ideals but no chain of $d + 2$ distinct proper prime ideals. Note that if K is a field, then $\dim K = 0$; conversely, if R is an integral domain and $\dim R = 0$, then R is a field.

7.2. Definition. *Let P be a proper prime ideal of R. The* **height** *of P, denoted by* $\operatorname{ht} P$, *is the Krull dimension of R_P. The* **depth** *of P, denoted by* $\operatorname{dpt} P$, *is the Krull dimension of R/P.*

Thus the height of P is the supremum of the lengths of all chains of distinct proper prime ideals of R having P as last term. The depth of P is the supremum of the lengths of all chains of distinct proper prime ideals of R having P as first term.

7.3. Definition. *Let A be a proper ideal of R. The* **height** *of A, denoted by* ht A, *and the* **dimension** *of A, denoted by* dim A, *are the infimum and supremum, respectively, of the values of* ht P *as P runs over all of the minimal prime divisors of A. The* **depth** *of A, denoted by* dpt A, *is the Krull dimension of R/A.*

Note that if A is a proper ideal of R, then ht $A \leq$ dim A.

For example, suppose that R is a Dedekind domain. Since every nonzero proper prime ideal of R is maximal, dim $R = 1$. If A is a nonzero proper ideal of R, then ht $A =$ dim $A = 1$ and dpt $A = 0$. Furthermore, ht $0 =$ dim $0 = 0$ and dpt $0 = 1$.

We shall now show that for a ring R, considered as an R-module, an interesting relation exists between (DCC), (ACC), and the property of having zero Krull dimension. In general, an R-module may satisfy one of (ACC) and (DCC) without satisfying the other (see Exercise 3 of Chapter I). However, if R is considered as an R-module, the two conditions are not independent.

7.4. Theorem. *A ring R satisfies* (DCC) *for ideals if and only if R is Noetherian and* dim $R = 0$.

Proof. Suppose that the zero ideal of R is the product of maximal ideals of R, say $0 = P_1 \cdots P_n$. Consider the chain of ideals of R,

$$0 = P_1 \cdots P_n \subseteq P_1 \cdots P_{n-1} \subseteq \cdots \subseteq P_1P_2 \subseteq P_1 \subset R.$$

There are no ideals strictly between P_1 and R. For $i = 1, \ldots, n-1$, $P_1 \cdots P_i/P_1 \cdots P_{i+1}$ is a vector space over the field R/P_{i+1}. If R satisfies (ACC), then each of these vector spaces satisfies (ACC) and so is finite dimensional. Thus each of these vector spaces has a composition series; this implies that the chain given above can be refined to a composition series of R, considered as an R-module. This, in turn, implies that R satisfies (DCC). In exactly the same way, we can show that if R satisfies (DCC), then it satisfies (ACC). The facts which we have used concerning composition series may be found in Exercise 4 of Chapter I.

Now suppose that R is Noetherian and that $\dim R = 0$. Then each proper prime ideal of R is a minimal prime divisor of 0 and it follows that R has only a finite number of proper prime ideals say $P_1, \ldots, P_k$, each of which is maximal. For some positive integer r we have

$$0 = (\text{Rad}(0))^r = (P_1 \cap \cdots \cap P_k)^r = (P_1 \cdots P_k)^r,$$

by Exercise 2 of Chapter II. Hence R satisfies (DCC).

Conversely, suppose that R satisfies (DCC). Let P be a proper prime ideal of R and let $a \in R \backslash P$. The set of ideals $\{P + (a^n) \mid n = 1, 2, \ldots\}$ has a minimal element, so that for some positive integer m we have $P + (a^{m+1}) = P + (a^m)$. Then $a^m = x + ra^{m+1}$ for some $x \in P$ and $r \in R$. Since $a^m(1 - ra) = x \in P$ and $a^m \notin P$, we have $1 - ra \in P$, that is, $1 \in P + (a)$. Hence $P + (a) = R$; this implies that P is maximal and proves that $\dim R = 0$.

Let $P_1, P_2, \ldots$ be distinct maximal ideals of R. Then $P_1 \supset P_1 \cap P_2 \supset P_1 \cap P_2 \cap P_3 \supset \cdots$. Since R satisfies (DCC) this descending chain cannot have infinitely many terms. Hence R has only finitely many maximal ideals, say $P_1, \ldots, P_k$. Let $A = P_1 \cdots P_k$; if we show that some power of A is the zero ideal then it will follow that R is Noetherian. Since R satisfies (DCC), there is a positive integer r such that $A^r = A^{r+1}$. Assume that $A^r \neq 0$ and let B be an ideal which is minimal with respect to the properties that $BA^r \neq 0$ and $B \subseteq A$. Note that the set of ideals with these properties is not empty, for A belongs to this set. If $P = 0 : BA^r$, then $P \neq R$ since $BA^r \neq 0$. Let $ab \in P$ and $b \notin P$. Then $abBA^r = 0$ but $bBA^r \neq 0$; hence $bB = B$ by our choice of B and $aBA^r = 0$, that is, $a \in P$. Thus P is a proper prime ideal of R. Therefore, $A \subseteq P$ and $BA^r = BA^{r+1} \subseteq BA^rP = 0$, which contradicts our choice of B. Thus we must conclude that $A^r = 0$.

In the remainder of this section we shall obtain results which show that a definite relationship exists between the number of generators of an ideal of a Noetherian ring and the heights of the minimal prime divisors of the ideal. Recall that $J(R)$ denotes the Jacobson radical of R (Exercise 7 of Chapter II).

7.5. Lemma. *Let R be a Noetherian ring and let $a \in J(R)$. Let A and B be ideals such that $B \subseteq A \subseteq B + (a)$ and $A:(a) = A$. Then $A = B$.*

Proof. First assume that $B = 0$. Since $A \subseteq (a)$ we have $A = (A:(a))(a) = Aa$. Then $A = Aa^n$ for all positive integers n, and consequently

$$A = \bigcap_{n=1}^{\infty} (a)^n A.$$

Since $a \in J(R)$, it follows from Exercise 7(d) of Chapter II that $A = 0$. Now drop the assumption that $B = 0$. If we set $R' = R/B$, $A' = A/B$, and $a' = a + B$, then $a' \in J(R')$, $A' \subseteq (a')$ and $A':(a') = A'$. Each of these facts is obvious except the last one. If $d + B \in A':(a')$, then $da + B \in A'$, so $da - c \in B$ for some $c \in A$. But then $da - c \in A$; hence $da \in A$ and $d \in A:(a) = A$. Thus $d + B \in A'$. From the first part of the proof we conclude that $A' = 0$. Therefore $A = B$.

7.6. Theorem (Krull's Principal Ideal Theorem). *Let R be a Noetherian integral domain and let a be a nonzero nonunit in R. If P is a minimal prime divisor of (a), then* $\operatorname{ht} P = 1$.

Proof. Since $\operatorname{ht} P = \operatorname{ht} PR_P$, we may assume that P is the unique maximal ideal of R. Let Q be a prime ideal of R such that $Q \subset P$; we must show that $Q = 0$. Since P is a minimal prime divisor of (a), we have $\dim(R/(a)) = 0$. Since $R/(a)$ is Noetherian, it follows from Theorem 7.4 that $R/(a)$ satisfies (DCC). Hence for some positive integer n we have $Q^{(k)} + (a) = Q^{(n)} + (a)$ for all $k \geq n$. Since $\operatorname{Rad}(Q^{(n)}) = Q$ and $a \notin Q$ [since P is a minimal prime divisor of (a)], we have $Q^{(n)}:(a) = Q^{(n)}$ by Exercise 6(j) of Chapter II. It now follows from Lemma 7.5 that $Q^{(k)} = Q^{(n)}$ for all $k \geq n$. Since R is an integral domain, $\bigcap_{n=1}^{\infty} Q^{(n)} = 0$ by Proposition 3.14. Therefore $Q^{(n)} = 0$ for large n, and since $Q^n \subseteq Q^{(n)}$ we have $Q^n = 0$. Thus $Q = 0$.

7.7. Theorem. *Let R be a Noetherian ring. If $(a_1, \ldots, a_r)$ is a proper ideal of R, then* $\dim(a_1, \ldots, a_r) \leq r$.

Proof. Let P be a minimal prime divisor of $(a_1, \ldots, a_r)$. We shall show that $\operatorname{ht} P \leq r$, and in doing so we may assume that P is the unique maximal ideal of R. We must show that if $P = P_0 \supset P_1 \supset \cdots \supset P_s$, where each P_i is a prime ideal of R, then $s \leq r$. We may replace R by R/P_s, and so assume that R is an integral domain. Furthermore we may assume that there are no prime ideals of R strictly between P and P_1. Since P is a minimal prime divisor of $(a_1, \ldots, a_r)$, we have $(a_1, \ldots, a_r) \nsubseteq P_1$, say $a_1 \notin P_1$. Then there is no prime ideal P' of R with $P_1 + (a_1) \subseteq P' \subset P$. Hence P is the unique prime divisor of $P_1 + (a_1)$, from which it follows that $P = \text{Rad}(P_1 + (a_1))$. Thus there is an integer t such that $a_i^t \in P_1 + (a_1)$ for $i = 1, \ldots, r$. Let $a_i^t = a_1 b_i + c_i$, $b_i \in R$, $c_i \in P_1$, and consider the ideal $(c_2, \ldots, c_r)$. Since $(c_2, \ldots, c_r) \subseteq P_1$, there is a minimal prime divisor P_1' of $(c_2, \ldots, c_r)$ with $P_1' \subseteq P_1$. Since $a_i \in \text{Rad}((a_1, c_2, \ldots, c_r))$ for $i = 1, \ldots, r$, P is the only prime divisor of $(a_1, c_2, \ldots, c_r)$. Hence in R/P_1', P/P_1' is a minimal prime divisor of the principal ideal generated by $a_1 + P_1'$. If $a_1 \in P_1'$ then $(a_1, \ldots, a_r) \subseteq P_1'$, which is not the case since $P_1' \subseteq P_1$. Hence $a_1 + P_1'$ is a nonzero element of R/P_1'; it is also a nonunit in R/P_1' since it is contained in P/P_1'. Therefore by Theorem 7.6, $\text{ht}\, P/P_1' = 1$, and we conclude that $P_1' = P_1$.

We can now prove the theorem by induction on r. If $r = 1$, the assertion follows from the result of the preceding paragraph. Assume $r > 1$ and that the assertion is true for any proper ideal generated by fewer than r elements. Then, since P_1 is a minimal prime divisor of $(c_2, \ldots, c_r)$, we have $s - 1 \leq r - 1$, that is, $s \leq r$.

7.8. Corollary. *The dimension, and therefore the height, of a proper ideal of a Noetherian ring is finite.*

This does not mean that the dimension of a Noetherian ring is finite. For, the ring may have a sequence $P_1, P_2, \ldots$ of proper prime ideals such that $\lim_{n\to\infty} \text{ht}\, P_n = \infty$ (see Exercise 3).

There is a result which is almost the converse of Theorem 7.7.

7.9. Theorem. *Let A be a proper ideal of a Noetherian ring R such that* $\text{ht}\, A = r \geq 1$. *Then there are elements $a_1, \ldots, a_r \in A$ such that*

$$\text{ht}(a_1, \ldots, a_i) = i \qquad \textit{for } i = 1, \ldots, r.$$

Proof. Let $P_1, \ldots, P_k$ be the minimal prime divisors of 0. Since $\text{ht}\, A \geq 1$, $A \nsubseteq P_i$ for $i = 1, \ldots, k$, and so by Exercise 5(c) of Chapter II there is an element $a_1 \in A$ with $a_1 \notin P_i$ for $i = 1, \ldots, k$. Let P be a minimal prime divisor of (a_1). By Theorem 7.7, $\text{ht}\, P \leq 1$. But $P_i \subset P$ for some i, so $\text{ht}\, P = 1$, and we conclude that $\text{ht}(a_1) = 1$. Now suppose that $1 < j \leq r$ and that we have found elements $a_1, \ldots, a_{j-1} \in A$ such that $\text{ht}(a_1, \ldots, a_i) = i$ for $i = 1, \ldots, j-1$. Let $Q_1, \ldots, Q_n$ be those minimal prime divisors of $(a_1, \ldots, a_{j-1})$ such that $\text{ht}\, Q_i = j - 1$ for $i = 1, \ldots, n$. If $A \subseteq Q_i$ for some i, then $\text{ht}\, A \leq j - 1 < r$, contrary to our assumption. Hence $A \nsubseteq Q_i$ for $i = 1, \ldots, n$, and so there is an element $a_j \in A$ with $a_j \notin Q_i$ for $i = 1, \ldots, n$. Let P' be a minimal prime divisor of $(a_1, \ldots, a_j)$. If no Q_i is contained in P', then some minimal prime divisor of $(a_1, \ldots, a_{j-1})$ of height greater than $j - 1$ is contained in P'; hence $\text{ht}\, P' \geq j$. If $Q_i \subset P'$ for some i, then $\text{ht}\, P' \geq \text{ht}\, Q_i + 1 = j$. By Theorem 7.7, $\text{ht}\, P' \leq j$. Hence $\text{ht}\, P' = j$ and therefore $\text{ht}\,(a_1, \ldots, a_j) = j$.

2 THE KRULL DIMENSION OF A POLYNOMIAL RING

Let R be a ring. In this section we shall begin an investigation of the relationship between the Krull dimension of R and that of the polynomial ring $R[X]$.

Let A be an ideal of R. Then the ideal $AR[X]$ of $R[X]$ consists of those polynomials of $R[X]$ having coefficients in A; hence $AR[X] \cap R = A$. Let $\phi: R \to R/A$ be the canonical homomorphism. Define $\psi: R[X] \to (R/A)[X]$ by

$$\psi(a_0 + a_1 X + \cdots + a_n X^n) = \phi(a_0) + \phi(a_1)X + \cdots + \phi(a_n)X^n.$$

The mapping ψ is a surjective homomorphism and its kernel is precisely $AR[X]$. Hence,

$$R[X]/AR[X] \cong (R/A)[X]$$

Two special cases are of particular interest. If P is a proper prime ideal of R, then $(R/P)[X]$ is an integral domain and we conclude that $PR[X]$ is a prime ideal of $R[X]$. If, in addition, P is a maximal ideal of R, then $(R/P)[X]$ is a Dedekind domain. Thus every nonzero proper prime ideal of $R[X]/PR[X]$ is maximal. It follows that dpt $PR[X]=1$.

7.10. Proposition. *Suppose that R is an integral domain. If P is a nonzero prime ideal of $R[X]$ such that $P \cap R = 0$, then for every $f(X) \in R[X]\backslash P$ we have*

$$(P+(f(X))) \cap R \neq 0.$$

Proof. Let K be the quotient field of R. If $S = R\backslash\{0\}$, then $S^{-1}R[X] = K[X]$. Since $P \cap R = 0$, $P \cap S$ is empty, and consequently $S^{-1}P$ is a nonzero proper prime ideal of $K[X]$. As such, it is a maximal ideal of $K[X]$. Furthermore $S^{-1}P \cap R[X] = P$. Therefore, if $f(X) \in R[X]\backslash P$, then $f(X) \notin S^{-1}P$; hence $f(X)$ and $S^{-1}P$ generate $K[X]$. Now $K[X]$ is a principal ideal domain, so $S^{-1}P$ is generated by some $g(X)/s$, where $g(X) \in P$ and $s \in S$. Hence there are elements $h(X)$, $k(X) \in R[X]$ and elements $u, v \in S$ such that $1 = (h(X)/u)(g(X)/s) + (k(X)/v)f(X)$. Then

$$uvs = vh(X)g(X) + usk(X)f(X) \in (P+(f(X))) \cap R \quad \text{and} \quad uvs \neq 0.$$

7.11. Corollary. *Suppose that R is an integral domain. Let P and P' be nonzero ideals of $R[X]$ such that $P \subset P'$. If P is a prime ideal, then $P' \cap R \neq 0$.*

Proof. If $P \cap R \neq 0$, then $P' \cap R \neq 0$. If $P \cap R = 0$ and if $f(X) \in P'\backslash P$, then

$$0 \neq (P+(f(X))) \cap R \subseteq P' \cap R.$$

7.12. Corollary. *Let P_1, P_2, P_3 be ideals of $R[X]$, with P_2 prime, such that $P_1 \subset P_2 \subset P_3$. Then $P_1 \cap R \subset P_3 \cap R$.*

Proof. If $P_1 \cap R \subset P_2 \cap R$, we have finished. Suppose that $P_1 \cap R = P_2 \cap R = P$, which is a prime ideal of R. By the remarks at the beginning of the section, $R[X]/PR[X]$ may be viewed as the polynomial ring in one indeterminate over the integral domain R/P. Since $PR[X] \subseteq P_1 \subset P_2$ we have $0 \neq P_2/PR[X] \subset P_3/PR[X]$. Hence, by Corollary 7.11, $P_3/PR[X] \cap R/P \neq 0$. Thus there is an element

$a \in R \backslash P$ and a polynomial $f(X) \in P_3$ such that $a - f(X) \in PR[X]$. If b is the constant term of $f(X)$, then $b \notin P$. However $b \in P_3 \cap R$, so $P_1 \cap R \subset P_3 \cap R$.

7.13. Theorem. *If* $\dim R = d$, *where d is an integer, then*

$$d + 1 \leq \dim R[X] \leq 2d + 1.$$

Proof. Let $P_0 \subset P_1 \subset \cdots \subset P_r$ be a chain of distinct proper prime ideals of $R[X]$. By Corollary 7.12, $P_0 \cap R \subset P_2 \cap R \subset \cdots$ is a chain of distinct proper prime ideals of R. Hence if s is the number of distinct terms in the chain $P_0 \cap R \subseteq P_1 \cap R \subseteq \cdots \subseteq P_r \cap R$, then we must have $s \geq (r+1)/2$. But $s \leq d+1$, so $(r+1)/2 \leq d+1$. Therefore $r \leq 2d+1$, and we conclude that $\dim R[X] \leq 2d+1$. On the other hand, if $Q_0 \subset Q_1 \subset \cdots \subset Q_d$ is a chain of $d+1$ distinct proper prime ideals of R, then

$$Q_0 R[X] \subset Q_1 R[X] \subset \cdots \subset Q_d R[X] \subset Q_d R[X] + (X)$$

is a chain of $d+2$ distinct proper prime ideals of $R[X]$. Therefore $\dim R[X] \geq d+1$.

There are large classes of rings for which the lower bound for the Krull dimension of $R[X]$ is realized. For example, this is true for Noetherian rings.

7.14. Theorem. *If R is a Noetherian ring of finite Krull dimension, then* $\dim R[X] = \dim R + 1$.

Proof. Let $\dim R = d$; we shall show that $\dim R[X] \leq d+1$. Let P be a proper prime ideal of $R[X]$. If we show that P is a minimal prime divisor of an ideal which is generated by $d+1$ or fewer elements, then $\operatorname{ht} P \leq d+1$ by Theorem 7.7, and the asserted inequality will follow. Let $P' = P \cap R$. Then P' is a proper prime ideal of R and so $\operatorname{ht} P' \leq d$. Hence by Theorem 7.9, there are elements $a_1, \ldots, a_d \in P'$ such that if $A = (a_1, \ldots, a_d)$, then $\operatorname{ht} A = \operatorname{ht} P'$. It follows that P' is a minimal prime divisor of A. Suppose that P_1 is a prime ideal of $R[X]$ such that $AR[X] \subseteq P_1 \subseteq P'R[X]$. Then $A \subseteq P_1 \cap R \subseteq P'$ and so $P_1 \cap R = P'$; hence $P_1 = P'R[X]$. Thus $P'R[X]$ is a minimal prime divisor of $AR[X]$. Since $AR[X] = a_1R[X] + \cdots + a_d R[X]$, we have ht $P'R[X] \leq d$. If $P'R[X] = P$, then ht $P \leq d$. Suppose, on the other hand, that $P'R[X] \subset P$. Let $f(X) \in P \backslash P'R[X]$, and let Q be a minimal prime divisor of $AR[X] + (f(X))$ such that $Q \subseteq P$. Since $A \subseteq Q \cap R \subseteq P'$ we must have $Q \cap R = P'$, and thus $P'R[X] \subset Q$, equality being ruled out by the fact that $f(X) \in Q$. Since $P'R[X] \cap R = Q \cap R = P \cap R$ it follows from Corollary 7.12 that $Q = P$. Hence P is a minimal prime divisor of $AR[X] + (f(X))$, which is generated by $d + 1$ elements.

7.15. Corollary. *If R is a Noetherian ring of finite Krull dimension, and if $X_1, \ldots, X_n$ are independent indeterminates, then*

$$\dim R[X_1, \cdots X_n] = \dim R + n.$$

3 VALUATIVE DIMENSION

In this section we shall show that if R is a Prüfer domain, then the conclusion of Corollary 7.15 holds for R. Since Prüfer domains are characterized in terms of their overrings, it seems natural to study the Krull dimension of Prüfer domains by studying the Krull dimension of their overrings. We shall be particularly interested in the Krull dimension of those overrings which are valuation rings, the **valuation overrings** of R. We note that a valuation ring has finite Krull dimension d if and only if it has rank d (see the remark on p. 112).

Throughout this section, R will be an integral domain.

7.16. Definition. *If there is a nonnegative integer k such that* $\dim V \leq k$ *for every valuation overring V of R, with equality for at least one V, then R has* **valuative dimension** *k, and we write* $\dim_v R = k$. *If there is no such k, we write* $\dim_v R = \infty$.

7.17. Proposition. *We always have* $\dim R \leq \dim_v R$.

Proof. Let $P_0 \subset P_1 \subset \cdots \subset P_r$ be a chain of distinct proper prime ideals of R. By Exercise 3(b) of Chapter V there is a valuation overring V of R and a chain $P_0' \subset P_1' \subset \cdots \subset P_r'$ of distinct proper prime ideals of V such that P_i' lies over P_i for $i = 0, 1, \ldots, r$. The assertion follows immediately from this fact.

7.18. Proposition. *If* $\dim_v R = k$ *and if* $X_1, \ldots, X_n$ *are independent indeterminates, then* $\dim_v R[X_1, \ldots, X_n] = k + n$.

Proof. Let V' be a valuation overring of $R[X_1, \ldots, X_n]$, and let V be the intersection of V' and the quotient field K of R. Then V is a valuation overring of R and so $\dim V \leq k$. Since $V' \cap K = V$ and $X_1, \ldots, X_n \in V'$, we have $\dim V' \leq k + n$ by Theorem 5.24. Thus, $\dim_v R[X_1, \ldots, X_n] \leq k + n$. Now there is a valuation overring W of R with $\dim W = k$, and by Theorem 5.24 there is a valuation overring W' of $R[X_1, \ldots, X_n]$ with $\dim W' = k + n$. Therefore, $\dim_v R[X_1, \ldots, X_n] = k + n$.

7.19. Proposition. *Suppose that* $\dim R = \dim_v R$. *If* $X_1, \ldots, X_n$ *are independent indeterminates, then* $\dim R[X_1, \ldots, X_n] = \dim R + n$.

Proof. It follows from the two preceding propositions that

$$\dim R[X_1, \ldots, X_n] \leq \dim_v R[X_1, \ldots, X_n] = \dim_v R + n = \dim R + n.$$

The reverse inequality follows from Theorem 7.13.

7.20. Theorem. *If R is a Prüfer domain and if* $X_1, \ldots, X_n$ *are independent indeterminates, then* $\dim R[X_1, \ldots, X_n] = \dim R + n$.

Proof. It is sufficient to show that $\dim R \geq \dim_v R$. Let V be a valuation overring of R and let P be the intersection of R and the maximal ideal of V. Then $R_P \subseteq V$. Since R is a Prüfer domain, R_P is a valuation ring. Hence $\dim V \leq \dim R_P = \operatorname{ht} P \leq \dim R$. Therefore, $\dim_v R \leq \dim R$.

The converse of this result is true under the additional hypothesis that $\dim R = 1$. In order to prove this we shall require two preliminary results.

7.21. Proposition. *Suppose that R is integrally closed and that R has a unique maximal ideal P. Let a belong to the quotient field of R and* *assume that* $a \notin R$ *and* $1/a \notin R$. *Then* $PR[a]$ *is a nonmaximal proper prime ideal of* $R[a]$.

Proof. Suppose that $PR[a] + aR[a] = R[a]$. Then there is a positive integer k, elements $b_1, \ldots, b_k \in R$, and an element $b_0 \in P$ such that $1 = b_0 + b_1 a + \cdots + b_k a^k$. Then $b_0 - 1$ is a unit in R and

$$(1/a)^k + (b_1/(b_0 - 1))(1/a)^{k-1} + \cdots + b_k/(b_0 - 1) = 0,$$

so that $1/a$ is integral over R. Since this is not the case, we conclude that $PR[a] + aR[a] \neq R[a]$.

Let $f(X)$ be a nonconstant monic polynomial in $R[X]$, and let $g(X) = f(X) - f(a)$. If $f(a) \in R$ then, since $g(X)$ is monic and $g(a) = 0$, a is integral over R, which is not true; hence $f(a) \notin R$. In particular, $f(a) \neq 0$. Suppose $1/f(a) \in R$. Then $1/f(a) \in P$ and so $1 \in f(a)P \subseteq PR[a]$, which is not true; hence $1/f(a) \notin R$. Therefore by what was proved above,

$$PR[f(a)] + f(a)R[f(a)] \neq R[f(a)].$$

Since $g(a) = 0$, $R[a]$ is integral over $R[f(a)]$. Hence by Exercise 3(c) of Chapter IV, P and $f(a)$ together generate a proper ideal of $R[a]$, that is,

$$PR[a] + f(a)R[a] \neq R[a].$$

Since $1 + f(X)$ is also nonconstant and monic this implies that

$$PR[a] + (1 + f(a))R[a] \neq R[a].$$

It follows immediately that $f(a) \notin PR[a]$.

To summarize, we have shown that $PR[a]$ is a nonmaximal proper ideal of $R[a]$, and that if $f(X)$ is any nonconstant monic polynomial in $R[X]$, then $f(a) \notin PR[a]$. To complete the proof we shall show that under the homomorphism $R[X] \to R[a]$ given by $h(X) \mapsto h(a)$, the inverse image of $PR[a]$ is $PR[X]$. It will follow that $PR[a]$ is a prime ideal of $R[a]$.

Let $h(X) \in R[X] \backslash PR[X]$; we shall show that $h(a) \notin PR[a]$. Write $h(X) = h_1(X) + h_2(X)$, where $h_1(X)$ has no coefficient in P and $h_2(X) \in PR[X]$. Then $h_1(X) \neq 0$ and its leading coefficient d is a unit in R. Then $h_1(X)/d$ is a nonconstant monic polynomial in $R[X]$

and so $h_1(a)/d \notin PR[a]$. Therefore, $h_1(a) \notin PR[a]$, and consequently $h(a) \notin PR[a]$.

7.22. Proposition. *If R is integrally closed and if* $\dim_v R = 1$, *then R is a Prüfer domain.*

Proof. If P is a maximal ideal of R, then R_P is integrally closed by Exercise 7(b) of Chapter IV, and $\dim_v R_P = 1$. Hence, in proving the assertion, we may assume that P is the unique maximal ideal of R and show that R is a valuation ring. Since R is not a field we have $\dim R = 1$ by Proposition 7.17; hence $\dim R[X] = 2$ by Proposition 7.19.

Assume that R is not a valuation ring. Then there is an element a in the quotient field of R such that $a \notin R$ and $1/a \notin R$. By Proposition 7.21, $\dim R[a] \geq 2$. Consider the homomorphism $R[X] \to R[a]$ given by $f(X) \mapsto f(a)$. If $a = b/s$, where $b, s \in R$, and if $f(X) = sX - b$, then $f(a) = 0$; hence the kernel of this homomorphism is a nonzero prime ideal of $R[X]$. Therefore, $\dim R[X] \geq 3$, which is a contradiction.

7.23. Theorem. *If* $\dim R = 1$ *and if X is an indeterminate, then* $\dim R[X] = 2$ *if and only if the integral closure of R is a Prüfer domain.*

Proof. Let R' be the integral closure of R. By Exercise 7, $\dim R' = \dim R$ and $\dim_v R' = \dim_v R$. If R' is a Prüfer domain, then $\dim R' = \dim_v R'$, and so $\dim R = \dim_v R$, and $\dim R[X] = 2$ by Proposition 7.19.

Conversely, suppose that $\dim R = 1$ and $\dim R[X] = 2$. We shall show that this implies that $\dim_v R = 1$; it follows then that $\dim_v R' = 1$, and that R' is a Prüfer domain by Proposition 7.22.

Suppose that $\dim_v R \geq 2$. Then there is a valuation overring V of R with $\dim V \geq 2$. Hence there are proper nonzero prime ideals P_1 and P_2 of V such that $P_1 \subset P_2$. Choose $a \in P_2 \backslash P_1$ and consider the ring $R[a]$. We have $P_1 \cap R[a] \subset P_2 \cap R[a] \neq R[a]$. If b/s is a nonzero element of P_1, where $b, s \in R$, then $b = s(b/s)$ is a nonzero element of $P_1 \cap R[a]$. Therefore, $\dim R[a] \geq 2$. Now, just as in the final paragraph of the proof of Proposition 7.22, this implies that $\dim R[X] \geq 3$, contrary to assumption.

EXERCISES

1. Heights of prime ideals.

Let R be a Noetherian ring and let P be a proper prime ideal of R.

(a) Show that $\operatorname{ht} P \leq n$ if and only if P is a minimal prime divisor of an ideal of R generated by n elements.

(b) Suppose that $\operatorname{ht} P \geq 2$. Show that there are infinitely many prime ideals of height one contained in P.

2. Dimension of certain rings.

Let R be a Noetherian ring and X an indeterminate.

(a) Show that $\dim R(X) = \dim R$.

(b) Let a be an element of some ring having R as a subring, and suppose that $f(a) = 0$ for some nonzero $f(X) \in R[X]$. Show that $\dim R[a] \leq \dim R$.

3. A Noetherian ring of infinite dimension.

(a) Let R be a ring such that for every maximal ideal P of R the ring R_P is Noetherian and such that each nonzero element of R is contained in only a finite number of maximal ideals. Show that R is a Noetherian ring.

(b) Let K be a field and let $R = K[X_1, X_2, \ldots]$ be the ring of polynomials in infinitely many indeterminates $X_1, X_2, \ldots$ over K. Let $n_1, n_2, \ldots$ be positive integers such that $0 < n_2 - n_1 < n_3 - n_2 < \cdots$. Let $P_i = (X_{n_i}, \ldots, X_{n_{i+1}})$. Show that P_i is a prime ideal of R and that $\operatorname{ht} P_i = n_{i+1} - n_i$.

(c) Let R be the ring of part (b) and let $S = R \backslash \bigcup_{i=1}^{\infty} P_i$. Show that $S^{-1}R$ is a Noetherian ring and that $\dim S^{-1}R = \infty$.

4. Local rings.

Recall that R is a local ring if it has a unique maximal ideal. Let R be a Noetherian local ring which is not a field and let P be its maximal ideal. Let $\dim R = d$, and note that d is finite since $d = \operatorname{ht} P$.

(a) Show that there exist d elements $a_1, \ldots, a_d \in P$ such that $(a_1, \ldots, a_d)$ is P-primary. Then $a_1, \ldots, a_d$ is called a **system of parameters** of R.

(b) Show that no set of fewer than d elements of P generates a P-primary ideal.

(c) Show that an element $a \in P$ belongs to a system of parameters of R if and only if $\dim R/(a) = \dim R - 1$.

(d) Show that each of the following is sufficient for an element $a \in P$ to belong to a system of parameters of R.

(1) $\dim R/(a) < \dim R$.

(2) a is not a zero-divisor of R.

(3) There is no proper prime ideal P' of R with $a \in P'$ and $\operatorname{dpt} P' = \dim R$.

(e) Let $a_1, \ldots, a_k \in R$ be such that $\operatorname{ht}(a_1, \ldots, a_k) = k$. Show that $\dim R/(a_1, \ldots, a_k) = \dim R - k$.

5. Modules over a local ring.
Let R be a Noetherian local ring, let P be its maximal ideal, and let M be a finitely generated R-module.

(a) Show that a subset $\{x_\alpha \mid \alpha \in I\}$ of M generates M if and only if $\{x_\alpha + PM \mid \alpha \in I\}$ generates M/PM as a vector space over R/P. Thus $\{x_1, \ldots, x_k\}$ is a minimal set of generators of M if and only if $\{x_1 + PM, \ldots, x_k + PM\}$ is a basis of M/PM over R/P.

(b) Let $\{x_1, \ldots, x_k\}$ be a minimal set of generators of M. Let $A = [a_{ij}]$ be a $k \times k$ matrix over R and let $y_i = \sum_{j=1}^{k} a_{ij} x_j$ for $i = 1, \ldots, k$. Show that $\{y_1, \ldots, y_k\}$ is a minimal set of generators of M if and only if $\det A \notin P$.

6. Regular local rings.
Let R be a Noetherian local ring which is not a field and let P be its maximal ideal. Then R is called **regular** if there is a system of parameters of R which generates P.

(a) Show that if k is the dimension of the vector space P/P^2 over R/P, then $k \geq \dim R$, with equality if and only if R is regular.

(b) Let $a_1, \ldots, a_n \in P$. Show that if R is regular, then these elements belong to a system of parameters which generate P if and only if $a_1 + P^2, \ldots, a_n + P^2$ are linearly independent elements of P/P^2. Show that this is the case if $\operatorname{ht}(a_1, \ldots, a_n) = n$.

(c) Let $a_1, \ldots, a_n \in P$ be such that $\operatorname{ht}(a_1, \ldots, a_n) = n$. Show that if $R/(a_1, \ldots, a_n)$ is regular, then R is regular.

(d) Suppose that R is regular and let A be a proper ideal of R. Show that R/A is regular if and only if A is generated by a subset of a system of parameters which generate P.

7. Dimension of integral extensions.
Let R' be integral over a subring R.
(a) Show that $\dim R' = \dim R$.
(b) Assume that R' is an integral domain and show that $\dim_v R' = \dim_v R$.
(c) Let R be an integral domain and let $X_1, \ldots, X_n$ be independent indeterminates. Show that if the integral closure of R is a Prüfer domain, then $\dim R[X_1, \ldots, X_n] = \dim R + n$.

8. Dimension of other extensions.
Let R be an integral domain with quotient field K and let $X_1, \ldots, X_n$ be independent indeterminates.
(a) Let $a_1, \ldots, a_n$ be distinct elements of K and let P be the kernel of the homomorphism $R[X_1, \ldots, X_n] \to R[a_1, \ldots, a_n]$ given by $f(X_1, \ldots, X_n) \mapsto f(a_1, \ldots, a_n)$. Show that $\text{ht } P = n$.
(b) Show that the following statements are equivalent.
(1) If T is an overring of R, then $\dim T \leq n$.
(2) If V is a valuation overring of R, then $\dim V \leq n$.
(3) If $a_1, \ldots, a_n \in K$, then $\dim R[a_1, \ldots, a_n] \leq n$.
(4) $\dim R[X_1, \ldots, X_n] \leq 2n$.
(c) Show that the following statements are equivalent.
(1) $\dim_v R = n$.
(2) If T is an overring of R, then $\dim T \leq n$, and $\dim T = n$ for at least one such T.
(3) If $a_1, \ldots, a_n \in K$, then $\dim R[a_1, \ldots, a_n] \leq n$, and $\dim R[a_1, \ldots, a_n] = n$ for some choice of $a_1, \ldots, a_n$.
(4) $\dim R[X_1, \ldots, X_n] = 2n$.

CHAPTER

VIII

Krull Domains

1 KRULL DOMAINS

Let R be a Dedekind domain which is not a field, and let K be the quotient field of R.

(I) *There is a family $\{v_\alpha \mid \alpha \in I\}$ of discrete rank one valuations on K such that if V_α is the valuation ring of v_α, then*

$$R = \bigcap_{\alpha \in I} V_\alpha .$$

In fact, let $\{P_\alpha \mid \alpha \in I\}$ be the family of maximal ideals of R. For each $\alpha \in I$, $V_\alpha = R_{P_\alpha}$ is a Noetherian valuation ring. The valuation v_α determined by V_α has rank one and is discrete by Theorem 5.18, and R is the intersection of the V_α by Exercise 5(a) of Chapter III.

(II) *For every $a \in K^*$, the set of α in I for which $v_\alpha(a) \neq 0$ is finite.*

For, let $a \in K^*$, then $a = b/c$ where $b, c \in R$, and both b and c are not zero. For each $\alpha \in I$, $v_\alpha(a) = v_\alpha(b) - v_\alpha(c)$, so it is sufficient to verify the assertion when $a \in R$. However if $a \in R$, then $v_\alpha(a) \neq 0$ if and only if $a \in P_\alpha$, and we know that a is contained in only a finite number of maximal ideals of R [see Exercise 11(a) of Chapter VI].

8.1. Definition. *An integral domain which is not a field is a* **Krull domain** *if it satisfies* (*I*) *and* (*II*).

We have shown that every Dedekind domain which is not a field is a Krull domain. It will be clear from results we shall obtain that there are Krull domains which are not Dedekind domains. Many of the assertions we shall make about Krull domains will also be true of fields. However, we choose to exclude fields from the class of Krull domains.

Let R be an integral domain with quotient field K, and let $\mathscr{F}(R)$ be the set of nonzero fractional ideals of R. If $A \in \mathscr{F}(R)$ we call the fractional ideal

$$[R\colon A] = \{a \mid a \in K \text{ and } Aa \subseteq R\}$$

the **quasi-inverse** of A. Two nonzero fractional ideals A and B are said to be **quasi-equal** if their quasi-inverses are equal; in this case we write $A \sim B$. Clearly, quasi-equality is an equivalence relation on the set $\mathscr{F}(R)$. We denote the set of equivalence classes of $\mathscr{F}(R)$ with respect to this relation by $\mathscr{D}(R)$. The elements of $\mathscr{D}(R)$ are called **divisors** of R. If $A \in \mathscr{F}(R)$, the divisor represented by A will be denoted by div A. If $A = Ra$ for some nonzero $a \in K$, we shall write div a instead of div Ra.

8.2. Proposition. *If $A, B \in \mathscr{F}(R)$, then $A \sim B$ if and only if A and B are contained in the same principal fractional ideals.*

Proof. This follows immediately from the fact that if $a \in K^*$, then $A \subseteq Ra$ if and only if $1/a \in [R\colon A]$.

8.3. Corollary. *Let $A \in \mathscr{F}(R)$ and set*

$$\bar{A} = \bigcap_{A \subseteq Ra} Ra.$$

Then $\bar{A} \in \mathscr{F}(R)$ and *$\bar{A} \sim A$.*

Proof. Certainly, $\bar{A}$ is a nonzero R-module. If $Ad \subseteq R$ where $d \in R$, $d \neq 0$, then $A \subseteq R(1/d)$; hence $\bar{A} \subseteq R(1/d)$ and so $\bar{A}d \subseteq R$. Thus $\bar{A}d \in \mathscr{F}(R)$ and it is clear that $\bar{A} \sim A$.

8.4. Corollary. *If $A \in \mathscr{F}(R)$, then, in the set of fractional ideals quasi-equal to A, there is one which contains all others, namely, $\bar{A}$.*

Note that if $A, B \in \mathscr{F}(R)$, then $A \sim B$ if and only if $\bar{A} = \bar{B}$.

8.5. Definition. *A fractional ideal A is* **divisorial** *if $A = \bar{A}$.*

Each divisor of R is represented by exactly one divisorial fractional ideal.

We can define a partial ordering on $\mathscr{D}(R)$ by

$$\operatorname{div} A \leq \operatorname{div} B \qquad \text{if} \qquad \bar{B} \subseteq \bar{A}.$$

8.6. Proposition. *If $A, B, C \in \mathscr{F}(R)$ and if $\operatorname{div} A \leq \operatorname{div} B$, then $\operatorname{div} AC \leq \operatorname{div} BC$.*

Proof. We shall show that $\overline{BC} \subseteq \overline{AC}$. By hypothesis, $\bar{B} \subseteq \bar{A}$; that is, $[R: A] \subseteq [R: B]$. Then

$$[R: AC] = [[R: A]: C] \subseteq [[R: B]: C] = [R: BC],$$

so that $\overline{BC} \subseteq \overline{AC}$.

If $A, B \in \mathscr{F}(R)$, we define the **sum** of the divisors $\operatorname{div} A$ and $\operatorname{div} B$ by

$$\operatorname{div} A + \operatorname{div} B = \operatorname{div} AB.$$

This is a well-defined operation on $\mathscr{D}(R)$ since

$$\begin{aligned}[R: AB] &= [[R: A]: B] = [[R: \bar{A}]: B] = [[R: B]: \bar{A}] \\ &= [[R: \bar{B}]: \bar{A}] = [R: \bar{A}\bar{B}]\end{aligned}$$

implies that $\operatorname{div} A + \operatorname{div} B = \operatorname{div} \bar{A} + \operatorname{div} \bar{B}$. With this operation, $\mathscr{D}(R)$ is a commutative semigroup with identity element $\operatorname{div} R$. It follows from Proposition 8.6 that if $A, B, C \in \mathscr{F}(R)$, and if $\operatorname{div} A \leq \operatorname{div} B$, then

$$\operatorname{div} A + \operatorname{div} C \leq \operatorname{div} B + \operatorname{div} C.$$

Thus, $\mathscr{D}(R)$ is a **partially ordered semigroup.** We shall show that, in fact, $\mathscr{D}(R)$ is a **lattice ordered semigroup**; this means that with respect to the given partial ordering, $\mathscr{D}(R)$ is a lattice; that is, every pair of elements in $\mathscr{D}(R)$ has a least upper bound and a greatest lower bound. Lattice ordered semigroups are also called **multiplicative lattices**.

Let $A, B \in \mathscr{F}(R)$. We shall show that $\operatorname{div}(A \cap B)$ is the least upper bound and $\operatorname{div}(A + B)$ the greatest lower bound of the subset $\{\operatorname{div} A, \operatorname{div} B\}$ of $\mathscr{D}(R)$. We may assume A and B are divisorial; then $A \cap B$ is divisorial (for, $\overline{A \cap B} = \bar{A} \cap \bar{B} = A \cap B$) and the assertion concerning $\operatorname{div}(A \cap B)$ follows immediately. Now suppose that $C \in \mathscr{F}(R)$ and that $\operatorname{div} C \leq \operatorname{div} A$ and $\operatorname{div} C \leq \operatorname{div} B$. Then $A \subseteq \bar{C}$ and $B \subseteq \bar{C}$; hence $A + B \subseteq \bar{C}$ and so $\overline{A + B} \subseteq \bar{C}$. Therefore $\operatorname{div} C \leq \operatorname{div}(A + B)$. It follows that $\operatorname{div}(A + B)$ is the greatest lower bound of $\{\operatorname{div} A, \operatorname{div} B\}$.

If R is a Dedekind domain, then $\mathscr{D}(R) = \mathscr{F}(R)$, and consequently $\mathscr{D}(R)$ is a group. We shall now determine a necessary and sufficient condition, stated in terms of R itself, for $\mathscr{D}(R)$ to be a group.

8.7. Proposition. *$\mathscr{D}(R)$ is a group if and only if R is completely integrally closed.*

Proof. Let K be the quotient field of R. Suppose that R is completely integrally closed. We shall show that for all divisorial fractional ideals A in $\mathscr{F}(R)$, $A[R: A]$ and R are quasi-equal. Since $A[R: A] \subseteq R$ it follows from Proposition 8.2 that it is sufficient to show that if $A[R: A] \subseteq Ra$, then $R \subseteq Ra$; that is, if $A[R: A] \subseteq Ra$, then $1/a \in R$. Now suppose that $A \subseteq Rb$ for some $b \in K$; then $1/b \in [R: A]$ and so $(1/b)A \subseteq Ra$, that is, $(1/a)A \subseteq Rb$. Thus $(1/a)A \subseteq \bar{A} = A$, and it follows that $(1/a)^n A \subseteq A$ for all positive integers n. Let $d \in R$, $d \neq 0$, be such that $dA \subseteq R$. Then $d(1/a)^n A \subseteq R$, and if $c \in A$, $c \neq 0$, we have $(dc)(1/a)^n \in R$ for all positive integers n. Since R is completely integrally closed, and since $dc \in R$ and $dc \neq 0$, this implies that $1/a \in R$ by Theorem 4.20.

Conversely, suppose that $\mathscr{D}(R)$ is a group. Let a be a nonzero element of K and suppose that there exists $d \in R$, $d \neq 0$, such that $da^n \in R$ for all positive integers n. Then $R[a]$ is a fractional ideal of R and $aR[a] \subseteq R[a]$. But then $\operatorname{div} R[a] \leq \operatorname{div} a + \operatorname{div} R[a]$, and since $\mathscr{D}(R)$ is a group this implies that $\operatorname{div} a \geq \operatorname{div} R$; that is, $a \in R$.

In Theorem 5.19, we showed that a valuation ring has rank one if and only if it is completely integrally closed. Clearly the intersection of completely integrally closed integral domains is completely integrally closed. Hence we immediately obtain:

8.8. Corollary. *If R is a Krull domain, then $\mathscr{D}(R)$ is a lattice ordered group.*

Let R be a Krull domain. Let $\{v_\alpha \mid \alpha \in I\}$ be a family of discrete rank one valuations on the quotient field K of R which satisfy (I) and (II). It is clear that if we replace any of the v_α by an equivalent valuation, then (I) and (II) continue to hold. Since each v_α is discrete, it is equivalent to a valuation on K having as its value group the additive group of integers. Such a discrete rank one valuation is said to be **normed**, and we shall assume henceforth that each v_α is normed. If $A \in \mathscr{F}(R)$ and if $\alpha \in I$, we set

$$v_\alpha(A) = \max\{v_\alpha(a) \mid A \subseteq Ra\}.$$

This maximum always exists, for if $c \in A$, $c \neq 0$, then $v_\alpha(c) \geq v_\alpha(a)$ whenever $A \subseteq Ra$.

8.9. Proposition. *If R is a Krull domain and if $A \in \mathscr{F}(R)$, then $v_\alpha(A) \neq 0$ for only a finite number of α in I.*

Proof. Let $d \in R$, $d \neq 0$, be such that $A \subseteq Rd$. Then $v_\alpha(A) \geq v_\alpha(d)$. If $c \in A$, $c \neq 0$, then $v_\alpha(c) \geq v_\alpha(A)$. By (II) there is only a finite number of $\alpha \in I$ for which either $v_\alpha(c) \neq 0$ or $v_\alpha(d) \neq 0$. The assertion follows.

For each $\alpha \in I$, let $Z_\alpha = Z$, the additive group of integers, and consider the group

$$Z^{(I)} = \bigoplus_{\alpha \in I} Z_\alpha .$$

For the Krull domain R we can define a mapping ϕ from $\mathscr{D}(R)$ into $Z^{(I)}$ by setting $\phi(\operatorname{div} A)_\alpha = v_\alpha(A)$ for all $\alpha \in I$. This is a well-defined mapping since it is clear from Proposition 8.2 that if A and B are quasi-equal elements of $\mathscr{F}(R)$, then $v_\alpha(A) = v_\alpha(B)$ for all $\alpha \in I$. It is a consequence of the next proposition that ϕ is one-to-one.

8.10. Proposition. *Let R be a Krull domain and let A and B be divisorial fractional ideals of R. Then $A \subseteq B$ if and only if $v_\alpha(A) \geq V_\alpha(B)$ for all $\alpha \in I$.*

Proof. We need prove only the sufficiency of the condition. Assume $v_\alpha(A) \geq v_\alpha(B)$ for all $\alpha \in I$. Let $a \in A$; then for each $\alpha \in I$ we have $v_\alpha(a) \geq v_\alpha(A)$. If $B \subseteq Rb$, then $v_\alpha(B) \geq v_\alpha(b)$, so $v_\alpha(a) \geq v_\alpha(b)$. Hence $v_\alpha(a/b) \geq 0$; that is, a/b is contained in the valuation ring of v_α. Since this is true for each $\alpha \in I$, it follows from (I) that $a/b \in R$. Hence $a \in Rb$. Therefore $a \in \bigcap_{B \subseteq Rb} Rb = B$. Thus $A \subseteq B$.

If $m, n \in Z^{(I)}$, we define $m \leq n$ if $m_\alpha \leq n_\alpha$ for all $\alpha \in I$. This is a partial ordering on $Z^{(I)}$, and if $m_1, m_2, n \in Z^{(I)}$ and $m_1 \leq m_2$, then $m_1 + n \leq m_2 + n$. In fact, with this ordering, $Z^{(I)}$ is a lattice ordered group; the least upper bound h and the greatest lower bound k of elements $m, n \in Z^{(I)}$ are given by

$$h_\alpha = \max\{m_\alpha, n_\alpha\} \qquad k_\alpha = \min\{m_\alpha, n_\alpha\}$$

for all $\alpha \in I$. An element $m \in Z^{(I)}$ is called **positive** if $m > 0$.

8.11. Proposition. *Every nonempty set of positive elements of $Z^{(I)}$ has a minimal element.*

Proof. Let S be a nonempty set of positive elements of $Z^{(I)}$ and let $n \in S$. Then $n_\alpha \geq 0$ for all $\alpha \in I$, and there is a finite subset J of I such that $n_\alpha > 0$ for $\alpha \in J$ and $n_\alpha = 0$ for $\alpha \in I \backslash J$. If $m \in S$ and $m \leq n$, then $m_\alpha = 0$ for $\alpha \in I \backslash J$ and $0 \leq m_\alpha \leq n_\alpha$ for $\alpha \in I$. Hence there is only a finite number of such m and among them there is a minimal element of S.

If R is a Krull domain and if we compose the mapping ϕ and the mapping $A \mapsto \operatorname{div} A$, where A is divisorial, we obtain a one-to-one mapping ψ from the set of divisorial fractional ideals of R into $Z^{(I)}$. If $A \subseteq B$, then $\psi(A) \geq \psi(B)$; in particular, if $A \subseteq R$, then $\psi(A) \geq 0$. Hence if $\mathscr{S}$ is a nonempty set of divisorial ideals of R, then $\{\psi(A) \mid A \in \mathscr{S}\}$ has a minimal element. Therefore $\mathscr{S}$ has a maximal element.

We have verified the necessity of the conditions in the following assertion, which is the principal result of this section.

8.12. Theorem. *An integral domain R which is not a field is a Krull domain if and only if*

(a) *R is completely integrally closed, and*
(b) *every nonempty set of divisorial ideals of R has a maximal element.*

Let R be an integral domain which is not a field for which (a) and (b) hold. By Proposition 8.7, $\mathscr{D}(R)$ is a lattice ordered group. We shall denote elements of $\mathscr{D}(R)$ by lower case German letters. If $\mathfrak{a}$ and $\mathfrak{b}$ are in $\mathscr{D}(R)$ we denote the least upper bound and greatest lower bound of the set $\{\mathfrak{a}, \mathfrak{b}\}$ by $\mathfrak{a} \cup \mathfrak{b}$ and $\mathfrak{a} \cap \mathfrak{b}$, respectively.

As a consequence of (b), every nonempty subset of positive elements of $\mathscr{D}(R)$ has a minimal element. Let $\{\mathfrak{p}_\alpha \mid \alpha \in I\}$ be the set of all minimal positive elements of $\mathscr{D}(R)$; this set is not empty since R is not a field. For each $\alpha \in I$ let P_α be the divisorial ideal of R such that $\mathfrak{p}_\alpha = \operatorname{div} P_\alpha$. Then $\{P_\alpha \mid \alpha \in I\}$ is the set of maximal divisorial proper ideals of R.

Let $\mathfrak{a} \in \mathscr{D}(R)$ and let $\mathfrak{b} = \mathfrak{a} \cup 0$; then $\mathfrak{a} = \mathfrak{b} - (\mathfrak{b} - \mathfrak{a})$, and we have $\mathfrak{b} \geq 0$ and $\mathfrak{b} - \mathfrak{a} \geq 0$. Thus, every element of $\mathscr{D}(R)$ can be written as the difference of two elements of $\mathscr{D}(R)$ each of which is greater than or equal to zero. We shall show that every element of $\mathscr{D}(R)$ can be written as a linear combination of a finite number of elements of the set $\{\mathfrak{p}_\alpha \mid \alpha \in I\}$ with integer coefficients. By the remark made above, it is sufficient to show that each positive element of $\mathscr{D}(R)$ can be written as a sum of minimal positive divisors. Suppose this is not true, and that $\mathfrak{a}$ is minimal among those positive divisors of R which cannot be so written. Then $\mathfrak{a}$ is not a minimal positive divisor of R and so there is an $\alpha \in I$ such that $0 < \mathfrak{p}_\alpha < \mathfrak{a}$. Then $0 < \mathfrak{a} - \mathfrak{p}_\alpha < \mathfrak{a}$ and so $\mathfrak{a} - \mathfrak{p}_\alpha$ can be written as a sum of minimal positive divisors. But $\mathfrak{a} = \mathfrak{p}_\alpha + (\mathfrak{a} - \mathfrak{p}_\alpha)$ and so we have contradicted our choice of $\mathfrak{a}$.

Thus if $\mathfrak{a} \in \mathscr{D}(R)$, we have

$$\mathfrak{a} = \sum_{\alpha \in I} n_\alpha \mathfrak{p}_\alpha,$$

where each n_α is an integer and $n_\alpha \neq 0$ for only a finite number of $\alpha \in I$. We shall show that this expression is unique.

First of all, suppose that $\mathfrak{a}$ and $\mathfrak{b}$ are positive elements of $\mathscr{D}(R)$ and that $\mathfrak{p}_\alpha \leq \mathfrak{a} + \mathfrak{b}$. Since $\mathfrak{p}_\alpha$ is a minimal positive element of $\mathscr{D}(R)$, either $\mathfrak{p}_\alpha \cap \mathfrak{a} = \mathfrak{p}_\alpha$ or $\mathfrak{p}_\alpha \cap \mathfrak{a} = 0$. In the former case, $\mathfrak{p}_\alpha \leq \mathfrak{a}$.

In the latter case, $(\mathfrak{p}_\alpha + \mathfrak{b}) \cap (\mathfrak{a} + \mathfrak{b}) = \mathfrak{b}$. To see this, note that $\mathfrak{b} \leq (\mathfrak{p}_\alpha + \mathfrak{b}) \cap (\mathfrak{a} + \mathfrak{b})$. If $\mathfrak{c} \leq \mathfrak{p}_\alpha + \mathfrak{b}$ and $\mathfrak{c} \leq \mathfrak{a} + \mathfrak{b}$, then $\mathfrak{c} - \mathfrak{b} \leq \mathfrak{p}_\alpha$ and $\mathfrak{c} - \mathfrak{b} \leq \mathfrak{a}$. Therefore $\mathfrak{c} - \mathfrak{b} \leq \mathfrak{p}_\alpha \cap \mathfrak{a} = 0$ and hence $\mathfrak{c} \leq \mathfrak{b}$. This proves that $\mathfrak{b} \geq (\mathfrak{p}_\alpha + \mathfrak{b}) \cap (\mathfrak{a} + \mathfrak{b})$. Since $\mathfrak{p}_\alpha \leq \mathfrak{p}_\alpha + \mathfrak{b}$ and $\mathfrak{p}_\alpha \leq \mathfrak{a} + \mathfrak{b}$, it follows that $\mathfrak{p}_\alpha \leq \mathfrak{b}$. By an induction argument we can show that if $\mathfrak{p}_\alpha \leq \mathfrak{a}_1 + \cdots + \mathfrak{a}_k$, where each $\mathfrak{a}_i$ is positive, then for some i, we have $\mathfrak{p}_\alpha \leq \mathfrak{a}_i$.

Now suppose that

$$\sum_{\alpha \in I} n_\alpha \mathfrak{p}_\alpha = \sum_{\alpha \in I} n_\alpha' \mathfrak{p}_\alpha$$

where each sum has only a finite number of nonzero terms. Assume $n_\alpha - n_\alpha' > 0$ for some $\alpha \in I$ and let

$$J = \{\alpha \mid \alpha \in I \quad \text{and} \quad n_\alpha - n_\alpha' > 0\}.$$

Then with $m_\alpha = n_\alpha - n_\alpha'$ whenever $\alpha \in J$ and $m_\alpha = n_\alpha' - n_\alpha$ otherwise, we have

$$\sum_{\alpha \in J} m_\alpha \mathfrak{p}_\alpha = \sum_{\beta \in I \setminus J} m_\beta \mathfrak{p}_\beta .$$

Since J is not empty, both sides of this equation are nonzero. By what we have shown in the preceding paragraph, it follows that if $\alpha \in J$, then $\mathfrak{p}_\alpha \leq \mathfrak{p}_\beta$ for some $\beta \in I \setminus J$. By the minimality of $\mathfrak{p}_\beta$ we conclude that $\mathfrak{p}_\alpha = \mathfrak{p}_\beta$, which is not true.

Let K be the quotient field of R. If $a \in K^*$, then

$$\operatorname{div} a = \sum_{\alpha \in I} n_\alpha \mathfrak{p}_\alpha ,$$

where each n_α is an integer uniquely determined by a, and $n_\alpha \neq 0$ for only a finite number of $\alpha \in I$. We set $v_\alpha(a) = n_\alpha$. Then v_α is a mapping from K^* into the additive group of integers, which we extend to all of K by setting $v_\alpha(0) = \infty$. If $b \in K^*$ and

$$\operatorname{div} b = \sum_{\alpha \in I} n_\alpha' \mathfrak{p}_\alpha ,$$

then

$$\operatorname{div} ab = \operatorname{div} a + \operatorname{div} b = \sum_{\alpha \in I} (n_\alpha + n_\alpha')\mathfrak{p}_\alpha .$$

Hence $v_\alpha(ab) = n_\alpha + n_\alpha' = v_\alpha(a) + v_\alpha(b)$. We know that $(\operatorname{div} a) \cap$

$(\operatorname{div} b) = \operatorname{div}(Ra + Rb)$, and $R(a + b) \subseteq Ra + Rb$, so that $\operatorname{div}(a + b) \geq \operatorname{div}(Ra + Rb)$. Furthermore

$$\Big(\sum_{\alpha \in I} n_\alpha \mathfrak{p}_\alpha\Big) \cap \Big(\sum_{\alpha \in I} n_\alpha' \mathfrak{p}_\alpha\Big) = \sum_{\alpha \in I} \Big(\min\{n_\alpha, n_\alpha'\}\Big)\mathfrak{p}_\alpha .$$

Hence either $a + b = 0$, or

$$\operatorname{div}(a + b) = \sum_{\alpha \in I} m_\alpha \mathfrak{p}_\alpha \geq \sum_{\alpha \in I} \Big(\min\{n_\alpha, n_\alpha'\}\Big)\mathfrak{p}_\alpha .$$

It follows that $m_\alpha \geq \min\{n_\alpha, n_\alpha'\}$ for each $\alpha \in I$. Therefore $v_\alpha(a + b) \geq \min\{v_\alpha(a), v_\alpha(b)\}$, even when $a + b = 0$. Thus (ii) and (iii) in the definition of valuation hold for v_α when $a, b \in K^*$. Since it is clear that they hold when either $a = 0$ or $b = 0$, we see that for each $\alpha \in I$, v_α is a valuation on K.

Consider the family of valuations $\{v_\alpha \mid \alpha \in I\}$. For each $\alpha \in I$, v_α has rank one and is discrete. It is immediate that (II) holds. If $a \in K^*$ and $v_\alpha(a) \geq 0$ for all $\alpha \in I$, then $\operatorname{div} a \geq 0$; hence $Ra \subseteq R$; that is, $a \in R$. Thus (I) also holds. Therefore, R is a Krull domain. This concludes the proof of Theorem 8.12.

Since an integrally closed Noetherian integral domain is completely integrally closed [Exercise 1(c) of Chapter IV], and the converse holds even for non-Noetherian domains, we have the

8.13. Corollary. *A Noetherian integral domain which is not a field is a Krull domain if and only if it is integrally closed.*

2 ESSENTIAL VALUATIONS

Let R be a Krull domain with quotient field K. We shall continue with the notation of Section 1. In particular, let $\{\mathfrak{p}_\alpha \mid \alpha \in I\}$ be the set of minimal positive elements of $\mathscr{D}(R)$, and for each $\alpha \in I$, let P_α be the divisorial ideal of R such that $\mathfrak{p}_\alpha = \operatorname{div} P_\alpha$. Let $\{v_\alpha \mid \alpha \in I\}$ be the family of valuations on K constructed as in Section 1; (I) and (II) hold for this family of valuations. The valuations in this family are called the **essential valuations** of R.

We have shown that if $\mathfrak{a} \in \mathscr{D}(R)$, then we can write uniquely

$$\mathfrak{a} = \sum_{\alpha \in I} n_\alpha \mathfrak{p}_\alpha,$$

where n_α is an integer for each $\alpha \in I$, and $n_\alpha \neq 0$ for only a finite number of $\alpha \in I$. If $\mathfrak{b} \in \mathscr{D}(R)$ and

$$\mathfrak{b} = \sum_{\alpha \in I} m_\alpha \mathfrak{p}_\alpha,$$

then $\mathfrak{a} \leq \mathfrak{b}$ if and only if $n_\alpha \leq m_\alpha$ for all $\alpha \in I$. Hence if A is the divisorial fractional ideal of R such that $\mathfrak{a} = \operatorname{div} A$, then $a \in A$ if and only if $v_\alpha(a) \geq n_\alpha$ for all $\alpha \in I$. Thus every divisorial ideal of R can be described by such a set of inequalities.

Conversely, let $\{n_\alpha \mid \alpha \in I\}$ be a set of integers only a finite number of which are not zero. Then

$$A = \{a \mid a \in K \quad \text{and} \quad v_\alpha(a) \geq n_\alpha \quad \text{for all} \quad \alpha \in I\}$$

is a divisorial fractional ideal of R; it is the one whose divisor is $\sum_{\alpha \in I} n_\alpha \mathfrak{p}_\alpha$.

If A is a fractional ideal of R, then by Proposition 8.10, $a \in \bar{A}$ if and only if $v_\alpha(a) \geq v_\alpha(\bar{A}) = v_\alpha(A)$ for all $\alpha \in I$. Therefore

$$\operatorname{div} A = \sum_{\alpha \in I} v_\alpha(A) \mathfrak{p}_\alpha.$$

If for each $\beta \in J$, A_β is a divisorial fractional ideal of R, and if $A = \sum_{\beta \in J} A_\beta$ is a fractional ideal, then $\operatorname{div} A$ is the greatest lower bound of the set $\{\operatorname{div} A_\beta \mid \beta \in J\}$. We leave the proof of this as an easy exercise for the reader.

If A is a fractional ideal of R, then $A = \sum_{a \in A,\ a \neq 0} Ra$. Hence $\operatorname{div} A$ is the greatest lower bound of the set $\{\operatorname{div} Ra \mid a \in A,\ a \neq 0\}$. Therefore

$$\operatorname{div} A = \sum_{\alpha \in I} (\min\{v_\alpha(a) \mid a \in A\}) \mathfrak{p}_\alpha.$$

On comparing the two expressions that we have for $\operatorname{div} A$, we conclude that for all $\alpha \in I$

$$v_\alpha(A) = \min\{v_\alpha(a) \mid a \in A\}.$$

8.14. Proposition. *Each of the essential valuations of R is normed.*

Proof. Let $\alpha \in I$; then $\mathfrak{p}_\alpha < 2\mathfrak{p}_\alpha$ and so, if A is the divisorial ideal of R such that $2\mathfrak{p}_\alpha = \operatorname{div} A$, we have $A \subset P_\alpha$. Let $a \in P_\alpha \backslash A$. Then $\mathfrak{p}_\alpha \leq \operatorname{div} a < 2\mathfrak{p}_\alpha$, so $1 \leq v_\alpha(a) < 2$. Therefore $v_\alpha(a) = 1$.

Let $\phi: \mathscr{D}(R) \to Z^{(I)}$ be defined by $\phi(\operatorname{div} A)_\alpha = v_\alpha(A)$ for all $\alpha \in I$. We have seen that ϕ is a one-to-one mapping.

8.15. Proposition. *The mapping ϕ is an order-preserving isomorphism from $\mathscr{D}(R)$ onto $Z^{(I)}$*

Proof. First we shall show that ϕ is a homomorphism by showing that for divisorial fractional ideals A and B, we have $v_\alpha(AB) = v_\alpha(A) + v_\alpha(B)$ for each $\alpha \in I$. Let $a, b \in K$ be such that

$$A \subseteq Ra, \qquad B \subseteq Rb, \qquad v_\alpha(A) = v_\alpha(a), \qquad \text{and} \qquad v_\alpha(B) = v_\alpha(b).$$

Then

$$AB \subseteq Rab, \qquad \text{and} \qquad v_\alpha(AB) \geq v_\alpha(ab) = v_\alpha(A) + v_\alpha(B).$$

On the other hand, let $c \in K$ be such that $AB \subseteq Rc$ and $v_\alpha(AB) = v_\alpha(c)$. If $x \in B$, $x \neq 0$, then $Ax \subseteq Rc$, or $A \subseteq R(x/c)$. Hence

$$v_\alpha(A) \geq v_\alpha(c/x) = v_\alpha(AB) - v_\alpha(x);$$

thus $v_\alpha(A) + v_\alpha(x) \geq v_\alpha(AB)$. By the remarks made above we may choose x so that $v_\alpha(B) = v_\alpha(x)$; then $v_\alpha(A) + v_\alpha(B) \geq v_\alpha(AB)$. It follows from Proposition 8.10 that ϕ is order-preserving. To show that ϕ is surjective we note simply that if $n \in Z^{(I)}$, then $\phi(\sum_{\alpha \in I} n_\alpha \mathfrak{p}_\alpha) = n$.

8.16. Proposition. *For each $\alpha \in I$, P_α is a prime ideal of R. Furthermore if V_α is the valuation ring of v_α, then $V_\alpha = R_{P_\alpha}$.*

Proof. If $a \in R$, then $v_\alpha(a) > 0$ if and only if $a \in P_\alpha$. Thus P_α is the intersection of R and the ideal of nonunits of V_α. Hence P_α is a prime ideal. To prove that $V_\alpha = R_{P_\alpha}$, we note that since V_α is a Noetherian valuation ring it is sufficient, by Exercise 1(a) of Chapter V, to show that $V_\alpha \subseteq R_{P_\alpha} \subset K$. Let $a \in V_\alpha$ and let $\alpha_1, \ldots, \alpha_k$ be those elements of I such that $v_{\alpha_i}(a) < 0$ for $i = 1, \ldots, k$. For each i, $\alpha_i \neq \alpha$, so there is an element $s_i \in R \backslash P_\alpha$ such that $v_{\alpha_i}(s_i) > 0$. Then we can choose n so large that $v_{\alpha_i}(as_i{}^n) \geq 0$ for $i = 1, \ldots, k$. If $s = (s_1 \cdots s_k)^n$ then $v_\beta(as) \geq 0$ for all $\beta \in I$. Hence $as = b \in R$ and $a = b/s \in R_{P_\alpha}$. Since $P_\alpha \neq 0$ we have $R_{P_\alpha} \neq K$.

8.17. Proposition. *Let P be a nonzero proper prime ideal of R. Then P is divisorial if and only if $P = P_\alpha$ for some $\alpha \in I$.*

Proof. Suppose that P is divisorial. Since

$$\operatorname{div} P = \sum_{\alpha \in I} n_\alpha \mathfrak{p}_\alpha = \sum_{\alpha \in I} n_\alpha \operatorname{div} P_\alpha = \sum_{\alpha \in I} \operatorname{div} P_\alpha^{n_\alpha} = \operatorname{div} \prod_{\alpha \in I} P_\alpha^{n_\alpha},$$

we have $\prod_{\alpha \in I} P_\alpha^{n_\alpha} \subseteq P$. Then $P_\alpha \subseteq P$ for some $\alpha \in I$, and since P_α is maximal among the divisorial proper ideals of R, we have $P = P_\alpha$.

Now we are able to describe completely the ideals P_α of R in terms of ideal-theoretic properties of R. An ideal of R is called a **minimal prime ideal** if it is minimal among the nonzero prime ideals of R.

8.18. Proposition. *The family $\{P_\alpha \mid \alpha \in I\}$ is precisely the family of minimal prime ideals of R.*

Proof. We have seen that P_α is prime and that R_{P_α} is a Noetherian valuation ring. Hence by Theorem 5.9, $P_\alpha R_{P_\alpha}$ is the only nonzero proper prime ideal of R_{P_α}. Then by Corollary 3.11, there are no prime ideals of R strictly between 0 and P_α. Then P_α is a minimal prime ideal of R. If P is a minimal prime ideal of R, then by Exercise 5 and Proposition 8.16, there is a subset J of I such that $R_P = \bigcap_{\alpha \in J} R_{P_\alpha}$. Hence if $\alpha \in J$, $R_P \subseteq R_{P_\alpha}$ and so $P_\alpha \subseteq P$. Since P is minimal, $P = P_\alpha$.

8.19. Theorem. *Let R be a Krull domain with quotient field K. Let K' be a finite extension of K and let R' be the integral closure of R in K'. Then R' is a Krull domain. A prime ideal P' of R' is a minimal prime ideal of R' if and only if $P' \cap R$ is a minimal prime ideal of R.*

Proof. Let $\{v_\alpha \mid \alpha \in I\}$ be the family of essential valuations of R. Let $\{v_\beta' \mid \beta \in J\}$ be the family of valuations on K' determined by the following conditions:

(i) Every v_β' is a prolongation of some v_α.
(ii) Every prolongation to K' of each v_α is equivalent to exactly one v_β'.

Let V_β' be the valuation ring of v_β' and set $R'' = \bigcap_{\beta \in J} V_\beta'$. Let $a \in R'$, $a \neq 0$; then there are elements $b_0, b_1, \ldots, b_{n-1} \in R$ such that

$$b_0 + b_1 a + \cdots + b_{n-1} a^{n-1} + a^n = 0.$$

Let $\beta \in J$ and suppose that $v_\beta'(a) < 0$. Then $v_\beta'(b_i a^i) > v_\beta'(a^n)$ for $i = 0, \ldots, n-1$, and consequently

$$v_\beta'(a^n) < \min\{v_\beta'(b_0), v_\beta'(b_1 a), \ldots, v_\beta'(b_{n-1}a^{n-1})\}$$
$$\leq v_\beta'(b_0 + b_1 a + \cdots + b_{n-1}a^{n-1}),$$

which is not true. Hence $v_\beta'(a) \geq 0$; that is, $a \in V_\beta'$. This proves that $R' \subseteq R''$.

On the other hand, suppose that $a \in R''$. Let $f(X)$ be the monic irreducible polynomial in $K[X]$ having a as a root. By the lemma proved below, $f(X)$ has coefficients in R. Hence $a \in R'$, and we have shown that $R' = R''$. In doing so, we have verified that (I) holds for R' and the family of valuations $\{v_\beta' \mid \beta \in J\}$.

Again let $a \in R'$, $a \neq 0$, and let

$$b_0 + b_1 a + \cdots + b_{n-1}a^{n-1} + a^n = 0,$$

where $b_i \in R$ for $i = 0, \ldots, n-1$. For all but a finite number of $\alpha \in I$, we have $v_\alpha(b_i) = 0$ for $i = 0, \ldots, n-1$. Choose such an $\alpha \in I$; we shall show that if v_β' is a prolongation of v_α, then $v_\beta'(a) = 0$. This will show that (II) holds for R' and the family of valuations $\{v_\beta' \mid \beta \in J\}$ because, by Theorem 6.25, each v_α has only a finite number of prolongations in the family $\{v_\beta' \mid \beta \in J\}$.

If $v_\beta'(a) > 0$, then

$$v_\beta'(b_0) < \min\{v_\beta'(b_1 a), \ldots, v_\beta'(b_{n-1}a^{n-1}), v_\beta'(a^n)\}$$
$$\leq v_\beta'(b_1 a + \cdots + b_{n-1}a^{n-1} + a^n),$$

and if $v_\beta'(a) < 0$, then

$$v_\beta'(a^n) < v_\beta'(b_0 + b_1 a + \cdots + b_{n-1}a^{n-1}),$$

both of which are not true. Thus we must have $v_\beta'(a) = 0$.

The final assertion of the theorem is a consequence of the lying-over theorem and the going-down theorem.

8.20. Lemma. *Let V be a valuation ring, K the quotient field of V, and v the valuation on K determined by V. Let K' be a finite extension of K and let $a \in K'$. Let*

$$f(X) = c_0 + c_1 X + \cdots + c_{n-1}X^{n-1} + X^n$$

be the monic irreducible polynomial in $K[X]$ having a as a root. If $v'(a) \geq 0$ for every prolongation v' of v to K', then $c_0, c_1, \ldots, c_{n-1} \in V$.

Proof. Let K'' be a finite normal extension of K with $K' \subseteq K''$. Let v'' be a prolongation of v to K''. Let $\sigma \in G(K''/K)$ and define a mapping $\bar{v}$ from K'' into the value group of v'' by $\bar{v}(b) = v''(\sigma(b))$ for all $b \in K''$. It is easily seen that $\bar{v}$ is a valuation on K'', and that its restriction to K' is a prolongation of v to K'. Hence $\bar{v}(a) \geq 0$; that is, $v''(\sigma(a)) \geq 0$. Now let $a_1, \ldots, a_n$ be the roots of $f(X)$ in K'', with repeated roots listed as often as their multiplicity. For each i there is an element $\sigma \in G(K''/K)$ such that $\sigma(a) = a_i$. Hence, $v''(a_i) \geq 0$ for $i = 1, \ldots, n$. Since each coefficient of $f(X)$ is an elementary symmetric function of $a_1, \ldots, a_n$, we have $v''(c_i) \geq 0$; that is, $v(c_i) \geq 0$ for $i = 0, \ldots, n-1$.

8.21. Theorem. *If R is a Krull domain, then the polynomial ring $R[X]$ is a Krull domain. A prime ideal P' of $R[X]$ is a minimal prime ideal of $R[X]$ if and only if either $P' = PR[X]$ for some minimal prime ideal P of R, or $P' = Q \cap R[X]$ where Q is a nonzero proper prime ideal of $K[X]$, where K is the quotient field of R.*

Proof. If $p(X)$ is a nonconstant monic irreducible polynomial in $K[X]$, let $v_{p(X)}$ be the $p(X)$-adic valuation on $K(X)$, and let $V_{p(X)}$ be the valuation ring of $v_{p(X)}$. Let $\{v_\alpha \mid \alpha \in I\}$ be the family of essential valuations of R. For each $\alpha \in I$, let v_α' be the extension of v_α to $K(X)$, and let V_α' be the valuation ring of v_α'. Set

$$R' = \left(\bigcap_{p(X)} V_{p(X)}\right) \cap \left(\bigcap_{\alpha \in I} V_\alpha'\right).$$

It is immediate that $R[X] \subseteq R'$. On the other hand, let $f(X) \in R'$, where $f(X)$ is a rational function over K. Since $f(X) \in V_{p(X)}$ for each nonconstant monic irreducible polynomial $p(X)$ in $K[X]$, it follows that $f(X)$ is a polynomial in $K[X]$. Then since $f(X) \in V_\alpha'$ for each $\alpha \in I$, every coefficient of $f(X)$ is in V_α for each $\alpha \in I$. Thus $f(X) \in R[X]$. We conclude that $R' = R[X]$, and that (I) holds for $R[X]$ and the family of valuations consisting of all of the $v_{p(X)}$ and all of the v_α'. It is easy to see that (II) holds also, and the details of this argument are left to the reader. Therefore $R[X]$ is a Krull domain.

By Exercise 7(b), the various $v_{p(X)}$ and v_α' form precisely the family of essential valuations of $R[X]$. Hence a prime ideal of $R[X]$ is a minimal prime ideal if and only if it is the intersection with $R[X]$ of the ideal of nonunits of some $V_{p(X)}$ or some V_α'. The ideal of nonunits of $V_{p(X)}$ is $QV_{p(X)}$, where Q is the principal ideal $(p(X))$ of $K[X]$ and

$$QV_{p(X)} \cap R[X] = Q \cap R[X].$$

Furthermore every nonzero proper prime ideal of $K[X]$ is of this form. Let P_α' be the ideal of nonunits of V_α'; we shall show that $P_\alpha' \cap R[X] = P_\alpha R[X]$. We have $P_\alpha' \cap R = P_\alpha$, so that $P_\alpha R[X] \subseteq P_\alpha' \cap R[X]$. If $f(X) \in P_\alpha' \cap R[X]$, then every coefficient of $f(X)$ lies in P_α; hence $f(X) \in P_\alpha R[X]$. Therefore, $P_\alpha' \cap R[X] = P_\alpha R[X]$. This completes the proof of the theorem.

8.22. Corollary. *If R is a Krull domain, then the polynomial ring $R[X_1, \ldots, X_k]$ in k indeterminates is a Krull domain.*

3 THE DIVISOR CLASS GROUP

Let R be a Krull domain with quotient field K. We have seen in the preceding section that the family of essential valuations of R is in one-to-one correspondence with the family of minimal prime ideals of R. Thus we may index the essential valuations of R with the minimal prime ideals of R. If P is a minimal prime ideal of R, then v_P is the essential valuation of R whose valuation ring is R_P (see Proposition 8.16).

Let $\mathscr{D}(R)$ be the group of divisors of R, and let $\mathscr{H}(R)$ be the subgroup of $\mathscr{D}(R)$ consisting of all divisors of R of the form div a where $a \in K$, $a \neq 0$.

8.23. Definition. *The factor group $\mathscr{D}(R)/\mathscr{H}(R)$ is the* **divisor class group of** R; we denote it by $\mathscr{C}(R)$.

Let R' be a second Krull domain containing R as a subring. If P is a minimal prime ideal of R and if P' is a minimal prime ideal of R', we shall write $P < P'$ if $v_{P'}$ is a prolongation of v_P to the field of quotients of R'. In this case, there is a positive integer $e(P'/P)$ such that $v_{P'}(a) = e(P'/P)v_P(a)$ for all $a \in K$ [see Exercise 8(b)]. The integer $e(P'/P)$ is called the **ramification index** of P' over P.

We shall assume that the following condition holds, and we shall show that under this assumption there is a homomorphism from $\mathscr{C}(R)$ into $\mathscr{C}(R')$:

(*) If P' is a minimal prime ideal of R', then either $P' \cap R = 0$ or $P' \cap R$ is a minimal prime ideal of R.

If P is a minimal prime ideal of R and if P' is a minimal prime ideal of R' such that $P \subseteq P'$, then $P' \cap R = P$. From this it follows that $P'R_{P'} \cap R_P = PR_P$. Hence, by Proposition 5.21, $v_{P'}$ is a prolongation of v_P. If $a \in P$, $a \neq 0$, then $a \in P'$ and so there is only a finite number of minimal prime ideals of R' containing P. Of course, there may be none at all.

If P is a minimal prime ideal of R, we set

$$\sigma(\operatorname{div} P) = \sum_{P < P'} e(P'/P) \operatorname{div} P'$$

if P is contained in at least one minimal prime ideal of R'; otherwise, we set $\sigma(\operatorname{div} P) = 0$. It follows from what we have already shown that $\mathscr{D}(R)$ is a free Abelian group, freely generated by $\{\operatorname{div} P \mid P$ is a minimal prime ideal of $R\}$. Hence we can extend σ by linearity to a homomorphism from $\mathscr{D}(R)$ into $\mathscr{D}(R')$, which we also denote by σ.

8.24. Theorem. *The homomorphism σ maps $\mathscr{H}(R)$ into $\mathscr{H}(R')$ and so induces a homomorphism $\sigma' : \mathscr{C}(R) \to \mathscr{C}(R')$.*

Proof. If $a \in K$, $a \neq 0$, then

$$\operatorname{div} a = \sum_P v_P(a) \operatorname{div} P$$

where the summation is over all of the minimal prime ideals of R. Hence

$$\begin{aligned}\sigma(\operatorname{div} a) &= \sum_P v_P(a)\, \sigma(\operatorname{div} P) \\ &= \sum_P v_P(a) \sum_{P < P'} e(P'/P) \operatorname{div} P' \\ &= \sum_{P'} \Big(\sum_{\substack{P \\ P < P'}} e(P'/P) v_P(a)\Big) \operatorname{div} P',\end{aligned}$$

where the first summation is over all P' such that $P < P'$ for some P. Given P', we have $P < P'$ if and only if $P' \cap R = P$, and so the inner summation is just $e(P'/P)v_P(a) = v_{P'}(a)$. If we do not have $P < P'$ for some P, then $P' \cap R = 0$ and $v_{P'}(c) = 0$ for all $c \in K$, $c \neq 0$. Therefore

$$\sigma(\text{div } a) = \sum_{P'} v_{P'}(a) \text{ div } P'$$

which is the divisor div a of R'.

Thus if (∗) holds, we are assured of the existence of the homomorphism σ'. The two propositions which follow give two important cases when (∗) does hold.

8.25. Proposition. *If R' is integral over R, then (∗) holds.*

Proof. Let P' be a minimal prime ideal of R'. By the lying-over theorem (Theorem 4.6), $P' \cap R \neq 0$. Let $P' \cap R = P_1$. If P_1 is not a minimal prime ideal of R, then there is a nonzero prime ideal P_2 of R such that $P_2 \subset P_1$. Then by the going-down theorem (Theorem 4.9) there is a prime ideal P'' of R' such that $P'' \cap R = P_2$ and $P'' \subset P'$. It follows that $P'' \neq 0$, and thus we have contradicted the fact that P' is minimal among the nonzero prime ideals of R'.

8.26. Proposition. *If R' is a flat R-module, then (∗) holds.*

Proof. Suppose that there is a minimal prime ideal P' of R' such that $P' \cap R \neq 0$ and $P' \cap R$ is not a minimal prime ideal of R. Let $a \in P' \cap R$, $a \neq 0$, and let $P_1, \ldots, P_k$ be those minimal prime ideals of R such that $v_{P_i}(a) > 0$ for $i = 1, \ldots, k$. We have $P' \cap R \nsubseteq P_i$ for $i = 1, \ldots, k$; hence, by Exercise 5(c) of Chapter II, there is an element $b \in P' \cap R$ such that $b \notin P_i$ for $i = 1, \ldots, k$. Then for each i, $v_{P_i}(b) = 0$, and consequently

$$v_{P_i}(ab) = v_{P_i}(a) = \max\{v_{P_i}(a), v_{P_i}(b)\}.$$

If Q is a minimal prime ideal of R different from $P_1, \ldots, P_k$, then $v_Q(a) = 0$ and $v_Q(ab) = v_Q(b) = \max\{v_Q(a), v_Q(b)\}$. Thus $\text{div}(Ra \cap Rb) =$ the least upper bound of $\{\text{div } a, \text{ div } b\} = \text{div } ab$. Then since both Rab and $Ra \cap Rb$ are divisorial, we have $Rab = Ra \cap Rb$. Since R' is a flat R-module it follows that $R'ab = R'a \cap R'b$ (see the remark following the proof). Thus we must have

$v_{P'}(ab) = \max\{v_{P'}(a), v_{P'}(b)\}$. However $a, b \in P'$, so that $v_{P'}(a) > 0$ and $v_{P'}(b) > 0$. Then $v_{P'}(ab) = v_{P'}(a) + v_{P'}(b) > \max\{v_{P'}(a), v_{P'}(b)\}$, which is a contradiction.

Remark. In this proof, we have used the fact that if A and B are ideals of R, then since R' is a flat R-module, we have $(A \cap B)R' = AR' \cap BR'$. This is a consequence of Exercise 10(c) of Chapter I, used in conjunction with Proposition 1.18.

Let S be a multiplicative system in R. By Exercise 5, $S^{-1}R$ is a Krull domain, and the minimal prime ideals of $S^{-1}R$ are those ideals of the form $S^{-1}P$ where P belongs to the set $\mathscr{T} = \{P \mid P$ is a minimal prime ideal of R and $P \cap S$ is empty$\}$. Let $\mathscr{H}_1$ be the subgroup of $\mathscr{D}(R)$ generated by $\{\operatorname{div} P \mid P \in \mathscr{T}\}$ and $\mathscr{H}_2$ the subgroup generated by $\{\operatorname{div} P \mid P$ is a minimal prime ideal of R and $P \cap S$ is not empty$\}$. Then $\mathscr{D}(R) = \mathscr{H}_1 \oplus \mathscr{H}_2$ and the mapping $\tau: \mathscr{D}(S^{-1}R) \to \mathscr{H}_1$ defined by $\tau(\operatorname{div} S^{-1}P) = \operatorname{div} P$ is well-defined and is an isomorphism (the reader should verify this). Each element of $\mathscr{D}(R)$ can be written uniquely in the form $h_1 + h_2$ where $h_1 \in \mathscr{H}_1$ and $h_2 \in \mathscr{H}_2$, and the mapping $h_1 + h_2 \mapsto h_i$ is a surjective homomorphism from $\mathscr{D}(R)$ onto $\mathscr{H}_i$. (In the notation of Section 3 of Chapter I, this is the homomorphism ϕ_i.) Since for each $P \in \mathscr{T}$, $e(S^{-1}P/P) = 1$, the composition of this homomorphism with τ^{-1} is precisely the homomorphism $\sigma: \mathscr{D}(R) \to \mathscr{D}(S^{-1}R)$.

8.27. Theorem. *Let S be as above. Then $\sigma': \mathscr{C}(R) \to \mathscr{C}(S^{-1}R)$ is defined and is surjective. Suppose there is a subset S' of R such that every element of S is a product of elements of S' and (a) is a prime ideal of R for every $a \in S'$. Then σ' is an isomorphism.*

Proof. First $S^{-1}R$ is a flat R-module by Theorem 3.3; hence σ' is defined by Proposition 8.26. Furthermore, σ' is surjective since σ is surjective. Now we have

$$\operatorname{Ker} \sigma' = (\mathscr{H}(R) + \mathscr{H}_2)/\mathscr{H}(R) \cong \mathscr{H}_2/(\mathscr{H}_2 \cap \mathscr{H}(R)).$$

We will show that if there is a subset S' of R with the stated property, then $\mathscr{H}_2 = \mathscr{H}_2 \cap \mathscr{H}(R)$. Consider a divisor $\operatorname{div} P$ where P is a minimal prime ideal of R and $P \cap S$ is not empty. If $a \in P \cap S$, then $a = a_1 \cdots a_k$ where a_i is a nonzero element of S' for $i = 1, \ldots, k$.

Then $(a_i) \subseteq P$ for some i; since (a_i) is a prime ideal this implies that $P = (a_i)$. Thus div $P \in \mathscr{H}(R)$, and we conclude that $\mathscr{H}_2 \subseteq \mathscr{H}(R)$.

8.28. Theorem. *If R is a Krull domain and X is an indeterminate, then $\sigma' : \mathscr{C}(R) \to \mathscr{C}(R[X])$ is defined and is an isomorphism.*

Proof. First we note that σ' is defined, since $R[X]$ is a flat R-module [Exercise 19(a) of Chapter II]. If S is the set of nonzero elements of R, then $S^{-1}(R[X]) = K[X]$, where K is the quotient field of R. Let ρ be the homomorphism from $\mathscr{D}(R[X])$ into $\mathscr{D}(K[X])$ defined as was σ above, and let $\rho' : \mathscr{C}(R[X]) \to \mathscr{C}(K[X])$ be the induced homomorphism. It is well known from elementary abstract algebra that every ideal of $K[X]$ is principal. Hence $\mathscr{C}(K[X]) = 0$ and Ker $\rho' = \mathscr{C}(R[X])$. Let $\mathscr{H}$ be the subgroup of $\mathscr{D}(R[X])$ generated by the set of all div P', where P' is a minimal prime ideal of $R[X]$ and $P' \cap S$ is not empty. Then as we have noted already,

$$\mathscr{C}(R[X]) = \text{Ker } \rho' = (\mathscr{H}(R[X]) + \mathscr{H})/\mathscr{H}(R[X])$$
$$\cong \mathscr{H}/(\mathscr{H} \cap \mathscr{H}(R[X])).$$

By Theorem 8.21, and its proof, it is clear that if P is a minimal prime ideal of R, and P' is a minimal prime ideal of $R[X]$, then $P < P'$ if and only if $P' = PR[X]$; in this case, $e(P'/P) = 1$. Thus, for every minimal prime ideal P of R, we have

$$\sigma(\text{div } P) = \text{div } PR[X].$$

We conclude that σ maps $\mathscr{D}(R)$ onto $\mathscr{H}$. We shall show that $\sigma(\mathscr{H}(R)) = \mathscr{H} \cap \mathscr{H}(R[X])$. Then σ induces an isomorphism

$$\bar{\sigma} : \mathscr{C}(R) \to \mathscr{H}/(\mathscr{H} \cap \mathscr{H}(R[X])).$$

The composite of $\bar{\sigma}$ with the isomorphism from $\mathscr{H}/(\mathscr{H} \cap \mathscr{H}(R[X]))$ onto $(\mathscr{H}(R[X]) + \mathscr{H})/\mathscr{H}$ is σ'.

It is clear that $\sigma(\mathscr{H}(R)) \subseteq \mathscr{H} \cap \mathscr{H}(R[X])$. Suppose that $f(X) \in R[X]$ and that $R[X]f(X) \in \mathscr{H}$. If we show that this ideal of $R[X]$ is generated by an element of R, we will be finished. Now $f(X)$ is contained in no minimal prime ideal of $R[X]$ other than the ones whose divisors generate $\mathscr{H}$. Hence by Theorem 8.21, $f(X)$ is contained in no proper prime ideal of $K[X]$. Thus $f(X)$ is a constant; that is, an element of R.

4 FACTORIAL RINGS

8.29. Definition. *An integral domain R is a* **factorial ring** *if there is a set S of nonzero nonunits of R such that every nonzero element of R can be written uniquely in the form $ua_1 \cdots a_k$, where u is a unit of R and $a_1, \ldots, a_k \in S$, except for the order in which the factors are written.*

A factorial ring is also called a **unique factorization domain.** Familiar examples are any field with S the empty set, the ring Z of integers with S the set of primes, and the ring of polynomials in one indeterminate over a field with S the set of monic irreducible polynomials of positive degree.

Let R be an integral domain. An element $p \in R$ is called a **prime** if (p) is a nonzero proper prime ideal of R. An element $a \in R$ is called **irreducible** if it is a nonzero nonunit and if, whenever $a = bc$ where $b, c \in R$, then either b or c is a unit of R. Every prime is irreducible. For, let p be a prime; since $(p) \neq 0$ we have $p \neq 0$ and since $(p) \neq R$, we conclude that p is not a unit. Suppose $p = bc$ where $b, c \in R$. Then one of b and c is in (p), say $b \in (p)$; write $b = pd$. Then $p = pdc$, and since R is an integral domain and $p \neq 0$, $dc = 1$; that is, c is a unit of R.

8.30. Proposition. *Let R be a factorial ring, and let S be a set whose existence is required in the definition. Then every irreducible element of R is prime, every element of S is prime, and every prime of R is the product of a unit of R and an element of S.*

Proof. Let $p \in S$ and let $ab \in (p)$, where $a, b \in R$. Then $ab = pc$, where $c \in R$. There are elements $p_1, \ldots, p_k, q_1, \ldots, q_m, r_1, \ldots, r_n \in S$ and units u, v, w of R such that

$$a = up_1 \cdots p_k, \qquad b = vq_1 \cdots q_m, \qquad \text{and} \qquad c = wr_1 \cdots r_n.$$

Then

$$uvp_1 \cdots p_k q_1 \cdots q_m = wpr_1 \cdots r_n.$$

By the uniqueness of such factorizations, p must be p_i for some i or q_j for some j. Hence either $a \in (p)$ or $b \in (p)$. Thus (p) is a prime ideal and, since p is a nonzero nonunit, p is a prime.

Next let p be an arbitrary prime of R. Then $p = up_1 \cdots p_k$, where $p_1, \ldots, p_k \in S$ and u is a unit of R. One of the factors must be in (p), and since $(p) \neq R$, some p_i must be in (p), say $p_1 \in (p)$. Then $p_1 = pd$, where $d \in R$. If d is a unit, then $p = d^{-1}p_1$, which is the assertion. Suppose d is not a unit. Then $d = vq_1 \cdots q_m$, where $q_1, \ldots, q_m \in S$ and v is a unit of R. Then $p_1 = uvp_1 \cdots p_k q_1 \cdots q_m$, which contradicts the uniqueness of such factorizations.

Finally, let a be an irreducible element of R and write $a = up_1 \cdots p_k$, where $p_1, \ldots, p_k \in S$ and u is a unit of R. If $k > 1$, then a is the product of two nonunits, p_1 and $up_2 \cdots p_k$. Hence we must have $k = 1$ and $a = up_1$.

8.31. Theorem. *An integral domain R which is not a field is a factorial ring if and only if R is a Krull domain and $\mathscr{C}(R) = 0$.*

Proof. Let R be a factorial ring and let S be a set whose existence is required in the definition. If $a \in R$, $a \neq 0$, we can write

$$a = u \prod_{p \in S} p^{v_p(a)},$$

where u is a unit of R, $v_p(a)$ is a nonnegative integer uniquely determined by a for each $p \in S$, and $v_p(a) = 0$ for all but a finite number of $p \in S$. If a and b are nonzero elements of R, then

$$ab = w \prod_{p \in S} p^{v_p(a) + v_p(b)}$$

where w is a unit of R; that is, $v_p(ab) = v_p(a) + v_p(b)$ for all $p \in S$. If we set $v_p(0) = \infty$, then this equality holds for all $a, b \in R$. If either $a = 0$, $b = 0$, or $a + b = 0$, then $v_p(a + b) \geq \min\{v_p(a), v_p(b)\}$. Suppose a, b and $a + b$ are all nonzero. Then

$$\begin{aligned} u \prod_{p \in S} p^{v_p(a+b)} &= a + b \\ &= v \prod_{p \in S} p^{v_p(a)} + w \prod_{p \in S} p^{v_p(b)} \\ &= c \prod_{p \in S} p^{\min\{v_p(a), v_p(b)\}} \qquad \text{for some} \quad c \in R; \end{aligned}$$

hence $v_p(a + b) \geq \min\{v_p(a), v_p(b)\}$ for all $p \in S$. Thus for each $p \in S$, there exists a valuation v_p on the field of quotients of R (see Exercise 11(d) of Chapter V) such that for all $a \in R$, $a \neq 0$,

$$a = u \prod_{p \in S} p^{v_p(a)},$$

where u is a unit of R. We leave it to the reader to verify that the valuation ring of v_p is R_P, where $P = (p)$, and that the family of valuations $\{v_p \mid p \in S\}$ satisfies (I) and (II). Therefore R is a Krull domain. Note that for each $p \in S$, $v_p(p) = 1$ and $v_p(q) = 0$ for $q \in S$, $q \neq p$.

To show $\mathscr{C}(R) = 0$, we shall show that every divisorial fractional ideal A' of R is principal. By Exercise 10(c), $A' = Ra' \cap Rb'$, where $a', b' \in K$. Let $d' \in R$, $d' \neq 0$, be such that $a = d'a'$ and $b = d'b'$ are in R. If $A = d'A'$, then $A = (a) \cap (b)$ and A' is principal if and only if A is principal. We shall prove that A is principal by induction on the number of elements of S occurring in the factorization of b. If b is a unit of R, then $(b) = R$ and $A = (a)$. Assume that b is a nonunit and that $(a) \cap (f)$ is principal whenever f is an element of R requiring fewer elements of S than b in its factorization. Since b is not a unit, we can write $b = pc$ where $p \in S$ and $c \in R$. By our induction assumption, $(a) \cap (c) = (d)$ for some $d \in R$; then $v_q(d) = \max\{v_q(a), v_q(c)\}$ for all $q \in S$. Hence $v_q(d) = v_q(pd) = \max\{v_q(a), v_q(b)\}$ for all $q \in S$, $q \neq p$. If $v_p(a) \leq v_p(c)$, then $v_p((a) \cap (pc)) = \max\{v_p(a), v_p(pc)\} = v_p(pc) = v_p(pd)$. If $v_p(a) > v_p(c)$, then $v_p(a) \geq v_p(pc)$, and $v_p((a) \cap (pc)) = v_p(a) = v_p(d)$. Thus $(a) \cap (b) = (pd)$ or $(a) \cap (b) = (d)$.

Conversely, suppose that R is a Krull domain and that $\mathscr{C}(R) = 0$. Then the minimal prime ideals of R are principal. Choose from each minimal prime ideal a generator and let S be the set of elements of R so chosen. Then the family of essential valuations of R can be indexed by S. If $p \in S$ and $(p) = P$, we write v_p for v_P. Let a be a nonzero element of R. Then $v_p(a) > 0$ for only a finite number of $p \in S$, say $p_1, \ldots, p_k$. If

$$u = a \prod_{i=1}^{k} p_i^{-v_{p_i}(a)},$$

then for $i = 1, \ldots, k$, $v_{p_i}(u) = v_{p_i}(a) - v_{p_i}(a) v_{p_i}(p_i) = 0$, and for $q \in S$, $q \neq p_1, \ldots, p_k$, we have also $v_q(u) = 0$. Thus u is a unit of R and

$$a = u \prod_{i=1}^{k} p_i^{v_{p_i}(a)}.$$

This factorization is unique; for suppose $q \in S$ and that q^n is the exact power of q occurring in some factorization of a as a product of a unit and elements of S. Then $v_q(a) = n$; so that if $n > 0$, we have $q = p_i$ for some i and $n = v_{p_i}(a)$. Therefore R is a factorial ring.

Combining the results of Theorems 8.27 and 8.31 we have

8.32. Corollary. *Let R be a Krull domain and let S be a multiplicative system in R. If R is a factorial ring then $S^{-1}R$ is a factorial ring. On the other hand, if S is generated by a set of primes, and if $S^{-1}R$ is a factorial ring, then R is a factorial ring.*

Combining the results of Theorems 8.28 and 8.31, we have:

8.33. Corollary. *If R is a Krull domain and X is an indeterminate, then $R[X]$ is a factorial ring if and only if R is a factorial ring.*

If R is a Krull domain, then it follows from Theorem 8.12 that R satisfies:

Maximum condition on principal ideals (MPI): every non-empty set of principal ideals of R has a maximal element.

8.34. Theorem. *If R is an integral domain which is not a field, then the following statements are equivalent:*

(1) *R is a factorial ring.*
(2) *R is a Krull domain and every divisorial ideal of R is principal.*
(3) *R is a Krull domain and every prime divisorial ideal of R is principal.*
(4) *R satisfies* (MPI) *and every irreducible element of R is a prime.*
(5) *R satisfies* (MPI) *and the intersection of two principal ideals of R is principal.*

Proof. (1) $\Rightarrow$ (2). Assume R is a factorial ring; then R is a Krull domain and $\mathscr{D}(R) = \mathscr{H}(R)$. Hence, if A is a divisorial ideal, then $\operatorname{div} A = \operatorname{div} a$ for some $a \in R$. Then $A = (a)$.

(2) $\Rightarrow$ (3). Clear.

(3) $\Rightarrow$ (1). Assume that R is a Krull domain and that every prime

divisorial ideal of R is principal. As we have seen, $\mathscr{D}(R)$ is generated by the elements div P, where P runs over the minimal prime ideals of R, and each such P is divisorial. By assumption, div $P \in \mathscr{H}(R)$. Hence $\mathscr{H}(R) = \mathscr{D}(R)$, and R is a factorial ring.

(2) $\Rightarrow$ (5). Clear.

(5) $\Rightarrow$ (4). Assume that R satisfies (MPI) and that the intersection of two principal ideals of R is principal. Let p be an irreducible element of R and suppose that $ab \in (p)$, where $a, b \in R$. If $a = 0$, then $a \in (p)$; assume $a \neq 0$. Let $(a) \cap (p) = (c)$ and set $d = ap/c$. Since $ap \in (c)$, $d \in R$. If $c = ae$, then $p = de$ and, since p is irreducible, either d or e is a unit of R. If e is a unit, then $a \in (c) \subseteq (p)$. Suppose d is a unit. Then $(p) = (e)$ and so $(c) = (a)(p) = (ap)$, that is, $(ap) = (a) \cap (p)$. Hence $ab \in (ap)$ and consequently $b \in (p)$. Therefore (p) is a prime ideal.

(4) $\Rightarrow$ (1). Suppose that R satisfies (MPI) and that every irreducible element of R is a prime. It follows from (MPI) that every nonzero nonunit of R can be written as a product of a finite number of irreducible elements of R. From each principal ideal which is generated by an irreducible element choose one generator, and let S be the set of elements so chosen. If $p \in R$ is irreducible, then there is a unique $p' \in S$ such that $p = up'$, where u is a unit of R. Hence every nonzero element of R can be written as a product of a unit and a finite number of elements of S. Suppose that $up_1 \cdots p_m = nq_1 \cdots q_n$, where u and v are units of R and $p_1, \ldots, p_m, q_1, \ldots, q_n \in S$. Then $up_1 \cdots p_m \in (q_1)$ and $u \notin (q_1)$, so some p_i is in (q_1), say $p_1 \in (q_1)$. It follows from the definition of S that $p_1 = q_1$, and so $up_2 \cdots p_m = vq_2 \cdots q_n$. We can then proceed with an argument by induction to show that $m = n$, $u = v$, and that the p_i and q_j can be so numbered that $p_i = q_i$ for $i = 1, \ldots, m$. Therefore S meets the requirements in the definition of factorial ring.

EXERCISES

1. Residuals of fractional ideals.

Let R be an integral domain.

(a) Show that if A and B are fractional ideals of R, then

$$[A : B] = \bigcap_{\substack{b \in B \\ b \neq 0}} A(1/b).$$

(b) Show that if A is a fractional ideal of R, then $\bar{A} = [R:[R:A]]$.

(c) Show that R is a completely integrally closed if and only if $[A:A] = R$ for every divisorial ideal A of R.

(d) Show that R is integrally closed if and only if $[A:A] = R$ for every finitely generated nonzero ideal A of R.

2. Primary decomposition in Krull domains.
Let R be a Krull domain.

(a) Let P be a minimal prime ideal of R, and let n be a positive integer. Show that

$$P^{(n)} = \{a \mid a \in R \quad \text{and} \quad v_P(a) \geq n\}.$$

(b) Let $a \in R$, $a \neq 0$. Show that there is only a finite number of minimal prime ideals of R which contain a, say $P_1, \ldots, P_k$. Show that

$$(a) = P_1^{(n_1)} \cap \cdots \cap P_k^{(n_k)},$$

where $n_i = v_{P_i}(a)$ for $i = 1, \ldots, k$, and that this is the unique reduced primary decomposition of the ideal (a).

3. Krull domains.

(a) Show that an integral domain R is a Krull domain if and only if it satisfies the following two conditions:

(i) if P is a minimal prime ideal of R, then R_P is a Noetherian valuation ring, and $\bigcap_P R_P = R$; and

(ii) if $a \in R$, $a \neq 0$, then a is contained in only a finite number of minimal prime ideals of R.

Now assume that R is a Krull domain.

(b) Show that every nonzero prime ideal P of R contains a minimal prime ideal of R, and if P is not itself minimal, then $[R:P] = R$.

(c) Show that if R is Noetherian and if A is a proper ideal of R, then A is divisorial if and only if each prime divisor of A is a minimal prime ideal of R.

(d) Suppose that R has a unique maximal prime ideal P. Show that R is a Noetherian valuation ring if and only if P is divisorial.

4. Intersections of Krull domains.

(a) Let K be a field. Let I be a set and for each $\alpha \in I$ let R_α be a Krull domain which is a subring of K. Let $R = \bigcap_{\alpha \in I} R_\alpha$ and assume that for each $a \in R$, $a \neq 0$, a is a unit in R_α for all but a finite number of $\alpha \in I$. Show that R is a Krull domain.

(b) Let R be a Krull domain with field of quotients K. Let K' be a subfield of K. Show that $K' \cap R$ is a Krull domain.

5. Krull domains and rings of quotients.

Let R be a Krull domain and let $\{v_\alpha \mid \alpha \in I\}$ be the family of essential valuations of R. Let V_α be the valuation ring of v_α. Let S be a multiplicative system in R and let $J = \{\alpha \mid \alpha \in I$ and $v_\alpha(s) = 0$ for all $s \in S\}$. Show that

$$S^{-1}R = \bigcap_{\alpha \in J} V_\alpha$$

and conclude that $S^{-1}R$ is a Krull domain.

6. An example.

Let R be the ring of holomorphic functions of a complex variable.

(a) Show that R is an integral domain (with respect to the usual addition and multiplication).

(b) For the complex number a, let $P_a = \{f \mid f \in R$ and $f(a) = 0\}$. Show that P_a is a minimal prime ideal of R.

(c) For each complex number a and for each nonzero $f \in R$, set $v_a(f) = n$ if $f(Z) = (Z - a)^n g(Z)$ where $g(a) \neq 0$. Define $v_a(0) = \infty$. Show that v_a, extended to the field of quotients K of R in the usual way [see Exercise 9(d) of Chapter V], is a valuation on K. Show that the valuation ring of v_a is R_{P_a}.

(d) Show that $\bigcap R_{P_a} = R$, where the intersection is over all complex numbers.

(e) Note that if f is the sine function, then $f \in P_{n\pi}$ for all integers n. Explain how this fact implies that R is not a Krull domain.

7. Essential valuations on polynomial rings.

(a) Let R be a Krull domain and let $\{v_\alpha \mid \alpha \in I\}$ be a family of (normed) valuations on the field of quotients of R satisfying (I) and (II). Assume that for each $\beta \in I$ we have

$$R \neq \bigcap_{\alpha \in I,\, \beta \neq \alpha} V_\alpha$$

where V_α is the valuation ring of v_α. Show that $\{v_\alpha \mid \alpha \in I\}$ is the family of essential valuations of R.

(b) Show that the family of valuations of $R[X]$ consisting of all $p(X)$-adic valuations and all extensions of essential valuations of R satisfy the assumption of (a).

8. Prolongations of valuations.

Let v be a valuation on a field K, let K' be an extension of K, and let v' be a valuation on K' which is a prolongation of v. Let $G_0 = \{v'(a) \mid a \in K^*\}$. Then G_0 is a subgroup of the value group G' of v'. Denote the index of G_0 in G' by $e(v'/v)$. Let $a_1', \ldots, a_k' \in K'^*$ be such that $v'(a_1'), \ldots, v'(a_k')$ represent distinct cosets of G_0 in G'.

(a) Show that $a_1', \ldots, a_k'$ are linearly independent over K. Conclude that $e(v'/v) \leq [K' : K]$.

(b) Assume that v and v' are normed rank one and discrete. Show that for every $a \in K$ we have $v(a) = e(v'/v)v'(a)$.

(c) Let P be the maximal ideal of the valuation ring V of v. Set $\bar{K} = V/P$; $\bar{K}$ is called the **residue field of** K **relative to** v. Define $\bar{K}'$ in like manner and show that we may regard $\bar{K}$ as a subfield of $\bar{K}'$. Show that $f(v'/v) = [\bar{K}' : \bar{K}]$ is finite if $[K' : K]$ is finite.

(d) Assume that v has rank one and is discrete. Let K' be a finite extension of K and let $v_1', \ldots, v_k'$ be the valuations on K' which are prolongations of v. Assume that for $i = 1, \ldots, k$, the residue field of K' relative to v_i' is a separable extension of $\bar{K}$. Show that

$$\sum_{i=1}^{k} e(v_i'/v) f(v_i'/v) \leq [K' : K].$$

9. The approximation theorem.
Let $v_1, \ldots, v_k$ be normed rank one discrete valuations on a field K such that v_i and v_j are not equivalent when $i \neq j$.

(a) Show that there is an element $a \in K$ such that $v_1(a) < 0$ and $v_i(a) > 0$ for $i = 2, \ldots, k$.

(b) Let n be an integer. Show that there is an element $a \in K$ such that $v_1(a - 1) > n$ and $v_i(a) > n$ for $i = 2, \ldots, k$.

(c) Let $a_1, \ldots, a_k \in K$. Show that if n is an integer then there is an element $a \in K$ such that $v_i(a - a_i) > n$ for $i = 1, \ldots, k$.

10. The approximation theorem for Krull domains.
Let R be a Krull domain and let $\{v_\alpha \mid \alpha \in I\}$ be the family of essential valuations of R. Let $\alpha_1, \ldots, \alpha_k$ be (distinct) elements of I and let $n_1, \ldots, n_k$ be integers.

(a) Show that there is an element a in the quotient field of R such that $v_{\alpha_i}(a) = n_i$ for $i = 1, \ldots, k$, and $v_\alpha(a) \geq 0$ for all $\alpha \in I \setminus \{\alpha_1, \ldots, \alpha_k\}$.

(b) Let A, B, and C be divisorial fractional ideals of R such that $A \subseteq B$. Show that $A = B \cap Ca$ for some a in the quotient field of R.

(c) Show that every divisorial fractional ideal of R is an intersection of two principal fractional ideals of R.

11. Integral domains which are almost Krull.
Let R be an integral domain: R is **almost Krull** if R_P is a Krull domain for each proper prime ideal P of R.

(a) Let R be almost Krull. Show that R is integrally closed, and therefore, if R is Noetherian, then it is a Krull domain.

(b) Show that if R is almost Krull, then $R = \cap R_P$, where P runs over the minimal prime ideals of R. Show that R is completely integrally closed.

(c) Show that R is almost Krull if and only if there exists a family $\{v_\alpha \mid \alpha \in I\}$ of discrete rank one valuations on the quotient field of R with the following properties:

(1) For each maximal ideal M of R there is a subset I_M of I such that $R_M = \bigcap_{\alpha \in I_M} V_\alpha$, where V_α is the valuation ring of v_α.

(2) For each $\alpha \in I$, $V_\alpha = R_{P_\alpha}$, where $P_\alpha = \{x \mid x \in R \text{ and } v_\alpha(x) > 0\}$.

(3) For each maximal ideal M of R and each nonzero x in the quotient field of R, $v_\alpha(x) \neq 0$ for only a finite number of $\alpha \in I_M$.

(d) If R is almost Krull, show that $R[X]$ is almost Krull.

(e) If R is almost Krull, show that the integral closure of R in a finite extension of the quotient field of R is almost Krull.

(f) Show that if R is almost Krull, and if every nonzero proper ideal of R is contained in only a finite number of maximal ideals, then R is a Krull domain.

12. Principal ideal domains.

(a) Let R be an integral domain such that every ideal of R is principal; R is called a **principal ideal domain**. Show that R is a factorial ring.

(b) Give an example of a factorial ring which is not a principal ideal domain.

(c) An integral domain R is called a **Euclidean ring** if there is a mapping ϕ from R into the set of nonnegative integers such that $\phi(ab) \geq \phi(a)$ and for nonzero elements $a, b \in R$ there are elements $q, r \in R$ such that $a = bq + r$ and either $r = 0$ or $\phi(r) < \phi(b)$. Show that a Euclidean ring is a principal ideal domain.

(d) Show that the ring of integers and a ring of polynomials over a field (in one indeterminate) are Euclidean rings. Show that the same is true of the ring of all complex numbers $a + bi$, where a and b are integers.

13. Factorial rings.

(a) Prove directly that (3) implies (4) in Theorem 8.34.

(b) Let R be a Krull domain. Show that if every divisorial ideal of R is invertible, then R_P is a factorial ring for every maximal ideal P of R.

(c) Assume that every divisorial ideal of R is finitely generated and that R_P is a factorial ring for every maximal ideal P of R. Show that every divisorial ideal of R is invertible.

14. Krull domains and the ideal transform.

Refer to Exercise 16 of Chapter VI for terminology. Let R be an integral domain with more than one maximal ideal.

(a) Prove that if R is a Krull domain, then $T((x))$ is a Krull domain for each nonunit x of R.

(b) Let $\{P_\alpha\}$ be the collection of all minimal prime ideals of R. Suppose that $T((x))$ is a Krull domain for each nonunit x of R. Prove the following assertions:

(1) For each α, R_{P_α} is a discrete rank one valuation ring.

(2) $R = \bigcap R_{P_\alpha}$.

(3) If y is a nonzero element of R, then y is contained in only a finite number of the P_α's. [Hint. Assume that y is a nonunit, and show that either y is contained in only a finite number of the P_α's or that y is contained in every maximal ideal of R: do this by considering two cases—whether or not there exists a nonzero nonunit in $R \backslash \bigcup Q_\beta$, where $\{Q_\beta\}$ is the collection of prime ideals of R which contain y.] Now conclude that R is a Krull domain.

CHAPTER

IX

Generalizations of Dedekind Domains

In this chapter we shall study several of the important classes of rings which contain the class of Dedekind domains. The procedure will be to choose a property of Dedekind domains and study the class of rings, or domains, which have the chosen property. The proofs of many of the assertions to be made are similar to those of results in earlier chapters; some will be given in the text but others left as exercises.

1 ALMOST DEDEKIND DOMAINS

9.1. Definition. *An integral domain R is an* **almost Dedekind domain** *if for each maximal ideal M of R, the ring R_M is a Dedekind domain.*

Thus, an almost Dedekind domain is a domain which is locally Dedekind. Note that if R_M is a Dedekind domain which is not a field, then it follows from the results of Chapter VI that R_M is a discrete rank one valuation ring. Hence, if R is an almost Dedekind domain which is not a field, its Krull dimension is one, and R is a Prüfer domain. In this section we shall obtain several characterizations of almost Dedekind domains.

9.2. Proposition. *Let R be an integral domain which is not a field. Then the following are equivalent:*

(1) *The integral closure of R is an almost Dedekind domain.*
(2) *Each nontrivial valuation ring which is an overring of R has rank one and is discrete.*

Proof. By Corollary 5.8, every valuation ring which is an overring of R is an overring of the integral closure of R. Hence, there is no loss of generality in assuming R is integrally closed. Thus (1) implies (2) without further ado. Suppose (2) holds. By Exercise 8(c) of Chapter VII and Theorem 7.23, R is a Prüfer domain of Krull dimension one. Hence, if M is a maximal ideal of R, then R_M is a valuation ring, which has rank one and is discrete. Thus, R is almost Dedekind.

9.3. Corollary. *An overring of an almost Dedekind domain is an almost Dedekind domain.*

Proof. Let R be an almost Dedekind domain and let T be an overring of R. Since R is a Prüfer domain, T is integrally closed by Theorem 6.13. Furthermore, each nontrivial valuation ring which is an overring of T is also an overring of R, and thus has rank one and is discrete. Therefore, T is almost Dedekind.

9.4. Theorem. *If R is an integral domain which is not a field, then the following statements are equivalent:*

(1) *R is an almost Dedekind domain.*
(2) *R has Krull dimension one and each primary ideal of R is a power of its radical.*
(3) *If A, B, and C are ideals of R with $A \neq 0$, and if $AB = AC$, then $B = C$.*
(4) *R is a Prüfer domain of Krull dimension one which has no idempotent maximal ideals (an ideal A of R is called* **idempotent** *if $A^2 = A$).*
(5) *R is a Prüfer domain and for each proper ideal A of R, $\bigcap_{n=1}^{\infty} A^n = 0$.*
(6) *Each ideal of R which has prime radical is a power of its radical.*

Proof. (1) $\Rightarrow$ (2). Assume that R is an almost Dedekind domain. Since $\dim R_M = 1$ for every maximal ideal M of R, $\dim R = 1$. Let Q be a primary ideal of R. If Q is either 0 or R, then Q is its own radical; hence, we assume $Q \neq 0$ and $Q \neq R$. Then $M = \text{Rad}(Q)$ is a maximal ideal of R and QR_M is MR_M-primary. Since R_M is a Noetherian valuation ring, $QR_M = M^n R_M$ for some integer n. Since M is a maximal ideal, M^n is M-primary by Exercise 6(e) of Chapter II. Therefore,

$$Q = QR_M \cap R = M^n R_M \cap R = M^n.$$

(2) $\Rightarrow$ (1). Assume (2) holds and let M be a maximal ideal of R. Then every proper nonzero ideal Q of R_M has MR_M as its radical; thus Q is MR_M-primary. Therefore, $Q \cap R$ is an M-primary ideal of R. Hence $Q \cap R = M^n$ for some integer n, and thus $Q = M^n R_M$. Therefore, the only proper nonzero ideals of R_M are powers of MR_M, and it follows immediately from this that R_M is a Noetherian valuation ring.

(1) $\Rightarrow$ (3). Assume that R is an almost Dedekind domain, and let A, B, and C be ideals of R such that $A \neq 0$ and $AB = AC$. If M is a maximal ideal of R then $(AR_M)(BR_M) = (AR_M)(CR_M)$ by Exercise 4(a) of Chapter III. Since R_M is a Dedekind domain, $BR_M = CR_M$. Since this is true for every maximal ideal M of R, it follows by Proposition 3.13 that $B = C$.

(3) $\Rightarrow$ (4). Assume (3) holds. By Theorem 6.6, R is a Prüfer domain. If M is a maximal ideal of R, and if $M^2 = M = MR$, then $M = R$; hence $M^2 \neq M$. It remains to show that $\dim R = 1$. Let P be a proper nonzero prime ideal of R and let $x \in R \backslash P$. Since

$$(P + (x))^3 = (P + (x))(P^2 + (x^2)),$$

we have by (3), $(P + (x))^2 = P^2 + (x^2)$. Thus $xP \subseteq P^2 + (x^2)$. If $y \in P$ then $xy = a + rx^2$ for some $a \in P^2$ and $r \in R$. Then $rx^2 \in P$, and since $x^2 \notin P$, we have $r \in P$. Hence, $xP \subseteq P^2 + (x^2)P$ and it follows by (3) that $x \in P + (x^2)$. Then $x = b + sx^2$ for some $b \in P$ and $s \in R$. Thus $x(1 - sx) \in P$, and since $x \notin P$, we have $1 - sx \in P$. Therefore, $R = P + (x)$. Since this is true for every $x \in R \backslash P$, P is a maximal ideal of R.

(4) $\Rightarrow$ (2). This implication follows at once from (4) of Proposition 6.9.

(4) $\Rightarrow$ (5). This implication follows at once from (1) of Proposition 6.9.

(5) $\Rightarrow$ (4). Assume (5) holds and let M be a maximal ideal of R. Then R_M is a valuation ring, and

$$0 = \bigcap_{n=1}^{\infty} M^n = \bigcap_{n=1}^{\infty} (M^n R_M \cap R) = \left(\bigcap_{n=1}^{\infty} M^n R_M \right) \cap R;$$

hence $\bigcap_{n=1}^{\infty} M^n R_M = 0$. Thus, by Theorem 5.10, R_M has no proper nonzero prime ideal other than MR_M. Therefore, dim $R_M = 1$, and since is this true for every maximal ideal of R, dim $R = 1$. If M is an idempotent maximal ideal of R, then $MR_M = \bigcap_{n=1}^{\infty} M^n R_M$, which is not true; hence $M^2 \neq M$.

(2) $\Rightarrow$ (6). Clear.

(6) $\Rightarrow$ (2). Assume (6) holds and let P be a proper nonzero prime ideal of R. First of all, we shall show that if P is a minimal prime divisor of a principal ideal (x), then P is a maximal ideal of R. Under this assumption, xR_P is PR_P-primary. Hence $xR_P \cap R$ is P-primary, and consequently $xR_P \cap R = P^n$ for some integer n. Thus $xR_P = (PR_P)^n$; this implies that PR_P is invertible and so $P^2 R_P \neq PR_P$. Hence

$$P^2 \subseteq P^{(2)} = P^2 R_P \cap R \subset P.$$

By (6), $P^{(2)}$ is a power of P, so that we must have $P^2 = P^{(2)}$. Let $a \in P \backslash P^2$ and $b \in R \backslash P$. Then $\text{Rad}(P^2 + (ab)) = P$, and so $P^2 + (ab) = P^2$ or $P^2 + (ab) = P$. Since P^2 is P-primary, $ab \notin P^2$. Hence $P^2 + (ab) = P$. Thus $a \in P^2 + (ab)$, so that $a = c + rab$ for some $c \in P^2$ and $r \in R$. Then $a(1 - rb) \in P^2$ and since P^2 is P-primary and $a \notin P^2$, we have $1 - rb \in P$. Thus, $R = P + (b)$, and since this is true for every $b \in R \backslash P$, P is a maximal ideal of R.

If a is an arbitrary nonzero element of P, then P contains a minimal prime divisor P' of (a). By what we have shown, P' is a maximal ideal of R; hence the same is true of P. Therefore, dim $R = 1$. A primary ideal of R has prime radical, so is a power of its radical. Thus, (2) holds.

We conclude this section by proving that the analog of Theorem 6.24 holds for almost Dedekind domains.

9.5. Theorem. *Let R be an integral domain with quotient field K. Let K' be a finite extension of K, and let R' be the integral closure of R in K'. If R is almost Dedekind, then R' is almost Dedekind.*

Proof. Let M be a nonzero proper prime ideal of R'. Let $P = M \cap R$ and let $S = R \backslash P$. By Proposition 4.5, $S^{-1}R'$ is integral over R_P. Since R' is integrally closed in K', $S^{-1}R'$ is integrally closed in K' by Exercise 7(b) of Chapter IV. Thus, $S^{-1}R'$ is the integral closure in K' of the discrete rank one valuation ring R_P. Hence, by Theorem 6.24, $S^{-1}R'$ is a Dedekind domain (since R_P is). Therefore, $(S^{-1}R')_{S^{-1}M}$ is a discrete rank one valuation ring. By Exercise 6(c) of Chapter III, $(S^{-1}R')_{S^{-1}M} = R_M'$. Therefore, R' is an almost Dedekind domain.

2 ZPI-RINGS

If R is a Dedekind domain, then every ideal of R can be written as a product of prime ideals of R. In this section we shall study the class of rings with this property.

9.6. Definition. *A ring R is a* **ZPI-ring** *if every ideal of R can be written as a product of prime ideals of R.*

The letters ZPI stand for "Zerlegung Primideale." Some authors call a ring a ZPI-ring if every nonzero ideal can be written as a product of prime ideals; they call the rings of Definition 9.6 general ZPI-rings.

If R is a ZPI-ring, and if P is a proper prime ideal of R, then R/P is a ZPI-ring which is also an integral domain. Hence R/P is a Dedekind domain and it follows that $\dim R/P \leq 1$. Therefore, $\dim R \leq 1$.

A ring is said to be **indecomposable** if it cannot be written as a direct sum of nontrivial rings. Clearly, every integral domain is indecomposable. We shall show that every ZPI-ring is a direct sum of indecomposable ZPI-rings, and we shall determine all indecomposable ZPI-rings.

9.7. Definition. *A ring R is a* **special primary ring** *if R has a unique maximal ideal M and if each proper ideal of R is a power of M.*

To determine all special primary rings we shall use the following lemma.

9.8. Lemma. *Let A be an ideal of a ring R such that there are no ideals of R strictly between A and A^2. Then, for every positive integer n, the only ideals between A and A^n are $A, A^2, A^3, \ldots, A^n$.*

Proof. Since the assertion is certainly true when A is idempotent, we shall assume that $A^2 \subset A$. Let $x \in A \backslash A^2$; then $A = A^2 + (x)$ and so

$$(A/(x))^2 = (A^2 + (x))/(x) = A/(x).$$

Therefore, for every positive integer m,

$$A/(x) = (A/(x))^m = (A^m + (x))/(x),$$

and consequently $A = A^m + (x)$. Then $A/A^m = (A^m + (x))/A^m$, and for every positive integer k,

$$(A^k + A^m)/A^m = (A/A^m)^k = (A^m + (x^k))/A^m.$$

Therefore $A^k = A^m + (x^k)$ for all positive integers k and m with $k \leq m$.

Now let $b \in A^n \backslash A^{n+1}$. Since $A^n = A^{n+1} + (x^n)$, we have $b - rx^n \in A^{n+1}$ for some $rx^n \notin A^{n+1}$. Then $rx \in A \backslash A^2$ and therefore $A = A^2 + (rx)$. Thus

$$\begin{aligned} A^n &= AA^{n-1} = (A^2 + (rx))(A^n + (x^{n-1})) \\ &= A^{n+2} + A^2(x^{n-1}) + A^n(rx) + (rx^n) \subseteq A^{n+1} + (rx^n) \\ &= A^{n+1} + (b). \end{aligned}$$

Therefore, $A^n = A^{n+1} + (b)$, and since this is true for every $b \in A^n \backslash A^{n+1}$, we conclude that there are no ideals of R strictly between A^n and A^{n+1}.

Now consider an ideal B of R such that $A^n \subseteq B \subseteq A$. Choose s to be the largest positive integer such that $B \subseteq A^s$. If $s \geq n$ then $B = A^n$. Suppose that $s < n$; then $A^s = A^n + (x^s)$. If $c \in B \backslash A^{s+1}$ then $c - rx^s \in A^n$ for some $r \in R$, and $rx^s \notin A^{s+1}$ since $c \notin A^{s+1}$ and $A^n \subseteq A^{s+1}$.

Hence $rx \in A \backslash A^2$ and so $A = A^2 + (rx)$. Therefore,

$$\begin{aligned} A^s = AA^{s-1} &= (A^2 + (rx))(A^{n-1} + (x^{s-1})) \\ &= A^{n+1} + A^2(x^{s-1}) + A^{n-1}(rx) + (rx^s) \\ &\subseteq A^n + A^{s+1} + (rx^s) = A^{s+1} + (rx^s). \end{aligned}$$

Thus, $A^s = A^{s+1} + (rx^s)$ and it follows that

$$\begin{aligned} (A/(rx^s))^s &= (A^s + (rx^s))/(rx^s) = (A^{s+1} + (rx^s))/(rx^s) \\ &= (A/(rx^s))^{s+1} = (A/(rx^s))^s(A/(rx^s)) \\ &= (A/(rx^s))^{s+2} = \cdots = (A/(rx^s))^n. \end{aligned}$$

Consequently,

$$A^s \subseteq A^s + (rx^s) = A^n + (rx^s) = A^n + (c) \subseteq B,$$

and we conclude that $B = A^s$.

As an immediate consequence of this lemma we have the following result.

9.9. Proposition. *If M is a maximal ideal of a ring R such that there are no ideals of R strictly between M and M^2, then for each positive integer n, R/M^n is a special primary ring.*

9.10. Theorem. *The following statements are equivalent:*

(1) *R is a ZPI-ring.*
(2) *R is a Noetherian ring such that for each maximal ideal M of R, there are no ideals of R strictly between M and M^2.*
(3) *R is a direct sum of a finite number of Dedekind domains and special primary rings.*

Proof. (1) $\Rightarrow$ (2). Let R be a ZPI-ring. We shall show that every prime ideal of R is finitely generated; by Exercise 14(c) of Chapter II, this implies that R is Noetherian. Let P be a proper prime ideal of R, and let A be an ideal of R such that $P \subset A$. If we use the fact that R/P is a Dedekind domain, it follows exactly as in the proof of Theorem 6.16 that for any element $a \in A \backslash P$ we have

$$P = P^2 + P(a) \subseteq PA \subseteq P.$$

Hence $P = PA$. Therefore, if

$$0 = P_1^{n1} \cdots P_k^{nk},$$

where $P_1, \ldots, P_k$ are distinct proper prime ideals of R and $n_1, \ldots, n_k$ are positive integers, then we may assume that there are no containment relations among $P_1, \ldots, P_k$. Thus, $P_1, \ldots, P_k$ are the minimal prime divisors of the ideal 0. We shall show first that these prime ideals are finitely generated; it is sufficient to show that P_1 is finitely generated.

If $P_1^2 \subset P_1$ choose $x \in P_1 \backslash (P_1^2 \cup P_2 \cup \cdots \cup P_k)$; if $P_1^2 = P_1$ choose $x \in P_1 \backslash (P_2 \cup \cdots \cup P_k)$. Let $(x) = Q_1 \cdots Q_m$, where each Q_i is a proper prime ideal of R. Since $Q_1 \cdots Q_m \subseteq P_1$ we have $Q_i \subseteq P_1$, and so $Q_i = P_1$, for some i, say $Q_1 = P_1$. If $m = 1$ we are finished; assume $m \geq 2$, and consider some Q_j where $j \geq 2$. If $P_1^2 \subset P_1$ then $Q_j \neq P_i$ for $i = 1, \ldots, k$, so Q_j is a maximal ideal of R and not a minimal prime divisor of the ideal 0. If $P_1^2 = P_1$ then $Q_j \neq P_i$ for $i = 2, \ldots, k$. However, we might have $Q_j = P_1$, in which case $Q_1 Q_j = Q_1$. Hence, we may eliminate every Q_j for which this is true, and assume that for $j = 2, \ldots, m$, Q_j is a maximal ideal of R and not a minimal prime divisor of the ideal 0. If $Q_1 \subset Q_j$, then $Q_1 = Q_1 Q_j$ and we may eliminate Q_j. Thus, we may assume $P_1 = Q_1 \nsubseteq Q_j$ for $j = 2, \ldots, m$. Choose $y \in P_1 \backslash (Q_2 \cup \cdots \cup Q_m)$. We shall show that $P_1 = (x, y)$.

Let $(x, y) = N_1 \cdots N_s$, where each N_i is a proper prime ideal of R. Then $N_i = P_1$ for some i, say $N_1 = P_1$. Suppose $s \geq 2$ and consider N_2. Since $x \in N_2$ we have $Q_j \subseteq N_2$ for some j. If $j \geq 2$ then $Q_j = N_2$, which is impossible since $y \notin Q_j$. Hence, $P_1 = Q_1 \subseteq N_2$. If $P_1 \subset N_2$ then $P_1 = P_1 N_2$. If $P_1 = N_2$ then $x \in P_1^2$, so $P_1^2 = P_1$ and $P_1 = P_1 N_2$. Thus, in either case, N_2 can be eliminated from the product $N_1 \cdots N_s$. Repeating this argument, we come finally to the conclusion that $(x, y) = P_1$.

Now let P be a prime ideal of R which is not a minimal prime divisor of the ideal 0. Then $P_i \subset P$ for some i, say $P_1 \subset P$. The ideal P/P_1 of the Dedekind domain R/P_1 is finitely generated. Since P_1 is finitely generated it follows that P is finitely generated. This completes the proof that R is Noetherian. Let A be an ideal of R such that $P^2 \subseteq A \subseteq P$. If we write A as a product of proper prime ideals, then each factor contains P, and so is equal to P since P is maximal. Hence $A = P^k$ for some positive integer k. If $k = 1$ then $A = P$; if $k \geq 2$ then $A = P^2$.

(2) $\Rightarrow$ (3). Suppose that (2) holds. Let M be a maximal ideal of R. If there exists a prime ideal of R properly contained in M, then by Exercise 6(e) and (f), this prime ideal is $P = \bigcap_{n=1}^{\infty} M^n$. Thus, $\dim R \leq 1$. Since M is maximal, all powers of M are M-primary. Since R is Noetherian, each M-primary ideal Q of R contains a power of M. Thus, by Lemma 9.8, $Q = M^k$ for some integer k. By Exercise 6(d), $P = PM$; thus, by Exercise 6(b), there exists $m \in M$ such that $p = pm$ for all $p \in P$. Therefore, if N is a P-primary ideal of R, then $p(1-m) = 0 \in N$, $1 - m \notin P$, and so $p \in N$ for all $p \in P$. Hence $P = N$, and we conclude that P is the only P-primary ideal of R.

Now let $0 = Q_1 \cap \cdots \cap Q_n$ be a reduced primary decomposition of the ideal 0; let Q_i be P_i-primary for $i = 1, \ldots, n$. We claim that $Q_1, \ldots, Q_n$ are comaximal. For, let $i \neq j$ and assume Q_i and Q_j are contained in the same maximal ideal M. Then $P_i \subseteq M$ and $P_j \subseteq M$; it follows from the fact that $P_i \neq P_j$ that we have, say, $P_i \subset P_j = M$. But then $P_i \subseteq Q_j$, so that $Q_i \subseteq Q_j$, and this contradicts the fact that the primary decomposition is reduced. By Exercise 2(d) of Chapter II we have

$$R \cong R/Q_1 \oplus \cdots \oplus R/Q_n.$$

For each i, if P_i is maximal then Q_i is a power of P_i, while if P_i is not maximal then $Q_i = P_i$. In the first case, R/Q_i is a special primary ring. In the second case, R/Q_i is a Dedekind domain.

(3) $\Rightarrow$ (1). Each Dedekind domain and each special primary ring is a ZPI-ring. Hence, the assertion follows from Exercise 6(g).

9.11. Corollary. *An indecomposable ZPI-ring is either a Dedekind domain or a special primary ring.*

3 MULTIPLICATION RINGS

9.12. Definition. *A ring R is a* **multiplication ring** *if $A \subseteq B$, where A and B are ideals of R, implies that there exists an ideal C of R such that $A = BC$.*

If R is a Dedekind domain, then R is a multiplication ring. The converse is also true when applied to integral domains.

9.13. Proposition. *If R is an integral domain which is a multiplication ring, then R is a Dedekind domain.*

Proof. Let A be a nonzero ideal of R and let $a \in A$, $a \neq 0$. Since $(a) \subseteq A$ there is an ideal C of R such that $(a) = AC$. Since (a) is invertible the same is true of A. Therefore, by Theorem 6.19, R is a Dedekind domain.

In this section we shall find some equivalent conditions for a ring to be a multiplication ring, and in the process of doing this, we shall establish some properties of multiplication rings. In particular, we shall show that R is a multiplication ring if and only if it satisfies the apparently weaker condition that $A \subseteq P$, where A is an ideal of R and P is a prime ideal of R, implies that there exists an ideal C of R such that $A = PC$. Temporarily, we need a name for such rings; we shall call them **weak multiplication rings.**

Let R be a weak multiplication ring. If P is a proper prime ideal of R, then R/P is a Dedekind domain (see Exercise 7). Hence $\dim R \leq 1$.

9.14. Proposition. *Let R be a weak multiplication ring. If P is a maximal ideal of R, then there are no ideals of R strictly between P and* P^2.

Proof. Let P be a maximal ideal of R, and assume that there is an ideal A of R such that $P^2 \subset A \subset P$. There exists an ideal C of R such that $A = PC$, and $C \nsubseteq P$. Let $c \in C \backslash P$; since $(c) \subseteq C$ we have $P(c) \subseteq PC = A$. Hence $P(P + (c)) = P^2 + P(c) \subseteq A$. Since P is maximal, $P + (c) = R$. Therefore, $P \subseteq A$, which contradicts the fact that $A \subset P$.

9.15. Proposition. *Let R be a weak multiplication ring. If M is a maximal ideal of R and if P is a prime ideal of R such that* $P \subset M$, *then* $P = \bigcap_{n=1}^{\infty} M^n$ *and* $P = MP$.

Proof. Since R/P is a Dedekind domain, $\bigcap_{n=1}^{\infty} (M/P)^n = 0$, and therefore $\bigcap_{n=1}^{\infty} M^n \subseteq P$. Since $P \subset M$ there is an ideal A of R such that

$P = MA$. Since $M \nsubseteq P$, A must be contained in P. Hence, $P = MA \subseteq MP \subseteq P$, and consequently

$$P = MP = M^2P = \cdots \subseteq \bigcap_{n=1}^{\infty} M^n.$$

Recall the following definition from Exercise 9 of Chapter III: if P is a minimal prime divisor of an ideal A of a ring R, then $AR_P \cap R$ is called the **isolated P-primary component** of A. It is the unique minimal P-primary ideal of R which contains A.

9.16. Definition. *Let A be an ideal of a ring R. The* **kernel** *of A is the intersection of the isolated P-primary components of A, as P runs over the minimal prime divisors of A.*

9.17. Proposition. *If R is a weak multiplication ring, then every ideal of R is equal to its kernel.*

Proof. Let A be an ideal of R and let A^* be the kernel of A. Suppose $A \neq A^*$ and let $a \in A^* \backslash A$. Consider the ideal $B = A : (a)$. Let M be a minimal prime divisor of B. Since $A \subseteq B$, M contains a minimal prime divisor P of A, by Lemma 2.15. If $P = M$ then $a \in AR_M \cap R$; hence, by Proposition 3.12, there exists $s \in R \backslash M$ such that $as \in A$. But then $s \in A : (a) \subseteq M$, a contradiction. Thus, we have $P \subset M$. Since dim $R \leq 1$, M must be maximal, and it follows from Proposition 9.15 that $P = \bigcap_{n=1}^{\infty} M^n$ and $P = MP$. Since $B \subseteq M$ there is an ideal C of R such that $B = MC$. If $C \subseteq B$ then

$$B = MB = M^2B = \cdots \subseteq \bigcap_{n=1}^{\infty} M^n = P;$$

this is not the case since M is a minimal prime divisor of B, and so we conclude that $C \nsubseteq B$. Thus, $C(a) \nsubseteq A$. Since $a \in A^* \subseteq P$, we have $C(a) \subseteq P$, and so there is an ideal D of R such that $C(a) = PD$. Hence,

$$C(a) = MPD = MC(a) = B(a) \subseteq A.$$

This contradiction implies that the assertion of the proposition is true.

9.18. Theorem. *If R is a weak multiplication ring, then every primary ideal of R is a power of its radical.*

Proof. Let Q be a primary ideal of a weak multiplication ring R, and let $P = \text{Rad}(Q)$. If $P = R$, then $Q = R$, so we assume that $P \neq R$.

We consider two cases. Assume first that P is a maximal ideal of R. If $P^n \neq P^{n+1}$ for each positive integer n, then Q is not contained in every power of P, since $\bigcap_{n=1}^{\infty} P^n$ is a prime ideal of R [by Exercise 7(d)] which is properly contained in P. Thus, there exists a positive integer k such that $Q \subseteq P^k$ and $Q \nsubseteq P^{k+1}$. By Exercise 7(c), there is an ideal A of R such that $Q = P^k A$ and $A \nsubseteq P$. If $A \neq R$ than any maximal ideal of R containing A would contain Q, and hence would equal P. Since this is impossible we have $A = R$ and $Q = P^k$.

Continuing to assume that P is maximal, we suppose that $P^n = P^{n+1}$ for some positive integer n. Suppose further that $Q \subseteq P^n$. If $a \in P^n$ then by Exercise 7(c), there is an ideal A of R such that $(a) = P^n A = P^{2n} A = P^n(a)$. Hence, by Exercise 6(b), there exists $p \in P$ such that $a = pa = p^2 a = \cdots$, and consequently $a \in Q$ since $p^s \in Q$ for some positive integer s. Therefore, $Q = P^n$. On the other hand, suppose that $Q \nsubseteq P^n$. Then there exists a positive integer k such that $Q \subseteq P^k$ and $Q \nsubseteq P^{k+1}$ and, just as in the preceding paragraph, we have $Q = P^k$.

Now assume that P is not a maximal ideal of R. We shall show that $Q = P$. Suppose that there exists $x \in P$ with $x \notin Q$. Let M be a maximal ideal of R containing P, and let $B = M(x) + Q$. If $x \in B$ then $x(1 - m) \in Q$ for some $m \in M$. Since $x \notin Q$, $1 - m \in P \subset M$, so $1 \in M$, which is not true. Hence $x \notin B$. By Proposition 9.17, B is equal to its kernel. Let $\{P_\alpha\}$ be the set of minimal prime divisors of B, and let Q_α be the isolated P_α-primary component of B. Since $x \notin B$ there is an index α such that $x \notin Q_\alpha$. If $p \in P$ then for some positive integer n, $p^n \in Q \subseteq B \subseteq P_\alpha$, so $p \in P_\alpha$; thus $P \subseteq P_\alpha$. However, if $y \in M \backslash P$ then $yx \in B \subseteq Q_\alpha$, so $y \in P_\alpha$. Therefore, $P \subset P_\alpha$, which contradicts the fact that P_α is a minimal prime divisor of B. Thus, we must have $Q = P$.

9.19. Proposition. *Let R be a weak multiplication ring. If P is a prime ideal of R which is not maximal, then $P = P^2$.*

Proof. Suppose that $P \neq P^2$ and let $a \in P \backslash P^2$. Let M be a maximal ideal of R such that $P \subset M$. Then P is the only minimal prime divisor of $(P^2 + (a))M$, and since each ideal of R is equal to its kernel, we have

$$\begin{aligned}(P^2 + (a))M &= (P^2 + (a))MR_P \cap R \\ &= (P^2 + (a))R_P \cap R \quad \text{since} \quad MR_P = R_P \\ &= P^2 + (a).\end{aligned}$$

Hence, $a = b + ac$ for some $b \in P^2$ and $c \in M$. Since P^2 is equal to its kernel, it is P-primary. Hence, $a(1 - c) \in P^2$ and $a \notin P^2$ imply that $1 - c \in P \subset M$. Thus $1 \in M$, which is not true.

9.20. Proposition. *Let R be a weak multiplication ring, let A be an ideal of R, and let P be a minimal prime divisor of A. Let P^n be the isolated P-primary component of A (such an n exists by Theorem 9.18). If $P^n \neq P^{n+1}$, then P does not contain the intersection of the remaining isolated primary components of A.*

Proof. Let $\{P_\alpha\}$ be the set of minimal prime divisors of A other than P. For each α, let $P_\alpha^{n_\alpha}$ be the isolated P_α-primary component of A. Let $B = \bigcap P_\alpha^{n_\alpha}$. By Proposition 9.17, $A = P^n \cap B$. Since $P^n \neq P^{n+1}$, we certainly have $P \neq P^2$; hence P is a maximal ideal of R. Thus, P^{n+1} is P-primary. If $A \subseteq P^{n+1}$, then

$$P^n = AR_P \cap R \subseteq P^{n+1}R_P \cap R = P^{n+1},$$

contrary to hypothesis; hence $A \nsubseteq P^{n+1}$. By Exercise 7(c) there is an ideal C of R such that $A = P^nC$ and $C \nsubseteq P$. For each α, $P^nC \subseteq P_\alpha^{n_\alpha}$ and $P^n \nsubseteq P_\alpha$; hence $C \subseteq P_\alpha^{n_\alpha}$. Thus, $C \subseteq B$, and since $C \nsubseteq P$, we conclude that $B \nsubseteq P$.

Of course, all of the results obtained under the hypothesis that R is a weak multiplication ring hold for multiplication rings. We shall now show that, in fact, every weak multiplication ring is a multiplication ring.

9.21. Theorem. *Let R be a ring. The following statements are equivalent:*

(1) *R is a multiplication ring.*
(2) *R is a weak multiplication ring.*
(3) *R satisfies the following conditions:*
  (i) *Each ideal of R is equal to its kernel.*
  (ii) *Each primary ideal of R is a power of its radical.*

(iii) *If P is a minimal prime divisor of an ideal A of R, if n is the least positive integer such that P^n is the isolated P-primary component of A, and if $P^n \neq P^{n+1}$, then P does not contain the intersection of the remaining isolated primary components of A.*

Proof. The only part of the theorem that requires proof is the assertion that (3) implies (1). Assume that (3) holds. Let A and B be ideals of R such that $A \subseteq B$. Let $\{P_\alpha\}$ be the set of those prime ideals of R which are minimal prime divisors of both A and B, $\{P_\beta'\}$ the set of those which are minimal prime divisors of B but not of A, and $\{P_\gamma''\}$ the set of those which are minimal prime divisors of A but not of B. Since A and B are equal to their kernels, we may write

$$B = (\bigcap P_\alpha^{h_\alpha}) \cap (\bigcap P_\beta'^{k_\beta})$$

and

$$A = (\bigcap P_\alpha^{m_\alpha}) \cap (\bigcap P_\gamma''^{n_\gamma}),$$

where the exponents h_α, k_β, m_α, and n_γ are chosen to be the least positive integers such that $P_\alpha^{h_\alpha}$ is the isolated P_α-primary component of B, $P_\beta'^{k_\beta}$ is isolated P_β'-primary component of B, $P_\alpha^{m_\alpha}$ is the isolated P_α-primary component of A, and $P_\gamma''^{n_\gamma}$ is the isolated P_γ''-primary component of A. For each α, we have $P_\alpha^{m_\alpha} \subseteq P_\alpha^{h_\alpha}$, so $h_\alpha \leq m_\alpha$. Let

$$C = (\bigcap P_\alpha^{m_\alpha - h_\alpha}) \cap (\bigcap P_\gamma''^{n_\gamma}).$$

Clearly, $A \subseteq C$. Let $x \in BC$; then $x = \sum_{i=1}^n b_i c_i$, where $b_i \in B$ and $c_i \in C$ for $i = 1, \ldots, n$. Therefore, $b_i \in P_\alpha^{h_\alpha}$, $c_i \in P_\alpha^{m_\alpha - h_\alpha}$, and $c_i \in P_\gamma''^{n_\gamma}$ for each i and each α and γ. Consequently, for each i, $b_i c_i \in P_\alpha^{m_\alpha}$ and $b_i c_i \in P_\gamma''^{n_\gamma}$. Thus $x \in P_\alpha^{m_\alpha}$ and $x \in P_\gamma''^{n_\gamma}$ for each α and γ and, as a result, $x \in A$. Therefore, $BC \subseteq A$.

Let P be a minimal prime divisor of BC. Then P contains either B or C. If $B \subseteq P$, then P is a minimal prime divisor of B. It contains A and so must contain a minimal prime divisor of A; hence, since $BC \subseteq A$, it must be a minimal prime divisor of A. Thus, $P = P_\alpha$ for some α. If $B \nsubseteq P$ then $C \subseteq P$ and P is a minimal prime divisor of both A and C. Thus, $P = P_\gamma''$ for some γ. Since BC is equal to its kernel,

$$BC = (\bigcap P_\alpha^{r_\alpha}) \cap (\bigcap P_\gamma''^{s_\gamma}),$$

where r_α and s_γ are the least positive integers such that $P_\alpha^{r_\alpha}$ is the isolated P_α-primary component of BC and $P_\gamma''^{s_\gamma}$ is the isolated P_γ''-primary component of BC. Since $BC \subseteq A$ we must have $m_\alpha \leq r_\alpha$ and $n_\gamma \leq s_\gamma$ for all α and γ.

For each γ, $A \subseteq C \subseteq P_\gamma''^{n_\gamma}$; hence

$$P_\gamma''^{n_\gamma} = CR_{P_\gamma''} \cap R.$$

Since $B \nsubseteq P_\gamma''$, this implies that

$$BCR_{P_\gamma''} \cap R = P_\gamma''^{n_\gamma};$$

that is, $P_\gamma''^{n_\gamma}$ is the isolated P_γ''-primary component of BC. Therefore, $n_\gamma = s_\gamma$.

If $P_\alpha^{m_\alpha} = P_\alpha^{m_\alpha+1}$, then $P_\alpha^{m_\alpha} = P_\alpha^{r_\alpha}$ and we conclude that $m_\alpha = r_\alpha$. Suppose $P_\alpha^{m_\alpha} \neq P_\alpha^{m_\alpha+1}$. Let

$$C' = \left(\bigcap_{\delta \neq \alpha} P_\delta^{m_\delta - h_\delta}\right) \cap \left(\bigcap P_\gamma''^{n_\gamma}\right),$$

$$B' = \left(\bigcap_{\delta \neq \alpha} P_\delta^{h_\delta}\right) \cap \left(\bigcap P_\beta'^{k_\beta}\right),$$

and

$$A' = \left(\bigcap_{\delta \neq \alpha} P_\delta^{m_\delta}\right) \cap \left(\bigcap P_\gamma''^{n_\gamma}\right).$$

By (iii), $A' \nsubseteq P_\alpha$, and by Proposition 9.19, P_α is a maximal ideal, so that $R = P_\alpha^{m_\alpha} + A'$. Thus,

$$A = P_\alpha^{m_\alpha} \cap A' = P_\alpha^{m_\alpha} A'.$$

Since $h_\alpha \leq m_\alpha$, we have $P_\alpha^{h_\alpha} \neq P_\alpha^{h_\alpha+1}$. Hence, $B' \nsubseteq P_\alpha$, $R = P_\alpha^{h_\alpha} + B'$, and

$$B = P_\alpha^{h_\alpha} \cap B' = P_\alpha^{h_\alpha} B'.$$

Since $A' \subseteq C'$ and $A' \nsubseteq P_\alpha$, we must have $C' \nsubseteq P_\alpha$. Consequently, if $m_\alpha - h_\alpha > 0$, then

$$R = P_\alpha^{m_\alpha - h_\alpha} + C' \qquad \text{and} \qquad C = P_\alpha^{m_\alpha - h_\alpha} \cap C' = P_\alpha^{m_\alpha - h_\alpha} C'.$$

Thus, if $m_\alpha - h_\alpha > 0$, we have $BC = P_\alpha^{m_\alpha} B'C'$. If $m_\alpha - h_\alpha = 0$, then $C = C'$ and the same equality holds. We note that $B'C' \nsubseteq P_\alpha$; thus

$P_\alpha^{m_\alpha}$ is the isolated P_α-primary component of BC. Since $m_\alpha \leq r_\alpha$, it follows that $m_\alpha = r_\alpha$.

We have shown that BC and A have the same kernels. Therefore, $A = BC$.

4 ALMOST MULTIPLICATION RINGS

If R is a Dedekind domain, then each ideal of R which has prime radical is a power of its radical. In this section we shall consider those rings which have this property.

9.22. Definition. *A ring R is an* **almost multiplication ring** *if each ideal of R which has prime radical is a power of its radical.*

The terminology stems from the fact that a ring is an almost multiplication ring if and only if it is locally a multiplication ring; we shall prove this later on in this section. Thus, the almost multiplication rings bear the same relation to the multiplication rings as the almost Dedekind domains bear to the Dedekind domains.

If P is a prime ideal of ring R and if $S = R\backslash P$, we shall denote 0_S, the S-component of the zero ideal, by $N(P)$; that is,

$$N(P) = \{x \mid x \in R \quad \text{and} \quad xs = 0 \quad \text{for some} \quad s \in R\backslash P\}.$$

9.23. Theorem. *If R is an almost multiplication ring, then for each proper prime ideal P of R, the ring R_P is a ZPI-ring.*

Proof. Let P be a proper prime ideal of R. If P is a minimal prime divisor of 0, then PR_P is the only prime ideal of R_P. Hence, by Exercise 5(g) of Chapter III, every proper ideal of R_P is a power of PR_P. Thus, R_P is a ZPI-ring.

Now assume that P is not a minimal prime divisor of 0. Let Q be a minimal prime divisor of 0 contained in P. Then, R/Q is an almost Dedekind domain by Theorem 9.4. Hence, P/Q is a maximal ideal of R/Q and therefore P is a maximal ideal of R. Thus $\dim R \leq 1$. Another consequence of the fact that R/Q is an almost Dedekind domain is that $\bigcap_{n=1}^{\infty}(P/Q)^n = 0$ (see Theorem 9.4), that is,

$\bigcap_{n=1}^{\infty} P^n \subseteq Q$. Suppose the containment is proper. Then there is a positive integer n such that $Q \subseteq P^n$ but $Q \nsubseteq P^{n+1}$. Then, since $\mathrm{Rad}(Q + P^{n+1}) = P$, we have $Q + P^{n+1} = P^n$; whence $(P/Q)^{n+1} = (P/Q)^n$. Since R/Q is an almost Dedekind domain, this is impossible. Thus $\bigcap_{n=1}^{\infty} P^n = Q$.

If Q' is any other minimal prime divisor of 0 contained in P, the above argument shows that $Q' = \bigcap_{n=1}^{\infty} P^n = Q$. Thus, R_P has only two proper prime ideals, PR_P and QR_P. Therefore, every ideal R_P has prime radical and consequently is a prime power. This proves that R_P is a ZPI-ring.

Our next objective is to prove the converse of Theorem 9.23. In doing so we shall establish some important results concerning the ideal structure of almost multiplication rings.

9.24. Proposition. *Suppose that R_P is a ZPI-ring for each proper prime ideal P of R. If P is a proper prime ideal of R such that $N(P)$ is not prime, then* $\mathrm{Rad}(N(P)) = P$.

Proof. As we have seen, R_P has only one minimal prime divisor of 0. Hence, R_P is either a Dedekind domain or a special primary ring; since $N(P)$ is not prime, R_P is not a domain, and so is a special primary ring. Thus, there is a positive integer n such that $P^n R_P = 0$. Hence, for each $p \in P$ there exists $s \in R \backslash P$ such that $p^n s = 0$, that is, $p \in \mathrm{Rad}(N(P))$. Therefore, $\mathrm{Rad}(N(P)) = P$.

9.25. Proposition. *Suppose that R_P is a ZPI-ring for each proper prime ideal P of R. If P is a proper prime ideal of R such that $N(P)$ is not prime, then*

(1) *P is maximal and minimal,*
(2) *R_P is a special primary ring, and*
(3) *each ideal with radical P is a power of P.*

Proof. We have noted already that R_P is a special primary ring. Thus, P is a minimal prime ideal of R. Let M be a maximal ideal of R containing P. Then, R_M is a Dedekind domain or a special primary ring. If R_M is a special primary ring, then M is minimal, so $M = P$, and P is maximal. Suppose R_M is a Dedekind domain; then $N(M)$ is prime and M and $N(M)$ are the only prime ideals of R contained in M. Hence, either $P = M$ or $P = N(M)$. If $P = N(M)$, then $P = N(P)$ which is not true since $N(P)$ is not prime. Hence $P = M$, and consequently P is maximal. Finally, if $\mathrm{Rad}(A) = P$ then A is P-primary and thus $AR_P = P^n R_P$ for some positive integer. Therefore, $A = P^n$.

9.26. Proposition. *Suppose that R_P is a ZPI-ring for each proper prime ideal P of R. If P is a proper prime ideal of R such that $N(P)$ is a prime ideal and $P \neq N(P)$, then*

(1) *R_P is a discrete rank one valuation ring,*
(2) *$N(P)$ is the only prime ideal of R properly contained in P,*
(3) *P is maximal,*
(4) *each ideal with radical P is a power of P, and*
(5) *$N(P) = \bigcap_{n=1}^{\infty} P^n$.*

Proof. Since $N(P)$ is a prime ideal, R_P is an integral domain; hence R_P is a discrete rank one valuation ring. Thus, PR_P is the only nonzero prime ideal of R_P [it is not zero since $P \neq N(P)$]. By the one-to-one correspondence between the proper prime ideals of R_P and the prime ideals of R contained in P, we see that $N(P)$ is the only prime ideal of R properly contained in P. Let M be a maximal ideal of R containing P. If $N(M)$ is not a prime ideal of R, then M is a minimal prime ideal of R by Proposition 9.25; in this case $P = M$ and since $P \neq N(P)$ we have a contradiction. Therefore, $N(M)$ is a prime ideal and is the only prime ideal of R properly contained in M. Thus, either $N(M) = N(P)$ or $N(M) = P$. Since $P \neq N(P)$, we must have $M = P$. Thus, P is maximal.

If $\mathrm{Rad}(A) = P$ then $AR_P = P^n R_P$ for some positive integer n; since P is maximal, this implies that $A = P^n$. It follows also from the fact that P is maximal that each power of P is P-primary, and consequently $N(P) \subseteq P^n$ for each positive integer n. Since R_P is a Noetherian valuation ring, $\bigcap_{n=1}^{\infty} P^n R_P = 0$. Therefore, $\bigcap_{n=1}^{\infty} P^n = N(P)$.

Now we are ready for the converse of Theorem 9.23.

9.27. Theorem. *If R_P is a ZPI-ring for each proper prime ideal P of R, then R is an almost multiplication ring.*

Proof. Suppose that R_P is a ZPI-ring for each proper prime ideal P of R. Let A be a proper ideal of R such that $\text{Rad}(A)=P$ is a prime ideal of R. By Propositions 9.25 and 9.26, A is a power of P if either $N(P)$ is not a prime ideal or if $N(P)$ is a prime ideal and $P \neq N(P)$. So, we assume that $P=N(P)$. Let Q be a proper prime ideal of R which contains A. Then $P \subseteq Q$, and as in the proof of Proposition 9.26, $N(Q)=N(P)=P$. Thus,

$$N(Q) \subseteq AR_Q \cap R \subseteq PR_Q \cap R = P,$$

from which it follows that $AR_Q \cap R = PR_Q \cap R$. Thus, by Exercise 5(b) of Chapter III, $A=P$. Therefore, in every case, A is a power of P.

Now we can easily prove our assertion that almost multiplication rings are rings which are locally multiplication rings.

9.28. Theorem. *A ring R is an almost multiplication ring if and only if R_P is a multiplication ring for each proper prime ideal P of R.*

Proof. If R is an almost multiplication ring, then each R_P is a multiplication ring by Theorem 9.23. On the other hand, let P be an arbitrary proper prime ideal and assume that R_P is a multiplication ring. Then, by Theorem 9.21, primary ideals in R_P are prime powers and each ideal in R_P is equal to its kernel. Thus, in R_P, any ideal with prime radical is primary. Since primary ideals are prime powers, this implies that R_P is an almost multiplication ring. Consequently, $R_P = (R_P)_{PR_P}$ is a ZPI-ring. Therefore, by Theorem 9.27, R is an almost multiplication ring.

9.29. Theorem. *A ring R is an almost multiplication ring if and only if whenever an ideal A of R is such that* $\text{Rad}(A)=P_1 \cdots P_s$, *where $P_1, \ldots, P_s$ are distinct prime ideals of R, then $A=P_1^{n_1} \cdots P_s^{n_s}$ for some positive integers $n_1, \ldots, n_s$.*

Proof. By the definition of almost multiplication ring, any ring which satisfies the condition stated in the theorem is an almost multiplication ring. Conversely, let R be an almost multiplication ring and let $\text{Rad}(A)=P_1 \cdots P_s$, where $P_1, \ldots, P_s$ are distinct prime ideals of R. First, suppose that $P_1, \ldots, P_s$ are comaximal and that each is a minimal prime divisor of A; then these prime ideals are all of the minimal prime divisors of A. By Exercise 5(g) of Chapter III, the isolated P_i-primary component of A is a power, $P_i^{n_i}$, of P_i for $i=1, \ldots, s$. Hence, by Exercise 9(b),

$$A = P_1^{n_1} \cap \cdots \cap P_s^{n_s} = P_1^{n_1} \cdots P_s^{n_s}.$$

Now suppose one of the prime factors of Rad(A), say P_1, is not a minimal prime divisor of A. Then, P_1 contains a minimal prime divisor Q of A. Since $P_1 \cdots P_s \subset Q$, one of these prime ideals, other than P_1, must equal Q, say $P_2 = Q$. Then $P_2 \subset P_1$. Note that this is the only way in which two prime factors of Rad(A) can fail to be comaximal. By Exercise 9(c), $P_1 P_2 = P_2$. If $A = P_2^{n_2} \cdots P_s^{n_s}$, then $A = P_1^{n_2} P_2^{n_2} \cdots P_s^{n_s}$.

EXERCISES

1. Almost Dedekind domains.

Let R be an integral domain.

(a) Show that R is an almost Dedekind domain if and only if R if a Prüfer domain with no idempotent nonzero proper prime ideals and for each such prime ideal P, R_P is completely integrally closed.

(b) Show that an almost Dedekind domain is integrally closed.

(c) Show that if P is a prime ideal of an almost Dedekind domain, then there are no ideals strictly between P and P^2.

(d) Show that in an almost Dedekind domain, the set of primary ideals with the same radical is totally ordered by inclusion.

2. An example.

(a) Let $\{R_n \mid n = 1, 2, \ldots\}$ be a sequence of almost Dedekind domains such that $R_n \subseteq R_{n+1}$ for all $n \geq 1$. Show that $R' = \bigcup_{n=1}^{\infty} R_n$ is an almost Dedekind domain if and only if it has no idempotent nonzero proper prime ideals.

(b) Let R be an almost Dedekind domain with quotient field K. Let K' be an algebraic extension of K and let R' be the integral closure of R in K'. Show that R' is almost Dedekind if and only if it has no idempotent nonzero proper prime ideals.

(c) Let K be the field of rational numbers and let K' be obtained by adjoining to K the pth roots of unity for every prime p. Let R' be the integral closure of Z in K'. Show that R' is an almost Dedekind domain.

(d) Show that the domain R' of (c) has no finitely generated nonzero proper prime ideals, and hence that R' is not a Dedekind domain.

(e) Let R be an almost Dedekind domain which is not a Dedekind domain. Then R has a maximal ideal P which is not finitely generated (by Exercise 14 of Chapter II). Show that if a is a nonzero element of P then $(a): P = (a)$, but $(a)R_P : PR_P \neq (a)R_P$.

(f) With R and P as in (e), show that PR_M is invertible for each maximal ideal M of R, but P is not invertible. (Thus, the assertion of Exercise 1(b) of Chapter VI is no longer true if the ideal is not required to be finitely generated.)

3. Factorization in almost Dedekind domains.
Let R be an almost Dedekind domain and let $\{M_\alpha \mid \alpha \in I\}$ be the set of maximal ideals of R. If A is a nonzero ideal of R, for each $\alpha \in I$, let $f_\alpha(A)$ be the smallest nonnegative integer k such that $AR_{M_\alpha} \cap R = M_\alpha{}^k$ (take $M_\alpha{}^0 = R$). For nonzero $x \in R$, write $f_\alpha(x)$ for $f_\alpha((x))$.

(a) Show that if A is a nonzero ideal of R, then

$$A = \bigcap_{\alpha \in I} M_\alpha^{f_\alpha(A)}.$$

(b) Show that if A is a nonzero ideal of R, then for each $\alpha \in I$ there is an integer $k \geq 0$ such that $A \subseteq M_\alpha{}^k$ and $A \nsubseteq M_\alpha^{k+1}$; show that $k = f_\alpha(A)$.

(c) Show that for nonzero $x, y \in R$ and for $\alpha \in I$,

$$f_\alpha(xy) = f_\alpha(x) + f_\alpha(y), \qquad \text{and} \qquad f_\alpha(x+y) \geq \min\{f_\alpha(x), f_\alpha(y)\}.$$

Hence, for each $\alpha \in I$, there is a valuation v_α on the quotient field of R such that $v_\alpha(x) = f_\alpha(x)$ for all nonzero $x \in R$. Show that R_{M_α} is the valuation ring of v_α.

(d) Let $x_\alpha \in M_\alpha \backslash M_\alpha{}^2$ and let y be a nonzero element of the quotient field of R. Show that there exist $a, b \in R \backslash M_\alpha$ such that

$$y = a x_\alpha^{v_\alpha(y)}/b.$$

4. Certain types of almost Dedekind domains.
Let R be an almost Dedekind domain.
(a) Show that R is a Dedekind domain if and only if every nonzero proper ideal of R is contained in only finitely many maximal ideals of R.
(b) Show that if R has only finitely many maximal ideals then R is a principal ideal domain.
(c) Show that R is a Dedekind domain if and only if R is a Krull domain.

5. Domains with property (#).
Let R be an integral domain and let Δ be the set of maximal ideals of R. The domain R has **property** (#) if for every pair of distinct subsets Δ_1 and Δ_2 of Δ we have

$$\bigcap_{P \in \Delta_1} R_P \neq \bigcap_{P \in \Delta_2} R_P .$$

(a) Show that R has property (#) if and only if for every $P \in \Delta$ we have

$$\bigcap_{\substack{Q \in \Delta \\ Q \neq P}} R_Q \nsubseteq R_P .$$

(b) Show that if for every $P \in \Delta$ we have

$$P \nsubseteq \bigcup_{\substack{Q \in \Delta \\ Q \neq P}} Q,$$

then R has property (#).
(c) Show that if R has the QR-property (see Exercise 12 of Chapter VI), then R has property (#) if and only if each maximal ideal of R is the radical of a principal ideal.
(d) Suppose R is a Prüfer domain and $\dim R = 1$. Show that if R has property (#), then every overring of R has property (#).
(e) Suppose that R is a Prüfer domain, that $\dim R = 1$, and that R has property (#). Let $\Delta = \{M_\alpha \mid \alpha \in I\}$. Show that for each $\alpha \in I$, M_α is the radical of an ideal with two generators. [Hint. For each $\beta \in I$ let v_β be the valuation on the quotient field of R determined by R_{M_β}. Show that there exists $a, b \in R$ such that

$$v_\alpha(a) < v_\alpha(b), \qquad \text{and} \qquad v_\beta(a) \geq v_\beta(b) \quad \text{for all} \quad \beta \neq \alpha.$$

Then $b/a = s/t$, where $s \in M_\alpha$ and $t \in R \backslash M_\alpha$. Let $J = \{\beta \mid \beta \in I \text{ and } s \in M_\beta\}$. Let $T = R \backslash \bigcup_{\beta \in J} M_\beta$ and set $R' = R_T$. Show that there is an element $t \in R$ such that

$$t \in M_\alpha R' \backslash \bigcup_{\substack{\beta \in I \\ \beta \neq \alpha}} M_\beta R'.$$

Then show that $M_\alpha = \text{Rad}((s, t))$.]

(f) Show that if an almost Dedekind domain R has property (#), then R is a Dedekind domain.

6. Some lemmas to Theorem 9.10.

Let R be a ring and let A be a finitely generated ideal of R.

(a) Let R be a subring of a ring R'. Let N be a finitely generated submodule of the R-module R' with k generators. Let B and C be ideals of R such that $BN \subseteq CN$. Show that for any element $b \in B^k$ there exists $c \in C$ such that $bx = cx$ for each $x \in N$. [Hint. First, show this for elements of the form $b_1 \cdots b_k$, where $b_i \in B$ for $i = 1, \ldots, k$, and then for sums of elements of this type.]

(b) Show that if B is an ideal of R and if $AB = A$, then there exists $b \in B$ such that $ab = a$ for all $a \in A$.

(c) Show that if A is idempotent then there is an element $a \in A$ such that $a^2 = a$ and $A = (a)$.

(d) Show that if $B = \bigcap_{n=1}^{\infty} A^n$, and if AB is an intersection of primary ideals of R, then $AB = B$.

(e) Assume R is Noetherian and let M be a maximal ideal of R such that there are no ideals of R strictly between M and M^2. Show that for each positive integer n we have $M^{n+1} \subset M^n$ if and only if there is a prime ideal P of R such that $P \subset M$.

(f) Let R and M be as in (e). Assume that for each positive integer n, $M^{n+1} \subset M^n$. Show that $\bigcap_{n=1}^{\infty} M^n$ is the unique prime ideal of R contained in M.

(g) Show that the direct sum of a finite number of ZPI-rings is a ZPI-ring.

7. Weak multiplication rings.
Let R be a weak multiplication ring.
(a) Show that for every ideal A of R, the ring R/A is a weak multiplication ring.
(b) Show that if R is an integral domain, then R is a Dedekind domain.
(c) Let M be a maximal ideal of R and let A be an ideal of R with $A \subseteq M^n$. Show that there is an ideal B of R such that $A = M^n B$; if $A \nsubseteq M^{n+1}$ then $B \nsubseteq M$.
(d) Let M be a maximal ideal of R such that $M^n \neq M^{n+1}$ for each positive integer n. Show that $\bigcap_{n=1}^{\infty} M^n$ is a prime ideal of R.

8. Multiplication rings.
(a) Show that every multiplication ring is a subring of a direct sum of Dedekind domains and special primary rings.
(b) Let R be the ring in Exercise 13 of Chapter II. Show that R is a multiplication ring which is not a direct sum of Dedekind domains and special primary rings.

9. Some properties of almost multiplication rings.
Let R be an almost multiplication ring.
(a) Show that for each proper primary ideal Q of R there exists a maximal ideal M of R such that either $Q = N(M)$ or $Q = M^n$ for some positive integer n.
(b) Show that each ideal of R is equal to its kernel.
(c) Let P be a prime ideal of R and let A be any ideal of R such that $P \subset A$. Show that $PA = P$.
(d) Show that if A, B, and C are ideals of R, with A regular, and $AB = AC$, then $B = C$.
(e) Let S be a multiplicative system in R. Show that $S^{-1}R$ is an almost multiplication ring.

10. Noetherian multiplication rings.
(a) Show that if R is a multiplication ring, then the following statements are equivalent:
(1) The zero ideal of R has finitely many minimal prime divisors.
(2) The zero ideal of R is an intersection of a finite number of primary ideals.
(3) R is Noetherian.

(b) Show that R is a Noetherian multiplication ring if and only if it is a ZPI-ring.

(c) Show that a Noetherian almost multiplication ring is a multiplication ring.

11. Almost Dedekind domains and the ideal transform.
Refer to Exercise 16 of Chapter VI for terminology.

(a) Prove that an integral domain R with more than one maximal ideal is an almost Dedekind domain if and only if $T((x))$ is an almost Dedekind domain for each nonunit x of R.

(b) Prove that an integral domain R with more than one maximal ideal is a Dedekind domain if and only if $T((x))$ is a Dedekind domain for each nonunit x of R. [Hint. Use (a) and Exercise 14 of Chapter VIII.]

(c) Show that the hypothesis concerning maximal ideals in (a) and (b) is necessary.

CHAPTER

X

Prüfer Rings

In this chapter we shall present in detail some extensions of the results concerning Prüfer domains to rings which are not integral domains. Under a restrictive assumption, this was done in a series of exercises in Chapter VI. In this chapter we shall show that no such restrictive assumption is necessary.

1 VALUATION PAIRS

In this section a valuation theory for rings is developed which, when restricted to fields, coincides with the theory of Chapter V.

10.1. Definition. *A* **valuation** *on a ring R is a mapping v from R onto an ordered Abelian group with ∞ adjoined, as in Section 3 of Chapter V, such that for all $a, b \in R$,*

(i) $v(ab) = v(a) + v(b)$, *and*
(ii) $v(a + b) \geq \min\{v(a), v(b)\}$.

It is evident that $v(1) = 0$ and $v(0) = \infty$ for every valuation v on R. It is important to note that we have required that v map R onto the ordered Abelian group; thus, $\{v(a) \mid a \in R\} \setminus \{v(0)\}$ is an ordered Abelian group.

10.2. Proposition. *Let v be a valuation on R and set*

$$R_v = \{x \mid x \in R \quad \text{and} \quad v(x) \geq 0\},$$
$$P_v = \{x \mid x \in R \quad \text{and} \quad v(x) > 0\},$$
$$A_v = \{x \mid x \in R \quad \text{and} \quad v(x) = \infty\}.$$

Then, R_v is a subring of R, P_v is a prime ideal of R_v, and A_v is a prime ideal of R_v. Furthermore, if $R_v \neq R$ and if A is an ideal of R such that $A \subseteq R_v$, then $A \subseteq A_v$.

Proof. That R_v is a subring of R and P_v is a prime ideal of R_v follow as in the case of a valuation on a field. Clearly, A_v is an ideal of R_v. If $a, b \in R_v$ and $ab \in A_v$, then $v(a) + v(b) = \infty$, so $v(a) = \infty$ or $v(b) = \infty$; hence A_v is a prime ideal. Suppose that $R_v \neq R$, that A is an ideal of R contained in R_v, and that $A \nsubseteq A_v$. Let $a \in A$ be such that $v(a) \neq \infty$. Then there exists $b \in R$ such that $v(b) = -v(a)$, and since $R_v \neq R$ there exists $c \in R$ such that $v(c) < 0$. Then $abc \in A$ but $v(abc) = v(c) < 0$, which contradicts $A \subseteq R_v$.

10.3. Definition. *A* **valuation pair** *of a ring R is a pair (A, P), where A is a subring of R and P is a proper prime ideal of A such that for every $x \in R \backslash A$ there exists $y \in P$ such that $xy \in A \backslash P$.*

If (A, P) is a valuation pair of the total quotient ring of A, then the ring A is called a **valuation ring**.

10.4. Theorem. *If v is a valuation on a ring R, then (R_v, P_v) is a valuation pair of R. If (A, P) is a valuation pair of R, then there exists a valuation v on R such that $A = R_v$ and $P = P_v$.*

Proof. Let v be a valuation on R. If $x \in R \backslash R_v$ then $v(x) < 0$ and there exists $y \in R$ such that $v(y) = -v(x) > 0$. Then $y \in P_v$ and $v(xy) = 0$, so that $xy \in R_v \backslash P_v$. Therefore, (R_v, P_v) is a valuation pair of R.

Conversely, suppose that (A, P) is a valuation pair of R. If $x \in R$ we set

$$[P : x]_R = \{z \mid z \in R \quad \text{and} \quad xz \in P\},$$

and we define a relation $\sim$ on R by setting $x \sim y$ if $[P : x]_R = [P : y]_R$.

Clearly, this is an equivalence relation on R; denote by $v(x)$ the equivalence class of an element x of R. On the set of these equivalence classes we define a binary operation by $v(x) + v(y) = v(xy)$. This is a well-defined operation, for if $x' \in v(x)$ and $y' \in v(y)$, then for any $z \in R$, $(xy)z \in P$ if and only if $(x'y')z \in P$. Thus, $v(xy) = v(x'y')$.

Let $v(R) = \{v(x) \mid x \in R\}$. We shall now show that $v(R)\backslash\{v(0)\}$ is a group with respect to the operation defined above, and that $v(1) = A\backslash P$. Clearly, $v(1)$ is an identity element of $v(R)$ with respect to this operation. If $x \notin A$, then $xy \in A\backslash P$ for some $y \in P$. Then $y \in [P:1]_R$ but $y \notin [P:x]_R$, so $x \nsim 1$. If $x \in P$, then $1 \in [P:x]_R$; however, $1 \notin [P:1]_R$, so $x \nsim 1$. This shows that $v(1) \subseteq A\backslash P$. Now suppose that $x \in A\backslash P$ and $x \notin v(1)$. Since $[P:1]_R = P$, we have $[P:1]_R \subset [P:x]_R$; hence there exists $y \notin P$ such that $xy \in P$. Since P is a prime ideal of A, this implies that $y \notin A$. Hence there exists $z \in P$ such that $yz \in A\backslash P$. Thus $x(yz) \in A\backslash P$, while at the same time $(xy)z \in P$. Since this is impossible we conclude that we do have $v(1) = A\backslash P$. Now let $x \notin v(0)$. Since $[P:0]_R = R$ we have $xy \notin P$ for some $y \in R$. If $xy \in A$, then $v(x) + v(y) = v(1)$, so $v(y) = -v(x)$. If $xy \notin A$, then $(xy)z \in A\backslash P$ for some $z \in P$. Then $v(x) + v(yz) = v(1)$, so $v(yz) = -v(x)$. This proves that $v(R)\backslash\{v(0)\}$ is a group.

Now define $\leq$ on $v(R)\backslash\{v(0)\}$ by $v(x) \leq v(y)$ if $[P:x]_R \subseteq [P:y]_R$. Then $v(x) < v(y)$ is equivalent to the existence of an element $z \in R$ such that $xz \in A\backslash P$ and $yz \in P$. Clearly, $\leq$ is a well-defined relation. We leave it to the reader to verify that $v(R)\backslash\{v(0)\}$, together with the relation $\leq$, is an ordered Abelian group. If we set $v(x) \leq v(0)$ for all $x \in R$, then $v(0)$ plays the role of ∞. To prove that the mapping $x \mapsto v(x)$ from R onto $v(R)$ is a valuation it is sufficient to verify (ii) of Definition 10.1. Let $x, y \in R$ and let $z \in R$ be such that $v(z) \leq v(x)$ and $v(z) \leq v(y)$. Then $[P:z]_R \subseteq [P:x]_R$ and $[P:z]_R \subseteq [P:y]_R$; hence $[P:z]_R \subseteq [P:x+y]_R$ and $v(z) \leq v(x+y)$. It is clear from the definition of $\leq$ that $A = R_v$ and $P = P_v$.

10.5. Corollary. *Let (A, P) be a valuation pair of R. Then the following statements hold:*

(1) *$R\backslash A$ is closed under multiplication.*
(2) *$R\backslash P$ is closed under multiplication.*
(3) *If $x \in R$ and $x^n \in A$, then $x \in A$.*
(4) *$A = \{x \mid x \in R$ and $xy \in P$ for all $y \in P\}$.*
(5) *If $A \neq R$, then $P = \{x \mid x \in A$ and $xy \in A$ for some $y \notin A\}$.*

For a ring R, let $\mathscr{T}$ be the set of all pairs (A, P), where A is a subring of R and P is a proper prime ideal of A. Define a partial ordering $\leqslant$ on $\mathscr{T}$ by

$$(A_1, P_1) \leqslant (A_2, P_2)$$

if $A_1 \subseteq A_2$ and $P_2 \cap A_1 = P_1$; if this is the case, we say that (A_2, P_2) **dominates** (A_1, P_1). Zorn's lemma guarantees maximal elements of $\mathscr{T}$ which dominate any given pair (A, P). It follows immediately from the lying-over theorem that, if (A, P) is a maximal pair in $\mathscr{T}$, then A is integrally closed in R.

The next theorem shows that the maximal elements of $\mathscr{T}$ play the same role here as in the valuation theory of fields; it is an extension of part of Theorem 5.7 (for the extension of the other part of Theorem 5.7, see Exercise 2).

10.6. Theorem. *Let R be a ring and let $\mathscr{T}$ be as above. Then (A, P) is a valuation pair of R if and only if (A, P) is a maximal element of $\mathscr{T}$.*

Proof. Let (A, P) be a valuation pair and let $(B, Q) \in \mathscr{T}$ be such that $(A, P) \leqslant (B, Q)$. Assume $A \subset B$. If $x \in B \backslash A$, then there exists $y \in P$ such that $xy \in A \backslash P$. However, $y \in Q$ and so $xy \in Q \cap A = P$, which is impossible. Thus $B = A$ and $Q = P$. Therefore, (A, P) is maximal in $\mathscr{T}$.

Conversely, suppose that (A, P) is a maximal element of $\mathscr{T}$. If $A = R$ then (A, P) is a valuation pair. Suppose that $A \neq R$ and let $x \in R \backslash A$. Let $B = A[x]$ and $Q = PB$. Then Q is an ideal of B and $P \subseteq Q \cap A$. If $P = Q \cap A$ then $A \backslash P$ is a multiplicative system in B which does not meet Q. Therefore, there exists a prime ideal M of B containing Q and which does not meet $A \backslash P$. Then $M \cap A = P$ and so $(A, P) \leqslant (B, M)$. This contradicts the maximality of (A, P) since $x \notin A$. Thus, we are forced to conclude that $P \subset Q \cap A$.

Let $a \in Q \cap A$, $a \notin P$. Then

$$a = \sum_{i=0}^{n} p_i x^i,$$

where $p_0, \ldots, p_n \in P$, and n is chosen as small as possible. Multiplying by p_n^{n-1}, we obtain

$$(xp_n)^n + \sum_{i=0}^{n-1} (xp_n)^i p_n^{n-1-i} p_i - a p_n^{n-1} = 0.$$

Hence, xp_n is integral over A, and therefore $xp_n \in A$. If $xp_n \in P$, then

$$a = (xp_n + p_{n-1})x^{n-1} + \sum_{i=0}^{n-2} p_i x^i,$$

which contradicts our choice of n (note that we cannot have $n = 0$ or $n = 1$). Hence, $xp_n \in A \backslash P$, and so we conclude that (A, P) is a valuation pair.

For the remainder of this section, let (A, P) be a fixed valuation pair of a ring R and let v be the valuation determined by this valuation pair. Let $G_v = v(R) \backslash \{v(0)\}$. As in Chapter V, we denote by G_v^* the ordered Abelian group G_v with ∞ adjoined.

10.7. Definition. *An ideal I of $R_v (= A)$ is v-**closed** if for all $x \in I$ and $y \in R$, $v(y) \geq v(x)$ implies that $y \in I$.*

Also recall that a subgroup H of an ordered Abelian group G is *isolated* if for all nonnegative $\alpha \in H$ and $\beta \in G$, $0 \leq \beta \leq \alpha$ implies that $\beta \in H$. We shall establish a correspondence between the v-closed proper prime ideals of R_v and the isolated subgroups of G_v.

10.8. Proposition. *Let ϕ be an order-preserving homomorphism from G_v into an ordered Abelian group, and set $\phi(\infty) = \infty$. Then:*

(1) *ϕv is a valuation on R.*
(2) Ker *ϕ is an isolated subgroup of G_v.*
(3) *$P_{\phi v}$ is a v-closed proper prime ideal of R_v.*

Proof. (1) Im ϕ is an ordered Abelian group, ϕv maps R onto (Im $\phi)^*$, and (i) and (ii) of Definition 10.1 hold for ϕv.

(2) Let α be a nonnegative element of Ker ϕ and let $0 \leq \beta \leq \alpha$. Then $0 \leq \phi(\beta) \leq \phi(\alpha) = 0$; hence $\phi(\beta) = 0$, that is, $\beta \in \operatorname{Ker} \phi$.

(3) If $x \in P_{\phi v}$, then $\phi(v(x)) > 0$, and so $v(x) > 0$; thus $P_{\phi v} \subseteq P_v$. If $x \in R_v$, then $v(x) \geq 0$, and so $\phi(v(x)) \geq 0$; thus $R_v \subseteq R_{\phi v}$. Therefore, $P_{\phi v}$ is a proper ideal of R_v, and since it is a prime ideal of $R_{\phi v}$, it is also a prime ideal of R_v. Now let $x \in P_{\phi v}$, $y \in R$, and $v(y) \geq v(x)$. Then $(\phi v)(y) \geq (\phi v)(x) > 0$ and consequently $y \in P_{\phi v}$. Thus, $P_{\phi v}$ is v-closed.

10.9. Proposition. *Let H be an isolated subgroup of G_v. Then there is a surjective order-preserving homomorphism ϕ from G_v onto an ordered Abelian group such that ϕv is a valuation on R and $H = \operatorname{Ker} \phi$.*

Proof. Let ϕ be the canonical homomorphism from G_v onto G_v/H. Set $\alpha + H \leq \beta + H$ if $\alpha \leq \beta$. Since H is isolated, this is a well-defined total ordering of G_v/H, and with respect to this ordering, G_v/H is an ordered Abelian group. The assertion of the proposition follows immediately.

If H and ϕ are as in Proposition 10.9, we set $v_H = \phi v$. Then we have

$$\begin{aligned} P_{v_H} &= \{x \mid x \in R \quad \text{and} \quad v_H(x) > 0\} \\ &= \{x \mid x \in R \quad \text{and} \quad v(x) + H > H\} \\ &= \{x \mid x \in R \quad \text{and} \quad v(x) > \alpha \quad \text{for all} \quad \alpha \in H\}. \end{aligned}$$

On the other hand,

$$\begin{aligned} H &= \{\alpha \mid \alpha \in G_v \quad \text{and} \quad \phi(\alpha) = \phi(-\alpha) = 0\} \\ &= \{\alpha \mid \alpha \in G_v \quad \text{and} \quad \max\{\alpha, -\alpha\} < v(x) \quad \text{for all} \quad x \in P_{v_H}\}. \end{aligned}$$

10.10. Theorem. *Let v be a valuation on R. Then there is a one-to-one order-reversing correspondence between the isolated subgroups H of G_v and the v-closed proper prime ideals Q of R_v. The correspondence is given by*

$$H \leftrightarrow P_{v_H}$$

and

$$Q \leftrightarrow \{\alpha \mid \alpha \in G_v \quad \textit{and} \quad \max\{\alpha, -\alpha\} < v(x) \quad \textit{for all} \quad x \in Q\}$$

Proof. In the light of the remarks made above, all that remains to be shown is that, given Q, the set $H = \{\alpha \mid \alpha \in G_v \text{ and } \max\{\alpha, -\alpha\} < v(x)$ for all $x \in Q\}$ is an isolated subgroup of G_v, and that the correspondence is order-reversing.

Let $\alpha, \beta \in H$ and suppose $\alpha = v(x)$ and $\beta = v(y)$. By definition of H, $-\alpha \in H$. In showing that $\alpha + \beta \in H$, it is enough to do so under the assumption that $\alpha + \beta \geq 0$. If $\alpha + \beta \notin H$, then $v(xy) = \alpha + \beta \geq$

$v(z)$ for some $z \in Q$, and this implies that $xy \in Q$ since Q is v-closed. Since Q is a prime ideal and $x \notin Q$ and $y \notin Q$, we must have $x \notin R_v$ or $y \notin R_v$. Suppose the former. Then there exists $u \in R_v$ such that $v(uxy) = \beta$, contrary to the fact that $\beta \in H$. Thus, $\alpha + \beta \in H$, and we have shown that H is a subgroup of G_v. Now suppose that $\alpha \geq 0$ and that $\gamma \in G_v$ is such that $0 \leq \gamma \leq \alpha$. If $x \in Q$, then $\max\{\gamma, -\gamma\} = \gamma \leq \alpha = \max\{\alpha, -\alpha\} < v(x)$; hence $\gamma \in H$. Therefore, H is an isolated subgroup of G_v.

If Q_1 and Q_2 are proper prime ideals of R_v such that $Q_1 \subseteq Q_2$, and if $\max\{\alpha, -\alpha\} < v(x)$ for all $x \in Q_2$, then $\max\{\alpha, -\alpha\} < v(x)$ for all $x \in Q_1$. Hence, the correspondence is order-reversing.

2 COUNTEREXAMPLES

Although valuation pairs of arbitrary rings possess many of the properties of valuation pairs of fields, they fail to possess others. In this section we shall discuss two examples of this fact.

The first example shows that if (A, P) is a valuation pair of a ring R, then P may not necessarily be a maximal ideal of A. In this example, the ring A is a valuation ring. A second example of this phenomenon is given in Exercise 3; however, in that exercise, A is not a valuation ring.

In this example, R is the ring of Exercise 23 of Chapter VI; all notation will be as in that exercise.

By Zorn's lemma, there is a valuation pair (A, P) of the total quotient ring K of R such that $R \subseteq A$ and $P \cap R = N$. Then K is the total quotient ring of A. We shall show that P is not a maximal ideal of A. Let P' be the ideal of A generated by P and x. Clearly, $P \subset P' \subseteq A$. If $P' = A$ then there exist $s \in A$ and $q \in P$ such that $1 = sx + q$. Let $q = a/b$, where $a, b \in R$ and b is regular. Write

$$a = f(y) + xg(x, y) + w, \qquad \text{and} \qquad b = f'(y) + xg'(x, y) + w',$$

where $f(Y), f'(Y), Xg(X, Y), Xg'(X, Y) \in F[X, Y]$ and $w, w' \in Q$. By Exercise 23(c) of Chapter VI, $f'(y) + xg'(x, y)$ is regular, and consequently $f'(y) + xg'(x, y) - w'$ is regular. If we multiply both a and b by this element, we see that we may assume that $w' = 0$.

If we multiply both sides of $1 = sx + q$ by bz_X we get $bz_X = az_X$;

hence $f'(y)z_X = f(y)z_X$. This implies that $(f'(Y) - f(Y))Z_X \in I$, which in turn implies that $f'(Y) = f(Y)$. Thus, $f'(y) = f(y)$, and we have

$$\frac{a}{b} = \frac{f(y) + xg(x, y) + w}{f(y) + xg'(x, y)}.$$

Since $f(y) + xg'(x, y)$ is regular, and since $xg'(x, y)z_X = 0$, we must have $f(Y) \neq 0$. Write $f(Y) = Y^m(r + h(Y))$ where r is a nonzero element of F and $h(Y) \in YF[Y]$. By Exercise 23(b) of Chapter VI, $g'(X, Y) = Y^m g^*(X, Y)$, where $g^*(X, Y) \in F[X, Y]$. Therefore,

$$r + h(y) + (xg(x, y) + w)y^{-m} = (r + h(y) + xg^*(x, y))a/b \in P.$$

Next, we show that $wy^{-m} \in P$. Let $g_1, \ldots, g_k$ be those elements of G such that $z_{g_1}, \ldots, z_{g_k}$ occur in the generation of w as an element of Q. Since $g_i(x, y) \in A \backslash P$ for $i = 1, \ldots, k$, we have $\prod_{i=1}^k g_i(x, y) \in A \backslash P$. But $(\prod_{i=1}^k g_i(x, y))wy^{-m} = 0 \in P$, so $wy^{-m} \in P$.

Since $h(Y) \in YF[Y]$, we have $h(y) \in P$; hence $r + xg(x, y)y^{-m} \in P$. Let $g(X, Y) = Y^p(r' + Yk(X, Y))$, where r' is a nonzero element of $F[X]$ and $k(X, Y) \in F[X, Y]$. Then

$$r + xy^{p-m}(r' + yk(x, y)) \in P.$$

If $p - m \geq 0$, then $r + xy^{p-m}(r' + yk(x, y)) \in P \cap R = N$, which is not true since $r \neq 0$. If $p - m < 0$, then $x(r' + yk(x, y)) \in N$, which is not true since $r' \neq 0$. Thus, we are forced to conclude that $P' \neq A$, and consequently that P is not a maximal ideal of A.

Another way in which valuation pairs of arbitrary rings fail to behave like valuation pairs of fields is the following. There exists a valuation pair (A, P) and an overring B of A such that there is no ideal Q of B with the property that $Q \subseteq P$ and (B, Q) is a valuation pair.

In fact, let (A, P) be the valuation pair of the example discussed above. Let p be a regular element of P and let $a \in A \backslash P$. Let $B = A[a/p]$ and suppose that there does exist an ideal Q of B such that $Q \subseteq P$ and (B, Q) is a valuation pair. Then $p \notin Q$, for if $p \in Q$, then $p(a/p) = a \in Q \subseteq P$. Also, $1/p \in B$. For, suppose $1/p \notin B$; then there exists $q \in Q$ such that $q/p \in B \backslash Q$, and $q = p(q/p)$. Let

$$1/p = b_0 + b_1(a/p) + \cdots + b_n(a/p)^n,$$

where $b_0, \ldots, b_n \in A$ and $b_n \neq 0$. Suppose $n \geq 2$. Multiplying by p^{n-1} gives $p^{n-2} = b_n a^{n-1}(a/p) + t$ for some $t \in A$. Since $p^{n-2} \in A$, we have $b_n a^{n-1}(a/p) \in A$. If v is the valuation determined by (A, P), then $v(a^{n-1}) = 0$, and since $v(b_n a^{n-1}(a/p)) \geq 0$, we must have $v(b_n a/p) \geq 0$, that is, $b_n a/p \in A$. Then we can write

$$1/p = (b_n a/p + b_{n-1})(a/p)^{n-1} + \cdots + b_0.$$

Assuming we have chosen n as small as possible, we have arrived at a contradiction. Thus, $n \leq 1$. If $n = 0$, then $1/p = b_0 \in A$, which is not true. Hence, $n = 1$ and we have $1/p = b_1 a/p + b_0$. Then, $1 = b_1 a + b_0 p$ and so $P + (a) = A$. Since P is not a maximal ideal of A, we can choose $a \in A \backslash P$ such that $P + (a) \neq A$. With this choice of a, B has the desired property.

3 LARGE QUOTIENT RINGS

10.11. Definition. *Let R be a ring with total quotient ring K, and let S be a multiplicative system in R. The* **large quotient ring with respect to** *S, denoted by $R_{[S]}$, is the set of all $x \in K$ such that $xs \in R$ for some $s \in S$. If S is the complement in R of a proper prime ideal P of R, we write $R_{[P]}$ for $R_{[S]}$.*

If P is a proper prime ideal of R, it is not necessarily true that $PR_{[P]}$ is a prime ideal of $R_{[P]}$. In order to study the relationship between the prime ideals of $R_{[P]}$ and those of R, the following operation is introduced.

10.12. Definition. *Let S be a multiplicative system in a ring R. If A is an ideal of R, set*

$$[A]R_{[S]} = \{x \mid x \in K \quad \textit{and} \quad xs \in A \quad \textit{for some} \quad s \in S\};$$

this is called the **extension** *of A to $R_{[S]}$.*

If P is a prime ideal of a ring R, and if Q is a primary ideal of R contained in P, then $[Q]R_{[P]}$ is a primary ideal of $R_{[P]}$ and $[Q]R_{[P]} \cap R = Q$. The details are left to the reader. However, it is not necessarily true that $[P]R_{[P]}$ is a maximal ideal of $R_{[P]}$ when P is not a maximal ideal of R. Furthermore, if A and B are ideals of R contained in P, it is not the case that $[AB]R_{[P]}$ must equal $([A]R_{[P]})([B]R_{[P]})$.

10.13. Definition. *Let R be a ring and P a proper prime ideal of R. The* **core** *$C(P)$ of P in R is the set of all $x \in R$ such that for each regular element $r \in R$, there exists an element $s \in R \backslash P$ such that $xs/r \in R$.*

Some properties of the core of a proper prime ideal are given in Exercise 9. The importance of this concept and the others which have been introduced stems from the following theorem.

10.14. Theorem. *Let P be a proper prime ideal of a ring R. The following statements are equivalent:*

(1) *$(R_{[P]}, [P]R_{[P]})$ is a valuation pair of the total quotient ring of R.*
(2) *For all $x, y \in R$, either both x and y are in $C(P)$ or there exist $a, b \in R$, not both in P, such that $ax = by$.*
(3) *If A and B are ideals of R, not both contained in $C(P)$, then either $AR_P \subseteq BR_P$ or $BR_P \subseteq AR_P$.*

Proof. Assume (1) holds and let v be the valuation on the total quotient ring of R which is determined by $(R_{[P]}, [P]R_{[P]})$. Let $x, y \in R$ and assume $v(x) \geq v(y)$. If $v(y) = \infty$, then

$$x, y \in C([P]R_{[P]}) \cap R = [C(P)]R_{[P]} \cap R = C(P),$$

where we have used various parts of Exercise 9. If $v(y) \neq \infty$, then there exists $t \in K$ such that $v(ty) = 0$ and $v(tx) \geq 0$; then $ty \in R_{[P]} \backslash [P]R_{[P]}$. Hence, there exist $u, v \in R \backslash P$ such that $xtu \in R$ and $ytuv \in R \backslash P$. Now let $a = ytuv$ and $b = uxvt$. Then a and b are not both in P, and $ax = by$. Thus, (2) holds.

To prove the converse, let y be an element of the total quotient ring of R which does not belong to $R_{[P]}$. There is a regular element $r \in R$ such that $ry \in R$. If $ry \in C(P)$, then there exists $s \in R \backslash P$ such that $sy = sry/r \in R$; this implies that $y \in R_{[P]}$, which is not the case. Hence, $ry \notin C(P)$ and so, assuming that (2) holds, there exist $a, b \in R$, not both in P, such that $ary = br$. Since r is regular, $ay = b$. Since $y \notin R_{[P]}$, we must have $a \in P$. Thus, $b \in R \backslash P$. Consequently

$a \in [P]R_{[P]}$, and $ay \in R_{[P]} \backslash [P]R_{[P]}$ since $[P]R_{[P]} \cap R = P$. Therefore, $(R_{[P]}, [P]R_{[P]})$ is a valuation pair.

Now assume that (2) holds, and let A and B be ideals of R such that $A \nsubseteq C(P)$. Let $a \in A$, $a \notin C(P)$. If $AR_P \nsubseteq BR_P$, we claim that we may choose a so that $as \notin B$ for all $s \in R \backslash P$. For, assume that this is not the case, that is, assume that there exists $q \in R \backslash P$ such that $aq \in B$. Since $AR_P \nsubseteq BR_P$, there exists $d \in A$ such that $ds \notin B$ for all $s \in R \backslash P$. Since $a \notin C(P)$, it follows from (2) that there are elements $t, z \in R$, not both in P, such that $at = dz$. If $t \in P$, then $z \in R \backslash P$ and consequently $zq \in R \backslash P$ and $d(zq) = aqt \in B$. Since this cannot happen, we must have $t \in R \backslash P$. Since $a \notin C(P)$, there exists a regular element $r \in R$ such that $au \notin (r)$ for all $u \in R \backslash P$. Hence $ab = rw$ for $w \in R$ implies that $b \in P$. Therefore if $du \in (r)$, then $duz = atu \in (r)$, so $tu \in P$. But $t \in R \backslash P$, so $u \in P$. It follows that $d \notin C(P)$. If we replace a by d we see that we may assume in the first place that $as \notin B$ for all $s \in R \backslash P$. Now apply (2) to a and an arbitrary $f \in B$. There exist $x, y \in R$, not both in P, such that $ax = fy$. We must have $x \in P$, so $y \in R \backslash P$. Hence, $fy/y = ax/y \in AR_P$. Therefore, $BR_P \subseteq AR_P$.

Finally, assume that (3) holds, and let x and y be elements of R not both in $C(P)$. Then either $xR_P \subseteq yR_P$ or $yR_P \subseteq xR_P$. If $xR_P \subseteq yR_P$, then there exists $a \in R \backslash P$ such that $ax = by$ for some $b \in R$. Otherwise, there exists $b \in R \backslash P$ such that $ax = by$ for some $a \in R$.

4 PRÜFER RINGS

10.15. Lemma. *Let R be a ring in which $(A + B)(A \cap B) = AB$ for all ideals A and B of R, at least one of which is regular. Let P be a proper prime ideal of R and let $a, b, c \in R$ with a regular. If $aR_P \subseteq bR_P$, then either $bR_P \subseteq cR_P$ or $cR_P \subseteq bR_P$.*

Proof. Let $x \in R$ and $y \in R \backslash P$ be such that $ay = bx$; x and y exist, since $aR_P \subseteq bR_P$. Now $(a, b, c)((a, b) \cap (c)) = (a, b)(c)$, so $bc = x_1 a + x_2 b + x_3 c$, where $x_i \in (a, b) \cap (c)$ for $i = 1, 2, 3$. If $x_3 = au + bv$, where $u, v \in R$, then $x_3 y = bxu + byv$. Therefore,

$$bc(y - (xu + yv)) = x_1 ay + x_2 by = (x_1 x + x_2 y)b.$$

If $z = xu + yv \notin P$, then $bR_P \subseteq cR_P$ because $zb = x_3 y \in (c)$. If $z \in P$ then $y - z \notin P$ since $y \notin P$. Furthermore, x_1 and x_2 are in (a, b), so

$$(cR_P)(bR_P) \subseteq ((a, b)R_P)(bR_P) \subseteq (bR_P)(bR_P).$$

However, bR_P is a regular principal ideal of R_P, and so $cR_P \subseteq bR_P$.

10.16. Lemma. *Let R be a ring such that if A, B, and C are ideals of R with A finitely generated and regular and if $AB = AC$, then $B = C$. Let a and b be elements of R and let P be a proper prime ideal of R. If there is a regular element $c \in R$ such that $cR_P \subseteq aR_P$, then either $aR_P \subseteq bR_P$ or $bR_P \subseteq aR_P$.*

Proof. Since $(ab)(a, b, c) \subseteq (a^2, b^2, ac)(a, b, c)$, we have $ab \in (a^2, b^2, ac)$. Then $cR_P \subseteq aR_P$ implies that $abR_P \subseteq (a^2, b^2)R_P$. Therefore, there exists $y \in R \backslash P$ such that $aby = xa^2 + zb^2$ for some $x, z \in R$. This implies that $(zb)(a, b, c) \subseteq (a, c)(a, b, c)$, so $zb \in (a, c)$. Again using the fact that $cR_P \subseteq aR_P$, we have $(zb)R_P \subseteq aR_P$. Hence, there exists $v \in R \backslash P$ such that $zbv = au$ for some $u \in R$. Now

$$abyv = xa^2v + zb^2v = xa^2v + abu,$$

and therefore $(a)(b)(yv - u) \subseteq (a)^2$. If $u \notin P$, then $aR_P \subseteq bR_P$. On the other hand, if $u \in P$, then $yv - u \notin P$ since $yv \notin P$. Thus $(aR_P)(bR_P) \subseteq (aR_P)(aR_P)$. Since aR_P is a regular principal ideal of R_P, this implies that $bR_P \subseteq aR_P$.

Now we are ready to define Prüfer ring and to prove that Prüfer rings are characterized by conditions analogous to those characterizing Prüfer domains.

10.17. Definition. *A ring R is a* **Prüfer ring** *if every finitely generated regular ideal of R is invertible.*

10.18. Theorem. *For a ring R, the following statements are equivalent:*

(1) *R is a Prüfer ring.*
(2) *For every maximal ideal P of R, $(R_{[P]}, [P]R_{[P]})$ is a valuation pair of the total quotient ring of R.*

(3) *Every overring of R is a flat R-module.*
(4) *Every overring of R is integrally closed.*
(5) $A(B \cap C) = AB \cap AC$ *for all ideals A, B, C of R, with B or C regular.*
(6) $(A + B)(A \cap B) = AB$ *for all ideals A, B of R, with A or B regular.*
(7) *Every regular ideal of R generated by two elements is invertible.*
(8) *If $AB = AC$, where A, B, C are ideals of R and A is finitely generated and regular, then $B = C$.*
(9) *R is integrally closed, and for any $a, b \in R$, at least one of which is regular, there exists an integer $n > 1$ such that $(a, b)^n = (a^n, b^n)$.*
(10) *If A and B are ideals of R, with B finitely generated and regular, and if $A \subseteq B$, then there is an ideal C of R such that $A = BC$.*
(11) $(A + B) : C = A : C + B : C$ *for all ideals A, B, C of R, with A regular and C finitely generated.*
(12) $A : (B \cap C) = A : B + A : C$ *for all ideals A, B, C of R, with C regular and B and C finitely generated.*
(13) $A \cap (B + C) = A \cap B + A \cap C$ *for all ideals A, B, C of R, at least one of which is regular.*

Proof. Let K be the total quotient ring of R.

(1) $\Rightarrow$ (2). Assume R is a Prüfer ring and let P be a maximal ideal of R. Let $x \in K \backslash R_{[P]}$; then there is a regular element $b \in R$ such that $xb \in R$. Then (b, xb) is invertible. Hence there exists a fractional ideal A of R such that $(b, xb)A = R$. For each $a \in A$, $ab \in P$, for if $ab \notin P$, then since $xba \in R$ we would have $x \in R_{[P]}$. Since $(b, xb)A = R$, we have $xba \in R \backslash P$ for some $a \in A$. For this element of A, $ab \in [P]R_{[P]}$ and $xba \in R_{[P]} \backslash [P]R_{[P]}$.

(2) $\Rightarrow$ (3). Assume (2) holds and let T be an overring of R. Let M be a maximal ideal of T and set $P = M \cap R$. Let Q be a maximal ideal of R containing P. Then $R_{[Q]} \subseteq R_{[P]} \subseteq T_{[M]}$. If $x \in K \backslash R_{[P]}$, then $x \notin R_{[Q]}$, so there exists $b \in Q$ such that $bx \in R \backslash Q$. If $b \notin P$, then $bx \in R$ would imply that $x \in R_{[P]}$, which is a contradiction. Now suppose that $x \in T_{[M]}$. Then there exists $t \in T \backslash M$ such that $xt \in T$. But $b \in P \subseteq M$ and $bx \in R \backslash P \subseteq T \backslash M$. Hence $bxt \in M \cap (T \backslash M)$, which is absurd. Thus $x \notin T_{[M]}$, and we conclude that $T_{[M]} = R_{[P]}$. Consequently, by Exercise 11(a), T is a flat R-module.

(3) $\Rightarrow$ (4). Assume (3) holds, let T be an overring of R, and let T^* be its integral closure. By (3), T^* is a flat R-module, so for any $x \in T^*$, $[R : x]_R T^* = T^*$ by Exercise 11(a). Therefore,

$$T^* \subseteq [T : x]_R T^* \subseteq [T : x]_T T^* \subseteq T^*,$$

so that $[T : x]_R T^* = T^*$. This means that T^* is a flat T-module, and using Exercise 11(b), we conclude that $T = T^*$.

(4) $\Rightarrow$ (2). Let P be a maximal ideal of R. Let $z \in K \backslash R_{[P]}$. Since $R_{[P]}[z^2]$ is integrally closed, $z \in R_{[P]}[z^2]$. Let

$$z = a_0 + a_1 z^2 + \cdots + a_n z^{2n}, \; a_i \in R_{[P]},$$

where $a_n \neq 0$ and n is chosen as small as possible. Note that we cannot have $n = 0$. Suppose $n > 1$. Multiplying by a_n^{2n-1}, we obtain

$$(a_n z)^{2n} + \cdots + a_n^{2n-3} a_1 (a_n z)^2 - a_n^{2n-2}(a_n z) + a_0 a_n^{2n-1} = 0,$$

so $a_n z \in R_{[P]}$. Then, multiplying the first equation by a_n^{n-1}, we have

$$(a_n z^2)^n + \cdots + a_n^{n-2} a_1 (a_n z^2) + (a_0 a_n^{n-1} - a_n^{n-2}(a_n z)) = 0$$

Since $R_{[P]}$ is integrally closed, this implies that $a_n z^2 \in R_{[P]}$. But then $a_{n-1} + a_n z^2 \in R_{[P]}$, and

$$z = a_0 + a_1 z^2 + \cdots + (a_{n-1} + a_n z^2) z^{2(n-1)},$$

which contradicts our choice of n. Thus, we must have $n = 1$, $z = a_0 + a_1 z^2$, and $a_1 z \in R_{[P]}$. Since $z \notin R_{[P]}$, we also have $a_1 \in [P]R_{[P]}$. Since $z(1 - a_1 z) = a_0 \in R_{[P]}$, there exists $t \in R \backslash P$ such that $tz(1 - a_1 z) \in R$. Also, there exists $s \in R \backslash P$ such that $sa_1 z \in R$. Now $z \notin R_{[P]}$, so $t(s - sa_1 z) \notin R \backslash P$; but this element is in R, so it must belong to P. Thus $1 - a_1 z \in [P]R_{[P]}$, and it follows that $a_1 z \in R_{[P]} \backslash [P]R_{[P]}$.

(2) $\Rightarrow$ (5), (11)–(13). Assume that (2) holds and let P be a proper prime ideal of R. If P is regular, then by Exercise 9(b), no regular ideal of R is contained in $C(P)$. Then by the equivalence of (2) and the third condition of Theorem 10.14, the equalities of (5) and (11)–(13) hold for the extensions of the ideals to R_P. The same is true when P is not regular; for, if P is not regular, then in each case at least one of the ideals has its extension to R_P equal to R_P.

(5) $\Rightarrow$ (6) $\Rightarrow$ (7). As in the proof of Theorem 6.6.

(7) $\Rightarrow$ (2). Same proof as proof that (1) implies (2).

(6) $\Rightarrow$ (1). Let A be a finitely generated regular ideal of R, and let

c be a regular element of A. Let $B=[(c):A]$. To prove that A is invertible, it is sufficient to show that $AB=(c)$; furthermore, since $AB \subseteq (c)$, all we have to do is to show that $cR_P \subseteq ABR_P$ for every proper prime ideal P of R. Let P be a proper prime ideal of R. By Lemma 10.15, there exist elements $a_1, \ldots, a_n \in A$ such that

$$A=(a_1, \ldots, a_n), \qquad a_k=c,$$

and

$$(a_1, \ldots, a_{k-1})R_P \subseteq a_k R_P \subseteq \cdots \subseteq a_n R_P.$$

For $i=1, \ldots, n-1$, let $x_i, y_i \in R$ be such that $y_i \notin P$, $a_i y_i = a_k x_i$ for $i=1, \ldots, k-1$, and $a_i y_i = a_{i+1} x_i$ for $i=k, \ldots, n-1$. Let $b=y_1 \cdots y_{k-1} x_k \cdots x_{n-1}$. Then $bA \subseteq (c)$, so $b \in B$; hence, $a_n b \in AB$. But $a_n b = y_1 \cdots y_{k-1} a_k$ and $y_1 \cdots y_{k-1} \notin P$. Therefore, $cR_P = a_k R_P \subseteq ABR_P$.

(1) $\Rightarrow$ (8). Clear.

(8) $\Rightarrow$ (5). Assume that (8) holds, and let A, B, C be ideals of R with B regular. Let P be a proper prime ideal of R. If $CR_P \nsubseteq BR_P$, then there exists $c \in C$ such that $cR_P \nsubseteq BR_P$. If b is a regular element of B, then by Lemma 10.16, either $bR_P \subseteq cR_P$ or $cR_P \subseteq bR_P$. By our choice of c we must have $bR_P \subseteq cR_P$. Then, if $x \in B$, it follows by Lemma 10.16 that $xR_P \subseteq cR_P$. Thus, $BR_P \subseteq CR_P$. We have shown that either $CR_P \subseteq BR_P$ or $BR_P \subseteq CR_P$. If the former holds, then $ACR_P \subseteq ABR_P$, and

$$\begin{aligned}(AB \cap AC)R_P &= ABR_P \cap ACR_P = (AR_P)(CR_P) \\ &= (AR_P)(BR_P \cap CR_P) = A(B \cap C)R_P.\end{aligned}$$

If the latter holds, we have likewise $(AB \cap AC)R_P = A(B \cap C)R_P$. Since P is an arbitrary proper prime ideal of R, we have $AB \cap AC = A(B \cap C)$.

(4) $\Rightarrow$ (9). Assume (4) holds and let a and b be elements of R, at least one of which is regular. By (4), R is integrally closed. Since (4) implies (8), and since $(a, b)(a, b)^2 = (a, b)(a^2, b^2)$, we have $(a, b)^2 = (a^2, b^2)$.

(9) $\Rightarrow$ (7). Suppose that $(a, b)^n = (a^n, b^n)$ for a pair of elements a and b of R with a regular, and some $n > 1$. Then $a^{n-1}b = xa^n + yb^n$ for some $x, y \in R$. Let m be the smallest integer greater than 1 for which $a^{m-1}b = xa^m + yb^m$ for some $x, y \in R$. Now assume that R is

integrally closed. Since $(yb/a)^m + xy^{m-1} - y^{m-2}(yb/a) = 0$, we have $yb/a = z \in R$. Then $a^{m-1}b = xa^m + zab^{m-1}$ and since a is regular, the minimality of m implies that $m = 2$. Hence $b = xa + zb$. Thus,

$$(a, b)(y, 1 - z) = (ay, by, a(1 - z), b(1 - z)) = (ay, az, a(1 - z), ax) = (a).$$

Therefore, (a, b) is invertible.

(1) $\Rightarrow$ (10). Take $C = B^{-1}A$.

(10) $\Rightarrow$ (1). Let A be a finitely generated regular ideal of R. If a is a regular element of A, then $(a) \subseteq A$, so there is an ideal B of R such that $(a) = AB$. Therefore, A is invertible.

(11)–(13) $\Rightarrow$ (2). In each case, we shall verify the second of the equivalent conditions of Theorem 10.14, and conclude that $(R_{[P]}, [P]R_{[P]})$ is a valuation pair for each regular maximal ideal P of R. Let P be such an ideal and let $x, y \in R$ with $x \notin C(P)$. Then there exists a regular element $r \in R$ such that $xs/r \notin R$ for all $s \in R \backslash P$.

Assume that (11) holds; then

$$\begin{aligned} R &= ((x) + (r)) : ((x) + (r)) \\ &= (x) : ((x) + (r)) + (r) : ((x) + (r)) \\ &= (x) : (r) + (r) : (x). \end{aligned}$$

Hence, $1 = s + t$, where $sr \in (x)$ and $tx \in (r)$. Then $tx/r \in R$ and consequently $t \in P$. Therefore, $s \notin P$. Now

$$R = ((y) + (x, r)) : ((y) + (x, r)) = (x, r) : (y) + (y) : (x, r).$$

Hence, $1 = u + v$, where $uy \in (x, r)$ and $vx \in (y)$. If $v \in R \backslash P$ then we have what we want. If $v \in P$ then $u \in R \backslash P$ and $suy \in (sx, sr) \subseteq (x)$ and $su \in R \backslash P$. Hence, the desired conclusion holds in this case also. Therefore, (11) implies (2).

The proof that (12) implies (2) is the same, for if (12) holds, then

$$R = ((x) \cap (r)) : ((x) \cap (r)) = (x) : (r) + (r) : (x)$$

and

$$R = ((x, r) \cap (y)) : ((x, r) \cap (y)) = (x, r) : (y) + (y) : (x, r).$$

Now assume that (13) holds. Since

$$\begin{aligned} (x) &= (x) \cap ((y, r) + (x - y)) = (x) \cap (y, r) + (x) \cap (y - x) \\ &= (x) \cap (y) + (x) \cap (r) + (x) \cap (y - x), \end{aligned}$$

we have $x = ux + vr + wx$, where $ux \in (y)$ and $wx = z(y - x)$ for some $z \in R$. Since $x(1 - u - w) = vr$, our choice of r implies that $1 - u - w \in P$. Thus, either $u \in R \backslash P$ or $w \in R \backslash P$. If $u \in R \backslash P$, the desired result follows from $ux \in (y)$. Suppose $w \in R \backslash P$ and consider $(w + z)x = zy$. Either $z \in R \backslash P$, or $z \in P$ in which case $w + z \in R \backslash P$. Thus, once more we draw the desired conclusion. Therefore, (12) implies (2).

This completes the proof of Theorem 10.18.

Because of the equivalence of (1) and (4) in Theorem 10.18, we have the following result.

10.19. Corollary. *Every overring of a Prüfer ring is a Prüfer ring.*

The next theorem shows that Prüfer rings have an ideal theory similar to that of Prüfer domains.

10.20. Theorem. *Let R be a Prüfer ring and T an overring of R. Let Δ be the set of regular prime ideals P of R such that $PT \neq T$. Then,*

(1) *for every regular maximal ideal M of T, $T_{[M]} = R_{[P]}$, where $P = M \cap R$, and $M = [P]R_{[P]} \cap T$,*
(2) *if P is a regular prime ideal of R, then $P \in \Delta$ if and only if $T \subseteq R_{[P]}$, and $T = \bigcap_{P \in \Delta} R_{[P]}$,*
(3) *$(A \cap R)T = A$ for every regular ideal A of T, and*
(4) *$\{PT \mid P \in \Delta\}$ is the set of regular prime ideals of T.*

Proof. By Corollary 10.19, T is a Prüfer ring. Let M be a regular maximal ideal of T. Then $(T_{[M]}, [M]T_{[M]})$ is a valuation pair. Since $R_{[P]} \subseteq T_{[M]}$, where $P = M \cap R$, and $R_{[P]}$ is a Prüfer ring, it follows from Exercise 12(b) that

$$T_{[M]} = (R_{[P]})_{[[Q]R_{[P]}]}$$

for some regular prime ideal $Q \subseteq P$ (prove this); in fact, Q is the regular prime ideal such that $[Q]R_{[P]} = [M]\, T_{[M]} \cap R_{[P]}$ [see Exercise 5(a)]. Then $T_{[M]} = R_{[Q]}$ and $[M]T_{[M]} = [Q]R_{[Q]}$. Thus,

$$Q = [Q]R_{[Q]} \cap R = [M]T_{[M]} \cap R = [M]T_{[M]} \cap T \cap R = M \cap R = P,$$

and so $T_{[M]} = R_{[P]}$ and $M = [M]T_{[M]} \cap T = [P]R_{[P]} \cap T$.

The first part of (2) follows from Exercise 11(a) since T, being an overring of R, is a flat R-module. By Exercise 6(c), $T = \bigcap T_{[M]}$, where M runs over the set of regular maximal ideals of T. Therefore,

$$T = \bigcap R_{[M \cap R]} = \bigcap_{P \in \Delta} R_{[P]}.$$

Now let A be a regular ideal of T. By Exercise 6(c),

$$A = \bigcap ([A]T_{[M]} \cap T),$$

where M runs over the set of regular maximal ideals of T. For each M, $T_{[M]} = R_{[P]}$ where $P = M \cap R$. If $x \in [A]T_{[M]}$, then there exists $s \in R \backslash P$ such that $xs \in A$. But $A \subseteq R_{[P]}$, so there exists $t \in R \backslash P$ such that $xst \in A \cap R$. Consequently,

$$x \in [A \cap R]R_{[P]} = [(A \cap R)T]T_{[M]}.$$

Since $(A \cap R)T \subseteq A$ we have, therefore,

$$[A]T_{[M]} = [(A \cap R)T]T_{[M]}.$$

The assertion of (3) now follows from another application of Exercise 6(c).

By (3) if Q is a regular prime ideal of T, then $Q = (Q \cap R)T$, and $Q \cap R \in \Delta$. On the other hand, if $P \in \Delta$, then $T \subseteq R_{[P]}$, and hence $[P]R_{[P]} \cap T$ is a prime ideal of T. Now, by (3) again,

$$[P]R_{[P]} \cap T = ([P]R_{[P]} \cap T \cap R)T = PT.$$

Thus, PT is a regular prime ideal of T.

If we turn now to the primary ideals of a Prüfer ring, we can recover most of the results which were obtained in the last part of Section 2 of Chapter VI.

Let R be a Prüfer ring and let P be a regular prime ideal of R such that P is not the only P-primary ideal of R. If there exist regular prime ideals of R properly contained in P, let P_0 be their union. Since $(R_{[P]}, [P]R_{[P]})$ is a valuation pair by Exercise 12(a), the set of ideals of R_P containing $C(P)R_P$ is totally ordered. Thus, P_0 is a prime ideal of R. If there are no regular prime ideals of R properly contained in P, set $P_0 = C(P)$. In either case, there are no regular prime ideals of R strictly between P_0 and P. Furthermore, P_0 is the intersection of the P-primary ideals of R.

Since $[P_0]R_{[P]}$ is a prime ideal of $R_{[P]}$, the ring $R_{[P]}/[P_0]R_{[P]}$ is a rank one valuation ring. Thus, by Exercise 6(b), $R_P/P_0 R_P$ is a rank one valuation ring. There exists a one-to-one correspondence between the P-primary ideals of R and the PR_P/P_0R_P-primary ideals of R_P/P_0R_P. This correspondence preserves residuals.

In Theorem 6.8 we proved that the product of P-primary ideals of a Prüfer domain is a P-primary ideal. Now we shall obtain the same result for regular prime ideals of a Prüfer ring.

10.21. Theorem. *Let P be a regular prime ideal of a Prüfer ring R. Every product of P-primary ideals of R is P-primary.*

Proof. We may assume that P is a proper prime ideal of R. Let M be a maximal ideal of R which contains P, and let Q_1 and Q_2 be P-primary ideals of R. It suffices to show that $Q_1Q_2R_M$ is PR_M-primary. Since $(R_{[M]}, [M]R_{[M]})$ is a valuation pair, $C(M)$ is contained in every regular ideal of R not meeting $R\backslash M$. Hence, $C(M) \subseteq Q_1Q_2$. Using various parts of Exercise 9, we can show that $C(M)$ is a prime ideal of R. Then, since the set of ideals of $R_M/C(M)R_M$ is totally ordered, this ring is a valuation ring. Thus, $Q_1Q_2R_M/C(M)R_M$ is $PR_M/C(M)R_M$-primary. Therefore $Q_1Q_2R_M$ is PR_M-primary.

We can now prove the following theorem in the same way that we proved a similar theorem in Section 2 of Chapter VI.

10.22. Theorem. *Let R be a Prüfer ring and let P be a regular prime ideal of R such that P is not the only P-primary ideal of R. If Q and Q_1 are P-primary ideals of R, then:*

(1) $\bigcap_{n=1}^{\infty} Q^n$ *is a prime ideal.*
(2) $Q^n = Q^{n+1}$ *for some $n \geq 1$ implies that $Q = Q^2 = P$.*
(3) $Q \subseteq Q_1 \subset P$ *implies that $Q_1{}^n \subseteq Q$ for some n.*
(4) $P^2 \subset P$ *implies that $Q = P^n$ for some n.*
(5) $Q \subset P$ *implies that $Q^2 \subset QP$.*
(6) $Q \subset Q_1$ *and $Q : Q_1 = Q$ imply that $Q_1 = P = P^2$.*

EXERCISES

1. The valuation determined by a valuation pair.
Let (A, P) be a valuation pair of a ring R, and let v be the valuation on R constructed in the proof of Theorem 10.4.

(a) Give the details of the proofs of the following facts used in the proof of Theorem 10.4:

(1) $\leq$ is well-defined;

(2) If $v(x) \neq v(y)$, then $v(x) < v(y)$ or $v(y) < v(x)$;

(3) If $v(x) < v(y)$ and $v(z) \neq v(0)$, then $v(xz) < v(yz)$.

(b) Let w be a valuation on R such that $A = R_w \neq R$ and $P = P_w$. Show that there is an order-preserving isomorphism $\phi : w(R)\backslash\{w(0)\} \to v(R)\backslash\{v(0)\}$ such that $\phi w = v$.

2. Maximal partial homomorphisms and valuation pairs.
Let R be a ring. A **partial homomorphism** of R is defined exactly the same way as a partial homomorphism of a field (see Section 1 of Chapter V). Let A be a subring of R and let P be a proper prime ideal of A. Show that (A, P) is a valuation pair of R if and only if there is a homomorphism ϕ from A into an algebraically closed field, with kernel P, and such that (ϕ, A) is a maximal partial homomorphism of R. [Hint. When (ϕ, A) is a maximal partial homomorphism and (A, P) is dominated by a maximal pair (B, Q), consider two cases: (i) For each $b \in B\backslash A$, there is a polynomial

$$f(X) = \sum_{i=0}^{n} a_i X^i \in A[X]$$

such that $f(b) = 0$ and $a_i \notin P$ for some $i = 0, 1, \ldots, n$; and (ii), the negation of (i).]

3. A counterexample.
Let K be the field of rational numbers and let $R = K[X]$. Let p be a rational prime and let A_p be the subring of K consisting of all rational numbers which can be written in the form m/n where p does not divide n.

(a) Show that if $A = A_p[X]$ and $P = pA$, then (A, P) is a valuation pair.

(b) Show that $B = P + AX$ is a proper prime ideal of A, and $P \subset B$.

4. v-closed prime ideals.
Let (R_v, P_v) be a valuation pair of a ring R, and let v be the valuation on R determined by this pair.

(a) Show that a prime ideal Q of R_v is v-closed if and only if $A_v \subseteq Q \subseteq P_v$.

(b) Show that the set of v-closed ideals of R_v is totally ordered.
(c) Show that the set of v-closed ideals of R_v is closed under arbitrary unions and intersections.
(d) Show that the set of v-closed prime ideals of R_v has the same property.
(e) Let Q be a v-closed prime ideal of R_v and let

$$H = \{\alpha \mid \alpha \in G_v \quad \text{and} \quad v(x) > \max\{\alpha, -\alpha\} \quad \text{for all} \quad x \in Q\}.$$

Show that $P_{v_H} = Q$ and $R_{v_H} = \{x \mid x \in R \text{ and } xQ \subseteq Q\}$.

5. Extensions of ideals to large quotient rings.
Let R be a ring and let P be a proper prime ideal of R.
(a) If Q is a primary ideal of R contained in P, show that $[Q]R_{[P]}$ is a primary ideal of $R_{[P]}$ and that $Q = [Q]R_{[P]} \cap R$.
(b) Show that every primary ideal of $R_{[P]}$ which is contained in $[P]R_{[P]}$ is of the form $[A]R_{[P]}$ for some ideal A of R.
(c) If P is a maximal ideal of R, show that $[P^n]R_{[P]} = ([P]R_{[P]})^n$ for each positive integer n.
(d) Suppose $(R_{[P]}, [P]R_{[P]})$ is a valuation pair and that there are no prime ideals of $R_{[P]}$ strictly between $C([P]R_{[P]})$ and $[P]R_{[P]}$. Show that if A and B are finitely generated regular ideals of R, then

$$[A]R_{[P]}[B]R_{[P]} = [AB]R_{[P]}.$$

(e) Show that it is not true in general that if S is a multiplicative system in R, and if A and B are ideals of R, then $[A]R_{[S]}[B]R_{[S]} = [AB]R_{[S]}$. Does it help to require that A and B be regular and/or finitely generated?

6. Large quotient rings.
Let R be a ring.
(a) Let S be a multiplicative system in R. Show that if R is integrally closed, then $R_{[S]}$ is integrally closed.
(b) Let P and Q be prime ideals of R with $Q \subseteq P$. Show that

$$R_{[P]}/[Q]R_{[P]} \cong R_P/QR_P.$$

[Hint. Use Proposition 3.15 and Theorem 3.17.]
(c) Let M run over the set of regular maximal ideals of R. Show that if A is a regular ideal of R, then

$$A = \bigcap ([A]R_{[M]} \cap R).$$

In particular, $R = \bigcap R_{[M]}$.

7. Valuation pairs and integral dependence.
Let R be a ring with total quotient ring K. Let x be an element of K which is not integral over R.

(a) If $z \in K$ is such that $zx \in R$, show that x is not integral over $R[z]$.

(b) If P is a proper prime ideal of R such that for all $y \in K$, $xy \in R$ implies that $y \in P$, show that x is not integral over $R_{[P]}$.

(c) If P is a proper prime ideal of R and if there exists $y \in R[x]$ such that, for some $w \in P$ we have $wy \in R \backslash P$, show that there exists a valuation pair (T, Q) of K such that $R \subseteq T$ and $x \notin T$.

(d) Suppose that for all valuation pairs (T, Q) of K with $R \subseteq T$ we have $x \in T$. Show that there exists an overring R' of R and a maximal ideal M of R' with the following properties:

(1) R' is integrally closed.
(2) Every regular nonunit of R' is in M.
(3) If $z \in K \backslash R'$, then x is integral over $R'[z]$.
(4) For all $y \in K$, $xy \in R'$ if and only if $y, xy \in M$.
(5) If $y \in K \backslash R'$ and $z \in R' \backslash M$, then $zy \in K \backslash R'$.
(6) If $y \in M$, then $x + y$ is a zero-divisor in K.

8. Valuation pairs and rings having few zero-divisors.
Let R be a ring which has few zero-divisors.

(a) Show that if P is a proper regular prime ideal of R, then $R_{[P]} = R_{S(P)}$ (see Exercise 10 of Chapter III).

(b) Show that R is integrally closed if and only if it is an intersection of valuation rings.

(c) Show that there exists a prime ideal P of R such that (R, P) is a valuation pair if and only if R is a quasi-valuation ring.

9. The core of a proper prime ideal.
Let P be a proper prime ideal of a ring R.

(a) Show that if P is regular, then $C(P) \subseteq P$.

(b) Show that if P is regular, then $C(P)$ consists entirely of zero-divisors.

(c) Show that if P is not regular, then $C(P) = R$ and $R_{[P]}$ is the total quotient ring of R.

(d) Show that $[C(P)]R_{[P]} = C([P]R_{[P]})$.

(e) Show that $[C(P)]R_{[P]} \cap R = C(P)$.
(f) If (R, P) is a valuation pair of the total quotient ring of R, show that $C(P) = A_v$, where v is the valuation determined by (R, P).

10. Large quotient rings and valuation pairs.
Let R be a ring and P a maximal ideal of R.
(a) Show that $R_{[P]} = (R_{S(P)})_{[PR_{S(P)}]}$.
(b) Show that $(R_{[P]}, [P]R_{[P]})$ is a valuation pair if and only if there is a prime ideal Q of $R_{S(P)}$ such that every regular nonunit of $R_{S(P)}$ is contained in Q, and $(R_{S(P)})_{[Q]}$ is a valuation ring.
(c) Show that $(R_{[P]}, [P]R_{(P)})$ is a valuation pair if and only if for every pair of ideals A and B of R, at least one of which is regular, either $AR_P \subseteq BR_P$ or $BR_P \subseteq AR_P$.

11. Flat overrings.
Let R be a ring and T an overring of R.
(a) Show that the following statements are equivalent:
  (1) T is a flat R-module.
  (2) For every $x \in T$, $[R: (x)]_R T = T$, where $[R: (x)]_R = \{a \mid a \in R \text{ and } ax \in R\}$.
  (3) For every prime ideal P of R, either $PT = T$ or $T \subseteq R_{[P]}$.
  (4) For every maximal ideal M of T, $R_{[M \cap R]} = T_{[M]}$.
(b) Show that if T is a flat R-module and is integral over R, then $T = R$.

12. Prüfer rings.
Let R be a Prüfer ring.
(a) Show that for every proper prime ideal P of R, $(R_{[P]}, [P]R_{[P]})$ is a valuation pair.
(b) Show that if (T, M) is a valuation pair of the total quotient ring of R such that $R \subseteq T$, then $T = R_{[M \cap R]}$.

13. Generators of an ideal.
Let R be a ring which satisfies the hypothesis of Lemma 10.15. Let A be a finitely generated ideal of R, let c be a regular element in A, and let P be a proper prime ideal of R. Show that there are elements $a_1, \ldots, a_n \in A$ such that $A = (a_1, \ldots, a_n)$, $a_k = c$ for some k, and

$$(a_1, \ldots, a_{k-1})R_P \subseteq a_k R_P \subseteq \cdots \subseteq a_n R_P.$$

14. A local characterization of Prüfer rings.
Let R be a ring and let K be its total quotient ring. Let P be a prime ideal of R and let $S = R\backslash P$. Then R_P can be considered as a subring of $S^{-1}K$.

(a) Prove that $(R_{[P]}, [P]R_{[P]})$ is a valuation pair of K if and only if (R_P, PR_P) is a valuation pair of $S^{-1}K$. [Hint. Define a mapping w on $S^{-1}K$ by $w(x/s) = v(x)$ for $x \in K$, $s \in S$, where v is the valuation of $(R_{[P]}, [P]R_{[P]})$.]

(b) Suppose R is a ring with unique maximal ideal P such that (R, P) is a valuation pair of K with valuation v. Show that $x \in K\backslash R$ implies that $x = 1/b$ where $b \in R$ and $v(b) < \infty$. Then show that every overring of R is of the form $S^{-1}R$ for some multiplicative system S in R.

(c) Let R be a Prüfer ring and let T be an overring of R. Let Q be a prime ideal of T and set $P = Q \cap R$. Prove that $R_P \cong T_Q$.

15. A generalization of Theorem 10.20.
Let R be a Prüfer ring and let T be an overring. Let Δ be the set of prime ideals of R such that $PT \neq T$. Prove the following assertions:

(a) If M is a maximal ideal of T, then $T_M = R_N$, where $N = M \cap R$ and the contraction of NR_N into T is M.

(b) $P \in \Delta$ if and only if $T \subseteq R_{[P]}$.

(c) $T = \bigcap_{P \in \Delta} R_{[P]}$.

(d) For any ideal A of T, $(A \cap R)T = A$. [Hint. Let Q be a prime ideal of T and let $P = Q \cap R$. Show that $AT_Q \cap R \subseteq (A \cap R)R_P \cap T$ (these are contractions and expansions) and use the fact that $A = \bigcap (AT_Q \cap T)$.]

(e) The set of prime ideals in T is precisely $\{PT \mid P \in \Delta\}$.

16. The ideal transform and large quotient rings.
Prove the following results about ideal transforms.

(a) If x is a regular nonunit of a ring R, then $T((x)) = R_{[N]}$, where $N = \{x^n \mid n = 1, 2, \ldots\}$.

(b) Let A be an ideal of a ring R and let R' be a ring such that $R \subseteq R' \subseteq T(A)$. Then there is a one-to-one correspondence between the set of prime ideals P of R not containing A and the set of prime ideals P' of R' not containing AR'. Furthermore, if P corresponds to P', then $P = P' \cap R$ and $R_{[P]} = R'_{[P']}$.

(c) Let a be a nonnilpotent element of R. Then $T((a)) = \bigcap R_{[P_\alpha]}$ where $\{P_\alpha\}$ the set of prime ideals of R such that $(a) \nsubseteq P_\alpha$.

(d) Let A be a finitely generated nonnilpotent ideal of R. Then $T(A) = \bigcap R_{[P_\alpha]}$ where $\{P_\alpha\}$ is the set of prime ideals of R which do not contain A. [Hint. Use the fact that if $A = (a_1, a_2, \ldots, a_n)$, then $T(A) = \bigcap_{i=1}^n T((a_i))$.]

(e) If no maximal ideal of R contains the collection $\{x_\alpha\}$ of regular nonunits of R, then $R = \bigcap T((x_\alpha))$.

(f) If no maximal ideal of R contains all of the regular nonunits of R, then R is a Prüfer ring if and only if $T((x))$ is a Prüfer ring for each regular nonunit x of R.

17. Prüfer valuation rings.
If (R, P) is a valuation pair and R is a Prüfer ring, we call (R, P) a **Prüfer valuation pair** and R a **Prüfer valuation ring**.

(a) Show that if R is a Prüfer ring, if A is a finitely generated regular ideal of R, and if P is a prime ideal of R, then $T(A) \subseteq R_{[P]}$ if and only if $A \nsubseteq P$.

(b) If M and N are regular prime ideals of a Prüfer ring R, prove that $N \subseteq M$ if and only if $R_{[M]} \subseteq R_{[N]}$.

(c) Prove that the following are equivalent:
   (1) (R, P) is a Prüfer valuation pair.
   (2) R is a Prüfer ring with a unique maximal regular ideal P.
   (3) (R, P) is a valuation pair where P is the unique maximal regular ideal of R.

(d) Let (R, P) be a Prüfer valuation pair, let V be an overring of R, and let M be a regular prime ideal of V. Prove that $M \subseteq P$.

(e) Prove that if (R, P) is a Prüfer valuation pair, then every overring of R is a Prüfer valuation ring.

18. Generalized transforms.
A collection $\mathscr{S}$ of ideals of a ring R is said to be multiplicatively closed if $A, B \in \mathscr{S}$ implies that $AB \in \mathscr{S}$. The $\mathscr{S}$-transform $R_{\mathscr{S}}$ of the ring R is the set

$$R_{\mathscr{S}} = \{x \in K \mid xA \subseteq R \quad \text{for some} \quad A \in \mathscr{S}\}.$$

If B is an ideal of R, denote by $B_{\mathscr{S}}$ the set

$$B_{\mathscr{S}} = \{x \in K \mid xA \subseteq B \quad \text{for some} \quad A \in \mathscr{S}\}.$$

(a) Show that large quotient rings and ideal transforms are generalized transforms.

(b) Let $\mathscr{S}$ be a multiplicatively closed collection of ideals of R. Let $\mathscr{F}$ be the set of all prime ideals of R such that $A \nsubseteq P$ for all $A \in \mathscr{S}$ and let $\mathscr{G}$ be the set of all prime ideals P' of $R_{\mathscr{S}}$ such that $AR_{\mathscr{S}} \nsubseteq P'$ for each $A \in \mathscr{S}$. Prove the following:
   (1) If $P \in \mathscr{F}$ and Q is a P-primary ideal of R, then $P_{\mathscr{S}}$ is a prime ideal of $R_{\mathscr{S}}$, $Q_{\mathscr{S}} \cap R = Q$, and $AR_{\mathscr{S}} \nsubseteq Q_{\mathscr{S}}$ for all $A \in \mathscr{S}$.
   (2) If $P' \in \mathscr{G}$ and Q' is a P'-primary ideal of $R_{\mathscr{S}}$, then $Q' = (Q' \cap R)_{\mathscr{S}}$ and $A \nsubseteq Q' \cap R$ for all $A \in \mathscr{S}$.
   (3) $P \mapsto P_{\mathscr{S}}$ is a one-to-one mapping from $\mathscr{F}$ onto $\mathscr{G}$.
   (4) For $P \in \mathscr{F}$, $R_{[P]} = (R_{\mathscr{S}})_{[P_{\mathscr{S}}]}$.

(c) Let $P \in \mathscr{F}$. Show that $Q \mapsto Q_{\mathscr{S}}$ is a one-to-one correspondence between the P-primary ideals of R and the $P_{\mathscr{S}}$-primary ideals of $R_{\mathscr{S}}$ provided any of the following conditions hold:
   (1) Each ideal in $\mathscr{S}$ is finitely-generated.
   (2) Each P-primary ideal of R contains a power of P.
   (3) For each $A \in \mathscr{S}$, $AR_{\mathscr{S}} = R_{\mathscr{S}}$.
   (4) P is a maximal ideal of R.

(d) Let T be an overring of R and prove that the following are equivalent:
   (1) T is a flat R-module.
   (2) There exists a multiplicatively closed collection $\mathscr{S}$ of ideals of R such that $T = R_{\mathscr{S}}$ and $AT = T$ for all $A \in \mathscr{S}$.
   (3) For each proper prime ideal M of T, $T_{[M]} = R_{[M \cap R]}$. [Hint. Use Exercise 11.]

Appendix: Decomposition of Ideals in Noncommutative Rings

It is natural to ask if the primary decomposition theory of commutative rings can be generalized to noncommutative rings. As it turns out, the uniqueness results can be generalized, but the existence theorem cannot. This leads to the search for an appropriate decomposition theory. We shall discuss such a theory, the tertiary ideal theory of Lesieur and Croisot.

Let R be a noncommutative ring, that is, a ring which is not necessarily commutative; we assume, of course, that R has a unity. We shall confine our attention to the ideals of R. Much of what we say can be extended to left ideals or right ideals of R, or to R-modules.

If H and K are subsets of R we set

$$HK = \{\text{finite sums } \sum a_i b_i \mid a_i \in H,\, b_i \in K\}.$$

If L is another subset of R, then $(HK)L = H(KL)$. If H is a left ideal of R, so is HK; if K is a right ideal of R, so is HK. If $H = \{a\}$, we will write aK for HK; if $K = \{b\}$, we write Hb for HK. Note that $(a) = RaR$.

A.1. Definition. *An ideal P of R is a* **prime ideal** *if $AB \subseteq P$, where A and B are ideals of R, implies that either $A \subseteq P$ or $B \subseteq P$.*

A.2. Proposition. *Let P be an ideal of R. Then P is a prime ideal if and only if $aRb \subseteq P$, where $a, b \in R$, implies that either $a \in P$ or $b \in P$.*

This assertion, and certain other assertions made in this appendix, will be left unproved. Their proofs may be considered as exercises.

A nonempty subset M of R is called an m-**system** if for each pair of elements $a, b \in M$, there is an element $x \in R$ such that $axb \in M$. A proper ideal P of R is prime if and only if $R \backslash P$ is an m-system.

A.3. Definition. *Let A be an ideal of R. The* **prime radical** *of A, denoted by* $\operatorname{Rad}(A)$, *is the set of all $a \in R$ such that every m-system of R which contains a meets A.*

If A is an ideal of R, then

(i) $A \subseteq \operatorname{Rad}(A)$,
(ii) $\operatorname{Rad}(\operatorname{Rad}(A)) = \operatorname{Rad}(A)$, and
(iii) if P is a prime ideal of R and if $A \subseteq P$, then $\operatorname{Rad}(A) \subseteq P$.

A.4. Proposition. *Let A be an ideal of R. Then* $\operatorname{Rad}(A)$ *is the intersection of all of the prime ideals of R that contain A.*

Proof. We must show that if $a \notin \operatorname{Rad}(A)$, then there is a prime ideal P of R such that $A \subseteq P$ and $a \notin P$. Since $a \notin \operatorname{Rad}(A)$ there is an m-system M such that $a \in M$ and $M \cap A$ is empty. By Zorn's lemma, there is an ideal P of R maximal with respect to the properties that $A \subseteq P$ and $M \cap P$ is empty. We shall show that P is prime. Let $bRc \subseteq P$ and suppose $b \notin P$ and $c \notin P$. Then $A \subseteq P \subset P + (b)$ and $A \subseteq P \subset P + (c)$. Consequently, there are elements $x, y \in M$ such that $x \in P + (b)$ and $y \in P + (c)$. For some $z \in R$, we have $xzy \in M$, and

$$\begin{aligned} xzy &\in (P + (b))z(P + (c)) \\ &= PzP + Pzc + bzP + (b)z(c) \\ &\subseteq P + RbRcR \subseteq P, \end{aligned}$$

which contradicts the fact that $M \cap P$ is empty. Thus, $bRc \subseteq P$ implies $b \in P$ or $c \in P$; hence P is a prime ideal.

A.5. Corollary. *If A is an ideal of R, then* $\operatorname{Rad}(A)$ *is an ideal of R.*

A.6. Definition. *An ideal Q of R is a* **primary ideal** *if $AB \subseteq Q$ and $B \nsubseteq Q$, where A and B are ideals of R, imply that $A \subseteq \text{Rad}(Q)$.*

A.7. Proposition. *Let Q be an ideal of R. Then Q is a primary ideal if and only if $aRb \subseteq Q$ and $b \notin Q$, where $a, b \in R$, imply that $a \in \text{Rad}(Q)$.*

We shall call R a **Noetherian ring** if the ascending chain condition holds for ideals of R. Note that if R is a left (or right) Noetherian ring, then it is a Noetherian ring.

A.8. Proposition. *Let R be a Noetherian ring and let A be an ideal of R. Then there is a positive integer n such that $(\text{Rad}(A))^n \subseteq A$.*

This assertion is a consequence of Exercise 2(a) and Proposition A.4.

A.9. Proposition. *Let R be a Noetherian ring and let Q be an ideal of R. Then Q is a primary ideal if and only if $AB \subseteq Q$ and $B \nsubseteq Q$, where A and B are ideals of R, imply that there is a positive integer n such that $A^n \subseteq Q$.*

This completes our preliminary list of definitions and simple results. If R is a commutative ring, the definitions of prime ideal and primary ideal are equivalent to the ones given in Chapter II. If R is commutative and Noetherian, then every ideal of R is an intersection of a finite number of primary ideals. The following example shows that this may not be the case when R is not commutative.

Let K be a field and let R be a 3-dimensional vector space over K. Let the elements x, y, z of R form a basis of R over K. We define a multiplication in R in the following manner. First of all, we multiply x, y, and z according to the table

	x	y	z
x	x	0	z
y	0	y	0
z	0	z	0

We extend this operation to all of R by requiring that the distributive laws hold. Thus,

$$(ax + by + cz)(dx + ey + fz) = adx + bey + (af + ce)z.$$

Then the associative law holds, so that R is a ring. The unity of R is $x + y$. The mapping $a \mapsto ax + ay$ is an injective homomorphism from K into R. We identify each element of K with its image under this homomorphism, and thus consider K as a subring of R.

Each ideal of R is a subspace of the vector space R. Therefore, R is Noetherian. However, the ideal 0 is not an intersection of primary ideals of R. For, let Q be a primary ideal of R. We have

$$y(ax + by + cz)z = 0$$

so that $yRz \subseteq Q$. Hence, either $z \in Q$ or $y^n \in Q$ for some positive integer n. But $y^n = y$ for all n, and if $y \in Q$ then $z = zy \in Q$. Therefore, in any case, $z \in Q$. Thus, no intersection of primary ideals of R can be the ideal 0.

If A and B are ideals of R, the **right residual of A by B** is defined to be the set

$$A \mathbin{.\cdot} B = \{x \mid x \in R \quad \text{and} \quad Bx \subseteq A\}.$$

The **left residual of A by B** is defined to be the set

$$A \mathbin{\cdot .} B = \{x \mid x \in R \quad \text{and} \quad xB \subseteq A\}.$$

Both $A \mathbin{.\cdot} B$ and $A \mathbin{\cdot .} B$ are ideals of R. If C is an ideal of R, then $BC \subseteq A$ if and only if $C \subseteq A \mathbin{.\cdot} B$, and $CB \subseteq A$ if and only if $C \subseteq A \mathbin{\cdot .} B$. The left and right residual operations have properties like those given in Proposition 2.3. Note that $(A \mathbin{.\cdot} B) \mathbin{.\cdot} C = A \mathbin{.\cdot} BC$ and $(A \mathbin{\cdot .} B) \mathbin{\cdot .} C = A \mathbin{\cdot .} CB$.

A.10. Theorem. *If R is a Noetherian ring, then the following statements are equivalent:*

(1) *Every ideal of R is an intersection of a finite number of primary ideals.*

(2) *If A and B are ideals of R, then there is a positive integer n such that $A^n \cap B \subseteq AB$.*

(3) *If A and B are ideals of R, then there is a positive integer n such that $A = (A + B^n) \cap (A \mathbin{.\cdot} B^n)$.*

Proof. (1) $\Rightarrow$ (2). Let A and B be ideals of R and assume that $AB = Q_1 \cap \cdots \cap Q_k$, where Q_i is a primary ideal of R for $i = 1, \ldots, k$. For each i, $AB \subseteq Q_i$; hence, either $B \subseteq Q_i$ or there is a positive integer n_i such that $A^{n_i} \subseteq Q_i$. If $B \subseteq Q_i$, set $n_i = 1$, and let $n = \max(n_1, \ldots, n_k)$. Then

$$A^n \cap B \subseteq Q_1 \cap \cdots \cap Q_k = AB.$$

(2) $\Rightarrow$ (3). Assume (2) holds and let A and B be ideals of R. We have $A \mathbin{.\cdot} B \subseteq A \mathbin{.\cdot} B^2 \subseteq \ldots$, and so there is a positive integer t such that $A \mathbin{.\cdot} B^t = A \mathbin{.\cdot} B^{t+1} = \cdots$. By (2), there is a positive integer s such that

$$B^{st} \cap (A \mathbin{.\cdot} B^t) \subseteq B^t(A \mathbin{.\cdot} B^t) \subseteq A.$$

If $n \geq st$, then $B^n \cap (A \mathbin{.\cdot} B^n) \subseteq B^{st} \cap (A \mathbin{.\cdot} B^t) \subseteq A$, since $A \mathbin{.\cdot} B^n = A \mathbin{.\cdot} B^t$. Since $A \subseteq A \mathbin{.\cdot} B^n$, we have by the modular law (Theorem 1.3),

$$(A + B^n) \cap (A \mathbin{.\cdot} B^n) = A + (B^n \cap (A \mathbin{.\cdot} B^n)) \subseteq A.$$

The reverse containment always holds, so that we have the asserted equality.

(3) $\Rightarrow$ (1). Assume (3) holds. Since R is Noetherian, every ideal of R is an intersection of a finite number of irreducible ideals; the proof is exactly the same as in the case of commutative rings (see Proposition 2.6). Thus, to prove (1) it is sufficient to show that an irreducible ideal Q of R is primary. Let $AB \subseteq Q$, where A and B are ideals of R, and assume that $A^n \nsubseteq Q$ for every positive integer n. By (3), there is a positive integer n such that $Q = (Q + A^n) \cap (Q \mathbin{.\cdot} A^n)$. Since $Q \subset Q + A^n$, and Q is irreducible, we must have $Q \mathbin{.\cdot} A^n = Q$. Since $B \subseteq Q \mathbin{.\cdot} A^n$ it follows that Q is primary.

Since primary decompositions may fail to exist in a Noetherian ring, it is natural to look for some other class of ideals which have some algebraic significance and such that every ideal of a Noetherian ring is an intersection of a finite number of ideals in this class of ideals. This leads us to the notions of the tertiary radical of an ideal and tertiary ideal.

Throughout the rest of the appendix, R is a Noetherian ring. Let A be an ideal of R. The set of all ideals B of R such that for an ideal C of R,

$$(A \mathbin{.\cdot} B) \cap C \subseteq A \quad \text{implies that} \quad C \subseteq A,$$

is not empty since A belongs to this set. Hence, this set has a maximal element B. Let B' be an arbitrary ideal of this set and suppose that $(A \mathbin{.\cdot} (B + B')) \cap C \subseteq A$. Then $(A \mathbin{.\cdot} B) \cap (A \mathbin{.\cdot} B') \cap C \subseteq A$; hence $(A \mathbin{.\cdot} B') \cap C \subseteq A$ and therefore $C \subseteq A$. Thus, $B + B'$ belongs to the set of ideals in question; hence $B + B' = B$, that is, $B' \subseteq B$.

A.11. Definition. *The unique maximal element of the set of ideals described in the preceding paragraph is the* **tertiary radical** *of* A; *it is denoted by* $\mathrm{Ter}(A)$. *An ideal* A *of* R *is a* **tertiary ideal** *if* $BC \subseteq A$ *and* $C \nsubseteq A$, *where* B *and* C *are ideals of* R, *imply that* $B \subseteq \mathrm{Ter}(A)$.

A.12. Theorem. *Every ideal of* R *is an intersection of a finite number of tertiary ideals of* R.

Proof. We shall show that an irreducible ideal A of R is tertiary. Let B and C be ideals of R such that $BC \subseteq A$ and $C \nsubseteq A$. Then, since $B(A + C) = BA + BC \subseteq A$, we have $A \subset A + C \subseteq A \mathbin{.\cdot} B$. Now, suppose that $(A \mathbin{.\cdot} B) \cap D \subseteq A$, where D is an ideal of R. Then, by the modular law.

$$A = ((A \mathbin{.\cdot} B) \cap D) + A = (A \mathbin{.\cdot} B) \cap (D + A).$$

Since A is irreducible, this implies that $D + A = A$, that is, $D \subseteq A$. Therefore, $B \subseteq \mathrm{Ter}(A)$.

We can now give one more necessary and sufficient for the existence of primary decompositions of ideals of R.

A.13. Theorem. *Every ideal of* R *is an intersection of a finite number of primary ideals of* R *if and only if every tertiary ideal of* R *is primary.*

Proof. The sufficiency of the condition is obvious in view of Theorem A.12. Conversely, suppose every ideal of R is an intersection of a finite number of primary ideals of R. Let A be an ideal of R. By Theorem A.10, there is a positive integer n such that

$$A = (A + (\mathrm{Ter}(A))^n) \cap (A \mathrel{.\cdot} (\mathrm{Ter}(A))^n).$$

Hence

$$(A \mathrel{.\cdot} (\mathrm{Ter}(A))) \cap (A + (\mathrm{Ter}(A))^n) \subseteq A,$$

and therefore, $A + (\mathrm{Ter}(A))^n \subseteq A$, that is, $(\mathrm{Ter}(A))^n \subseteq A$. It follows, by Proposition A.4, that $\mathrm{Ter}(A) \subseteq \mathrm{Rad}(A)$. Now suppose that A is tertiary and that $BC \subseteq A$ and $C \nsubseteq A$, where B and C are ideals of R. Then $B \subseteq \mathrm{Ter}(A)$, so $B \subseteq \mathrm{Rad}(A)$. Therefore, A is primary.

It is sometimes convenient to have an elementwise description of the tertiary radical of an ideal; we now give such a description.

A.14. Proposition. *If A is an ideal of R then*

$$\mathrm{Ter}(A) = \{a \mid a \in R \text{ and if } b \notin A \text{ then there is an element } c \in (b) \text{ such that } c \notin A \text{ and } (a)(c) \subseteq A\}.$$

Proof. Let a belong to the set on the right and suppose that $(A \mathrel{.\cdot} (a)) \cap C \subseteq A$, where C is some ideal of R. If $C \nsubseteq A$ then there is an element $b \in C$ with $b \notin A$. Hence, there is an element $c \in (b)$ such that $c \notin A$ and $(a)(c) \subseteq A$. Then $c \in (A \mathrel{.\cdot} (a)) \cap C \subseteq A$, which contradicts the fact that $c \notin A$. Thus we do have $C \subseteq A$; we conclude that $a \in \mathrm{Ter}(A)$.

Conversely, let $a \in \mathrm{Ter}(A)$; then $(a) \subseteq \mathrm{Ter}(A)$. If $A = R$, then a belongs to the set on the right. If $A \neq R$, choose $b \in R$ so that $b \notin A$. Then $(b) \nsubseteq A$ and consequently, $(A \mathrel{.\cdot} (a)) \cap (b) \nsubseteq A$. Choose $c \in (A \mathrel{.\cdot} (a)) \cap (b)$ so that $c \notin A$. Then $c \in (b)$ and $(a)(c) \subseteq A$. Therefore, a belongs to the set on the right.

Let A be an ideal of R. A **tertiary decomposition** of A is an expression of A as an intersection of a finite number of tertiary ideals of R.

A.15. Proposition. *Let A be an ideal of R and let $A = T_1 \cap \cdots \cap T_k$ be a tertiary decomposition of A such that no T_i contains the intersection of the remaining T_j. Then,*

$$\mathrm{Ter}(A) = (\mathrm{Ter}(T_1)) \cap \cdots \cap (\mathrm{Ter}(T_k)).$$

Proof. Let a belong to the set on the right and let $b \notin A$. Then $b \notin T_i$ for some i, say $b \notin T_1$. Since $a \in \mathrm{Ter}(T_1)$, there is an element $b_1 \in (b)$ such that $b_1 \notin T_1$ and $(a)(b_1) \subseteq T_1$. If $b_1 \in T_2$, then $b_1 \notin T_1 \cap T_2$ and $(a)(b_1) \in T_1 \cap T_2$. If $b_1 \notin T_2$, then, since $a \in \mathrm{Ter}(T_2)$, there is an element $c_1 \in (b_1)$ such that $c_1 \notin T_2$ and $(a)(c_1) \subseteq T_2$. Then $c_1 \notin T_1 \cap T_2$ and $(a)(c_1) \subseteq T_1 \cap T_2$. Let $b_2 = b_1$ if $b_1 \in T_2$ and $b_2 = c_1$ if $b_1 \notin T_2$. Then $b_2 \in (b)$, $b_2 \notin T_1 \cap T_2$, and $(a)(b_2) \subseteq T_1 \cap T_2$. If we repeat this argument several times we will obtain an element $b_k \in (b)$ such that $b_k \notin T_1 \cap \cdots \cap T_k = A$ and $(a)(b_k) \subseteq A$. Therefore, $a \in \mathrm{Ter}(A)$.

Conversely, let $a \in \mathrm{Ter}(A)$. Let $b \in T_2 \cap \cdots \cap T_k$ and $b \notin T_1$ (if $k = 1$ the assertion is trivially true). Since $b \notin A$, there is an element $c \in (b)$ such that $c \notin A$ and $(a)(c) \subseteq A$. If $c \in T_1$, then $c \in T_1 \cap (b) \subseteq T_1 \cap \cdots \cap T_k = A$. Hence, $(c) \notin T_1$ but $(a)(c) \subseteq T_1$. Since T_1 is tertiary, this implies that $(a) \subseteq \mathrm{Ter}(T_1)$. If we apply this argument to each of $T_1, \ldots, T_k$, we conclude that $a \in (\mathrm{Ter}(T_1)) \cap \cdots \cap (\mathrm{Ter}(T_k))$.

A.16. Corollary. *Let T_1 and T_2 be tertiary ideals of R with $\mathrm{Ter}(T_1) = \mathrm{Ter}(T_2)$. Then $T = T_1 \cap T_2$ is a tertiary ideal of R with $\mathrm{Ter}(T) = \mathrm{Ter}(T_1)$.*

Proof. By Proposition A.15, we have $\mathrm{Ter}(T) = (\mathrm{Ter}(T_1)) \cap (\mathrm{Ter}(T_2)) = \mathrm{Ter}(T_1)$. Suppose that $BC \subseteq T$ and $C \nsubseteq T$, where B and C are ideals of R. Then $BC \subseteq T_1 \cap T_2$, and either $C \nsubseteq T_1$ or $C \nsubseteq T_2$. In either case, $B \subseteq \mathrm{Ter}(T)$. Therefore, T is a tertiary ideal.

A tertiary decomposition $A = T_1 \cap \cdots \cap T_k$ of an ideal A of R is called **reduced** if:

(i) no T_i contains the intersection of the remaining T_j, and
(ii) $\mathrm{Ter}(T_i) \neq \mathrm{Ter}(T_j)$ for $i \neq j$.

Because of Corollary A.16, it is evident that every ideal of R has a reduced tertiary decomposition. We shall now obtain a uniqueness result for such decompositions.

A.17. Lemma. *Let A be an ideal of R and suppose that $A = T \cap B = T' \cap B'$, where T and T' are tertiary ideals of R, and B and B' are ideals of R. If $\mathrm{Ter}(T) \neq \mathrm{Ter}(T')$, then $A = B \cap B'$.*

Proof. Since $\text{Ter}(T) \neq \text{Ter}(T')$, we may assume that $\text{Ter}(T) \nsubseteq \text{Ter}(T')$. Then, from $(\text{Ter}(T))(T' \mathbin{.\cdot} (\text{Ter}(T))) \subseteq T'$, it follows that $T' \mathbin{.\cdot} (\text{Ter}(T)) \subseteq T'$. Thus $T' \mathbin{.\cdot} (\text{Ter}(T)) = T'$. Then

$$\begin{aligned}(T \mathbin{.\cdot} (\text{Ter}(T))) &\cap (B \mathbin{.\cdot} (\text{Ter}(T)))\\ &= (T \cap B) \mathbin{.\cdot} (\text{Ter}(T)) = (T' \cap B') \mathbin{.\cdot} (\text{Ter}(T))\\ &= (T' \mathbin{.\cdot} (\text{Ter}(T))) \cap (B' \mathbin{.\cdot} (\text{Ter}(T))) = T' \cap (B' \mathbin{.\cdot} (\text{Ter}(T))).\end{aligned}$$

If we intersect each side of this equality with $B \cap B'$, and use the fact that $B \cap B' \subseteq B \mathbin{.\cdot} (\text{Ter}(T))$ and $B \cap B' \subseteq B' \mathbin{.\cdot} (\text{Ter}(T))$, we obtain

$$\begin{aligned}(T \mathbin{.\cdot} (\text{Ter}(T))) \cap (B \cap B') &= T' \cap (B \cap B')\\ &= (T' \cap B') \cap B = A \cap B = A \subseteq T.\end{aligned}$$

Therefore, $B \cap B' \subseteq T$ and we have

$$A = T \cap (B \cap B') = B \cap B'.$$

A.18. Theorem. *Let A be an ideal of R and let*

$$A = T_1 \cap \cdots \cap T_m = T_1' \cap \cdots \cap T_n'$$

be two reduced tertiary decompositions of A. Then, $m = n$ and the T_i and T_i' can be so numbered that $\text{Ter}(T_i) = \text{Ter}(T_i')$ *for $i = 1, \ldots, n$.*

Proof. The assertion is obviously true if $A = R$; hence we assume that $A \neq R$. Since the decompositions are reduced, this implies that $T_i \neq R$ for $i = 1, \ldots, m$ and $T_i' \neq R$ for $i = 1, \ldots, n$. Suppose that $n = 1$. If $m = 1$, then $A = T_1 = T_1'$ and $\text{Ter}(T_1) = \text{Ter}(T_1')$. Suppose that $m > 1$. If $\text{Ter}(T_1) \neq \text{Ter}(T_1')$, then by Lemma A.17 we have $A = T_2 \cap \cdots \cap T_m$ (take $B = T_2 \cap \cdots \cap T_m$ and $B' = R$) which contradicts our assumption that the tertiary decompositions are reduced. Thus $\text{Ter}(T_1) = \text{Ter}(T_1')$. By the same argument, $\text{Ter}(T_2) = \text{Ter}(T_1')$, again a contradiction. Thus, $n = 1$ if and only if $m = 1$.

Now suppose that $n > 1$. Suppose further that $\text{Ter}(T_1) \neq \text{Ter}(T_i')$ for $i = 1, \ldots, n$. Then, by Lemma A.17,

$$\begin{aligned}A &= T_2 \cap \cdots \cap T_m \cap T_2' \cap \cdots \cap T_n'\\ &= T_2 \cap \cdots \cap T_m \cap T_3' \cap \cdots \cap T_n'\\ &= \cdots\\ &= T_2 \cap \cdots \cap T_m \cap T_n'\\ &= T_2 \cap \cdots \cap T_m,\end{aligned}$$

once more a contradiction. Thus $\mathrm{Ter}(T_1) = \mathrm{Ter}(T_i')$ for some i. By the same argument, for each $j = 2, \ldots, m$, $\mathrm{Ter}(T_j) = \mathrm{Ter}(T_i')$ for some i, and for each $j = 1, \ldots, n$, $\mathrm{Ter}(T_j') = \mathrm{Ter}(T_i)$ for some i. Hence, $m = n$, and the assertion is proved.

The remainder of the appendix will consist of a proof that the tertiary radical of a tertiary ideal is a prime ideal. Several exercises expand on the ideas used in this proof.

Let A be a proper ideal of R. If B is an ideal of R such that $B \nsubseteq A$, then $A \cdot . B$ is called a **proper left residual** of A.

A.19. Proposition. *An ideal C of R is a proper left residual of a proper ideal A of R if and only if $A \cdot . (A . \cdot C) = C$ and $A \subset A . \cdot C$.*

Proof. The sufficiency of the condition is clear. Conversely, suppose that C is a proper left residual of A and let $C = A \cdot . B$ where B is an ideal of R and $B \nsubseteq A$. Then $CB \subseteq A$ and so $B \subseteq A . \cdot C$. Hence, $A \cdot . (A . \cdot C) \subseteq A \cdot . B = C$. On the other hand, since $C(A . \cdot C) \subseteq A$, we have $C \subseteq A \cdot .(A . \cdot C)$. Therefore, $A \cdot . (A . \cdot C) = C$. If $A = A . \cdot C$, then $C = R$ and $B \subseteq A$, contrary to fact. Hence, $A \subset A . \cdot C$.

A.20. Proposition. *If A is a proper ideal of R, then A has a prime proper left residual.*

Proof. Since $A \neq R$, the set of proper left residuals of A is not empty. Hence, this set has a maximal element P. Let $P = A \cdot . B$, where B is an ideal of R and $B \nsubseteq A$. Suppose that $CD \subseteq P$, where C and D are ideals of R, and suppose that $D \nsubseteq P$. We have $CDB \subseteq PB \subseteq A$, so $C \subseteq A \cdot . DB$. Since $DB \subseteq B$, we have $A \cdot . B \subseteq A \cdot . DB$, and it follows from the maximality of $A \cdot . B$ that $P = A \cdot . DB$. Thus, $C \subseteq P$. Therefore, P is a prime proper left residual of A.

A.21. Proposition. *Let A be a proper ideal of R. Then $A \subset A . \cdot B$ if and only if $B \subseteq P$ for some prime proper left residual P of A.*

Proof. Suppose $A \subset A . \cdot B$. Since $B(A . \cdot B) \subseteq A$, we have $B \subseteq A \cdot . (A . \cdot B)$; this is a proper left residual of A. It is contained in a maximal proper left residual of A, which is prime by the proof of

Proposition A.20. Conversely, suppose $B \subseteq P = A \cdot . C$, where C is an ideal of R and $C \nsubseteq A$, and P is a prime ideal of R. From $(A \cdot . C)C \subseteq A$, it follows that $C \subseteq A . \cdot (A \cdot . C) = A . \cdot P \subseteq A . \cdot B$. If $A = A . \cdot B$, then $C \subseteq A$, which is not true. Hence, $A \subset A . \cdot B$.

A.22. Definition. *Let A be a proper ideal of R. An ideal C of R is an* **essential left residual** *of A if there is an ideal B of R with $A \subset B$ such that $C = A \cdot . B$ and such that $A \subset B' \subseteq B$, where B' is an ideal of R, implies that $A \cdot . B' = A \cdot . B$.*

A.23. Proposition. *Let A be a proper ideal of R. Every maximal proper left residual of A is an essential left residual of A. Every essential left residual of A is a prime proper left residual of A.*

Proof. Let C be a maximal proper left residual of A. By Proposition A.19, $C = A \cdot . B$, where B is an ideal of R such that $A \subset B$. Suppose that $A \subset B' \subseteq B$, where B' is an ideal of R. Then $A \cdot . B \subseteq A \cdot . B'$, so we must have $A \cdot . B' = A \cdot . B$.

Now suppose that P is an essential left residual of A; let $P = A \cdot . B$ where B is an ideal of R satisfying the condition in Definition A.22. Let $CD \subseteq P$, where C and D are ideals of R and $D \nsubseteq P$. Then $CDB \subseteq PB \subseteq A$, and so $C \subseteq A \cdot . DB$. Since $DB \nsubseteq A$ and $DB \subseteq B$, it follows that $A \subset A + DB \subseteq B$. Consequently,

$$P = A \cdot . B = A \cdot . (A + DB) = A \cdot . DB.$$

Thus, $C \subseteq P$. Therefore, P is a prime ideal of R.

A.24. Theorem. *A proper ideal T of R is tertiary if and only if* $\mathrm{Ter}(T)$ *is the unique essential left residual of T.*

Proof. Let C be an essential left residual of T. Let $C = T \cdot . B$, where B is an ideal of R satisfying the condition of Definition A.22. We have

$$\begin{aligned} \mathrm{Ter}(T) &\subseteq T \cdot . (T . \cdot (\mathrm{Ter}(T))) \\ &\subseteq T \cdot . ((T . \cdot (\mathrm{Ter}(T))) \cap B) \end{aligned}$$

and

$$T \subseteq (T \,.\!\cdot\, (\mathrm{Ter}(T))) \cap B \subseteq B.$$

If $T = (T \,.\!\cdot\, (\mathrm{Ter}(T))) \cap B$, then from the definition of $\mathrm{Ter}(T)$ it follows that $B = T$, which is not true. Hence, $T \subset (T \,.\!\cdot\, (\mathrm{Ter}(T))) \cap B$, and consequently

$$\mathrm{Ter}(T) \subseteq T \cdot\!. \,((T \,.\!\cdot\, (\mathrm{Ter}(T))) \cap B) = T \cdot\!. \, B = C.$$

Note that in this part of the proof we have not used the fact that T is tertiary.

Now, assume that T is tertiary. By Propositions A.23 and A.19, $T \subset T \,.\!\cdot\, C$. Since $C(T \,.\!\cdot\, C) \subseteq T$, it follows that $C \subseteq \mathrm{Ter}(T)$. Therefore, $\mathrm{Ter}(T) = C$, and C is the unique essential left residual of T.

Conversely, suppose that $\mathrm{Ter}(T)$ is the unique essential left residual of T. Suppose also that $BC \subseteq T$ and $B \nsubseteq \mathrm{Ter}(T)$, where B and C are ideals of R. By Proposition A.23, B is not contained in any prime proper left residual of T. Hence $T \,.\!\cdot\, B = T$ by Proposition A.21. But $C \subseteq T \,.\!\cdot\, B$, so $C \subseteq T$. Therefore, T is a tertiary ideal of R.

A.25. Corollary. *If T is a tertiary ideal of R, then* $\mathrm{Ter}(T)$ *is a prime ideal of R.*

EXERCISES

1. Prime ideals and the prime radical.
Let R be a ring and let P and A be ideals of R.

(a) Show that the following statements are equivalent:

(1) P is a prime ideal of R.

(2) $(a)(b) \subseteq P$, where $a, b \in R$, implies that either $a \in P$ or $b \in P$.

(3) $BC \subseteq P$, where B and C are left ideals of R, implies that either $B \subseteq P$ or $C \subseteq P$.

(4) $BC \subseteq P$, where B and C are right ideals of R, implies that either $B \subseteq P$ or $C \subseteq P$.

(b) Define **minimal prime divisor** of A in exactly the same way as in the commutative case. Show that every prime ideal of R which contains A contains a minimal prime divisor of A; conclude that $\mathrm{Rad}(A)$ is the intersection of all of the minimal prime divisors of A.

(c) Show that if $a \in \text{Rad}(A)$, then $a^n \in A$ for some positive integer n.

(d) Show that if $a \in R$, then $\{a^{2^n} \mid n \text{ is a positive integer}\}$ is an m-system. Show by example that even when R is commutative, this set need not be multiplicatively closed.

2. Noetherian rings.
Let R be a Noetherian ring.

(a) Show that an ideal A of R contains a product of a finite number of prime ideals of R, each of which contains A.

(b) Show that an ideal A of R has only a finite number of minimal prime divisors.

(c) Suppose that the equivalent conditions of Theorem A.10 hold. Show that if B is an ideal of R and if $A = \bigcap_{n=1}^{\infty} B^n$, then $A = B^m A$ for every positive integer m.

(d) Again, suppose that the equivalent conditions of Theorem A.10 hold. Let $P_1, \ldots, P_k$ be the minimal prime divisors of an ideal A of R. Show that for every large positive integer, n,

$$A = (A + P_1{}^n) \cap \cdots \cap (A + P_k{}^n).$$

3. Residuals.
Let R be a ring and let A, B, and C be ideals of R.

(a) Show that $(A \cdot . B) . \cdot C = (A . \cdot C) \cdot . B$.

(b) Show that $B = A \cdot . (A . \cdot B)$(or $B = A . \cdot (A \cdot . B)$) if and only if $B = A \cdot . D$(or $B = A . \cdot D$) for some ideal D of R.

(c) Assume $A \neq R$ and let P be a prime proper left residual of A. Show that if $B \not\subseteq P$, then P is a prime proper left residual of $A . \cdot B$. (The definition of proper left residual does not really require that R be Noetherian, nor does this result.)

4. Noetherian rings, continued.
Let R be a Noetherian ring.

(a) Show that a proper ideal of R has only a finite number of prime proper left residuals.

(b) Show that a proper ideal Q of R is primary if and only if Q has a unique prime proper left residual P; show that if this is the case, then P is the unique minimal prime divisor of Q, and $\text{Rad}(Q) = P$. Show that every primary ideal of R is a tertiary ideal of R.

(c) Let Q be a proper ideal of R, and let P be an ideal of R such that $P \subseteq \text{Rad}(Q)$ and if $AB \subseteq Q$ and $B \nsubseteq Q$, where A and B are ideals of R, then $A \subseteq P$. Show that P is prime, Q is primary, and $\text{Rad}(Q) = P$.

(d) Show that if $Q_1, \ldots, Q_k$ are primary ideals of R with the same prime radical, then $Q_1 \cap \cdots \cap Q_k$ is also primary with the same prime radical.

(e) Define **reduced primary decomposition** of an ideal A of R in the same way as in the commutative case. Assume A is an intersection of a finite number of primary ideals of R. Show that A has a reduced primary decomposition, and state and prove a result like Corollary 2.27 for A.

5. The tertiary radical and tertiary ideals.
Let R be a Noetherian ring.

(a) Let A be a proper ideal of R. Show that $\text{Ter}(A)$ is the intersection of the essential left residuals of A.

(b) Let T be a proper ideal of R, and let P be an ideal of R such that $P \subseteq \text{Ter}(T)$ and if $AB \subseteq T$ and $B \nsubseteq T$, where A and B are ideals of R, then $A \subseteq P$. Show that P is prime, T is tertiary, and $\text{Ter}(T) = P$.

(c) Give an alternate proof of Corollary A.16, without using Proposition A.15, but using part (b) of this exercise.

(d) Show that an ideal T of R is tertiary if and only if $aRb \subseteq T$ and $b \notin T$, where $a, b \in R$, imply that $a \in \text{Ter}(T)$.

Bibliography

The list of books which follows contains most of the recent books on the theory of rings and several books on related topics such as the theory of fields and algebraic number theory. Expositions of many aspects of the theory of rings may be found also in the many books on abstract algebra which are available.

Following the list of books is a lengthy list of papers, virtually all of which deal directly with some topic mentioned in the text. The book of Krull contains a list of the older papers on commutative rings.

The book of Lesieur and Croisot contains a more detailed exposition of tertiary decomposition theory than that given in the appendix to this book. The list of papers contains several which are concerned with noncommutative rings.

BOOKS

Atiyah, M. F., and **MacDonald, I. G.,** "Introduction to Commutative Algebra." Addison-Wesley, Reading, Massachusetts, 1969.

Bourbaki, N., "Élements de Mathématique, Algébre," Chaps. 4 and 5 (Polynomes et fractions rationnelles; Corps commutatifs). Hermann, Paris, 1959.

Bourbaki, N., "Élements of Mathématique, Algébre Commutative," Chaps. 1 and 2 (Modules Plats; Localisation). Hermann, Paris, 1961.

Bourbaki, N., "Élements de Mathématique, Algébre Commutative," Chaps. 3 and 4 (Graduations, filtrations, et topologies; Idéaux premier associés et décomposition primaire). Hermann, Paris, 1961.

Bourbaki, N., "Élements de Mathématique, Algébre," 3rd ed., Chap. 2 (Algébre linéaire). Hermann, Paris, 1962.

Bourbaki, N., "Élements de Mathématique, Algébre Commutative," Chaps. 5 and 6 (Entiers; Valuations). Hermann, Paris, 1964.

Bourbaki, N., "Élements de Mathématique, Algébre," Chaps. 6 and 7 (Groupes et corps ordonnés; Modules sur les anneaux principaux). Hermann, Paris, 1964.

Bourbaki, N., "Élements de Mathématique, Algébre Commutative," Chap. 7 (Diviseurs). Hermann, Paris, 1965.

Budach, L., "Quotientenfunktoren und Erweiterungstheorie." VEB Deutsch. Verlag Wiss., Berlin, 1967.

Burton, D. A., "A First Course in Rings and Ideals." Addison-Wesley, Reading, Massachusetts, 1970.

Chevalley, C., "Introduction to the Theory of Algebraic Functions of One Variable." Amer. Math. Soc., Providence, Rhode Island, 1950.

Curtis, C. W., and **Reiner, I.**, "Representation Theory of Finite Groups and Associative Algebras." Wiley (Interscience), New York, 1962.

Dedekind, R., "Gesammelte mathematische Werke." Friedr. Vieweg und Sohn, Braunschweig, 1932.

Deuring, M., "Algebren," 2nd ed. Springer-Verlag, Berlin, 1968.

Dirichlet, P. G. L., and **Dedekind, R.**, "Verlesungen uber Zahlentheorie, vierte Auflage." Friedr. Vieweg und Sohn, Braunschweig, 1894.

Endler, O., "Bewertungstheorie unter Benutzung einer Vorlesung von W. Krull." Bonner math. Schriften, Bonn, 1963.

Faith, C., "Algebra: Categories, Rings, and Modules." Saunders, Philadelphia, Pennsylvania, in preparation.

Gilmer, R. W., "Multiplicative Ideal Theory," Part I, Queens Papers on Pure and Applied Mathematics, No. 12. Queens Univ. Press, Kingston, Ontario, 1968.

Gilmer, R. W., "Multiplicative Ideal Theory," Part II, Queens Papers on Pure and Applied Mathematics, No. 12. Queens Univ. Press, Kingston, Ontario, 1968.

Gray, M., "A Radical Approach to Algebra." Addison-Wesley, Reading, Massachusetts, 1970.

Jacobson, N., "Lectures in Abstract Algebra," Vol. III. Van Nostrand Reinhold, Princeton, New Jersey, 1964.

Jaffard, P., "Les Systèmes d'Idéaux." Dunod, Paris, 1960.

Jaffard, P., "Théorie de la Dimension dans les Anneaux de Polynomes." Gauthier-Villars, Paris, 1960.

Kaplansky, I., "Infinite Abelian Groups." Univ. of Michigan Press, Ann Arbor, Michigan, 1954.

Kaplansky, I., "Commutative Rings." Allyn and Bacon, Boston, Massachusetts, 1969.

Krull, W., "Idealtheorie, zweite Auflage." Springer-Verlag, Berlin, 1968.

Lang, S., "Introduction to Algebraic Geometry." Wiley (Interscience), New York, 1958.

Lesieur, L., and **Croisot, R.**, "Algèbre Noethérienne Non Commutative." Mem. Sci. Math., Fasc. CLIV. Gauthier-Villars, Paris, 1963.

Macaulay, F. S., "Algebraic Theory of Modular Systems." Cambridge Univ. Press, London and New York, 1916.

McCarthy, P. J., "Algebraic Extensions of Fields." Ginn (Blaisdell), Boston, Massachusetts, 1966.

McCoy, N. H., "Rings and Ideals." Open Court, LaSalle, Illinois, 1948.

Mann, H. B., "Introduction to Algebraic Number Theory." Ohio State Univ. Press, Columbus, Ohio, 1955.

Nagata, M., "Local Rings." Wiley (Interscience), New York, 1962.

Northcott, D. G., "Ideal Theory." Cambridge Univ. Press, London and New York, 1960.

Northcott, D. G., "Lessons on Rings, Modules, and Multiplicities." Cambridge Univ. Press, London and New York, 1968.

Ribenboim, P., "Théorie des Groupes Ordonnés." Univ. Press, Bahia Blanca, Argentina, 1959.

Ribenboim, P., "Théorie des Valuations." Univ. of Montreal Press, Montreal, 1964.

Ribenboim, P., "Rings and Modules." Wiley (Interscience), New York, 1969.

Roggenkamp, K. H., and **Huber-Dyson, V.**, "Lattices over Orders I." Springer-Verlag, Berlin, 1970.

Samuel, P., "Anneaux Factoriels." Sociedade de Matematica de Sao Paulo, Sao Paulo, 1963.

Samuel, P., "Unique Factorization Domains." Tata Institute of Fundamental Research, Bombay, 1964.

Schenkman, E., "Theory of Groups." Van Nostrand-Reinhold, Princeton, New Jersey, 1965.

Schilling, O. F. G., "The Theory of Valuations." Amer. Math. Soc., New York, 1950.

van der Waerden, B. L., "Modern Algebra," 2nd English ed., Vol. II. Ungar, New York, 1950.

van der Waerden, B. L., "Modern Algebra," 2nd English ed., Vol. I. Ungar, New York, 1953.

Zariski, O., and **Samuel, P.**, "Commutative Algebra," Vol. I. Van Nostrand-Reinhold, Princeton, New Jersey, 1958.

Zariski, O., and **Samuel, P.**, "Commutative Algebra," Vol. II. Van Nostrand-Reinhold, Princeton, New Jersey, 1960.

Papers

Abhayankar, S. S.

1955 Splitting of valuations in extensions of local domains II. *Proc. Nat. Acad. Sci. U.S.A. 41*, 220–223.

1956 Two notes on formal power series. *Proc. Amer. Math. Soc. 7*, 903–905.

Abhayankar, S. S., and **Zariski, O.**

1955 Splitting of valuations in extensions of local domains. *Proc. Nat. Acad. Sci. U.S.A. 41*, 220–223.

Akiba, T.

1964 Remarks on generalized quotient rings. *Proc. Japan Acad. 40*, 801–806.

1965 Remarks on generalized rings of quotients II. *J. Math. Kyoto Univ. 5*, 39–44.

1969 Remarks on generalized rings of quotients III. *J. Math. Kyoto Univ. 9*, 205–212.

Akizuki, Y.

1935a Einige Bemerkungen über primäre Integritätsbereiche mit Teilerkettensatz. *Proc. Phys.-Math. Soc. Japan Ser. III 17*, 327–336.

1935b Teilerkettensatz and Vielfachenkettensatz. *Proc. Phys.-Math. Soc. Japan Ser. III 17*, 337–345.

1937 Zur Idealtheorie Ringbereiche mit dem Teilerkettensatz. *Proc. Imp. Acad. Tokyo 13*, 53–55.

1938a Zur Idealtheorie der einartigen Ringbereiche mit dem Teilerkettensatz II. *Japan J. Math. 14*, 177–187.

1938b Zur Idealtheorie der einartigen Ringbereiche mit dem Teilerkettensatz III. *Japan J. Math. 15*, 1–11.

Alling, N. L.

1965 Rings of continuous integer-valued functions and non-standard arithmetic. *Trans. Amer. Math. Soc. 118*, 498–525.

Amitsur, S. A.

1948 On unique factorization in rings. *Riveon Lematematika 2*, 28–29.

1956 Radicals of polynomial rings. *Canad. J. Math. 8*, 355–361.

Arezzo, M., and **Greco, S.**

1967 Sul gruppo delle classi di ideali. *Ann. Scuola Norm. Sup. Pisa 21*, 459–483.

Arnold, J. T.

1969a On the dimension theory of overrings of an integral domain. *Trans. Amer. Math. Soc. 138*, 313–326.

1969b On the ideal theory of the Kronecker function ring and the domain $D(X)$. *Canad. J. Math. 21*, 558–563.

Arnold, J. T., and **Brewer, J.**

1971a Commutative rings which are locally Noetherian. *J. Math. Kyoto Univ. 11*, 45–49.

1971b Kronecker function rings and flat $D[X]$-modules. *Proc. Amer. Math. Soc. 27*, 326–330.

On flat overrings, ideal transforms, and generalized transforms of a commutative ring. *J. Algebra* (to appear).

Arnold, J. T., and **Gilmer, R. W.**

1967 Idempotent ideals and unions of nets of Prüfer domains. *J. Sci. Hiroshima Univ. Ser. A-I 31*, 131–145.

1970 On the contents of polynomials. *Proc. Amer. Math. Soc. 24*, 556–562.

Artin, E., and **Tate, J. T.**

1951 A note on finite ring extensions. *J. Math. Soc. Japan 3*, 74–77.

Artin, E., and **Whaples, G.**

1945 Axiomatic characterization of fields by the product formula for valuations. *Bull. Amer. Math. Soc. 51*, 469–492.

1946 A note on axiomatic characterizations of fields. *Bull. Amer. Math. Soc. 52*, 245–247.

Asano, K.

1951 Über kommutative Ringe, in denen jedes Ideal als Produkt von Primidealen darstellbar ist. *J. Math. Soc. Japan 3*, 82–90.

Asano, K. and **Nakayama, T.**

1940 A remark on the arithmetic of a subfield. *Proc. Imp. Acad. Tokyo 16*, 529–531.

Aubert, K. E.

1954 Some charactérizations of valuation rings. *Duke Math. J. 21*, 517–525.

1958 Charactérizations des anneaux de valuation à l'aide de la theorie des r-ideaux. Sem. P. Dubreil et C. Pisot *10*, Nr. 10.

Auslander, M., and **Buchsbaum, D.**

1959 Unique factorization in regular local rings. *Proc. Nat. Acad. Sci. U.S.A. 45*, 733–734.

Banaschewski, B.

1961 On the components of ideals in commutative rings. *Arch. Math.* (*Basel*) *12*, 22–29.

Bandyopadhyay, S. P.

1969 Valuations in groups and rings. *Czechoslovak Math. J. 19*, 275–276.

Bang, C. M.

1970 Countably generated modules over complete discrete valuation rings. *J. Algebra 14*, 552–560.

Barnes, W. E.

1956 Primal ideals and isolated components in noncommutative rings. *Trans. Amer. Math. Soc. 82*, 1–16.

Barnes, W. E., and **Cunnea, W. M.**

1965 Ideal decompositions in Noetherian rings. *Canad. J. Math. 17*, 178–184.

1967 On primary representations of ideals in noncommutative rings. *Math. Ann. 173*, 233–237.

Behrens, E. A.

1956 Zur additiven Idealtheorie in nichtassoziativen Ringen. *Math. Z. 64*, 149–182.

Berger, R.

1962 Über eine Klasse ubergabelter lokaler Ringe. *Math. Ann. 146*, 98–102.

Besserre, A.

1962a Un analogue de la clôture algébrique pour les anneaux. *C. R. Acad. Sci. Paris Ser. A-B 254*, 400–401.

1962b Sur quelques propriétés d'un couple d'anneaux. *C. R. Acad. Sci. Paris Ser. A-B 254*, 4407–4409.

Bialynichi-Birula, A., **Browkin, J.**, and **Schinzel, A.**

1959 On the representation of fields as finite unions of subfields. *Colloq. Math. 7*, 31–32.

Birkhoff, G.

1934 Ideals in algebraic rings. *Proc. Nat. Acad. Sci. U.S.A. 20*, 571–573.

Boccioni, D.

1967 Alcume asservazioui sugli anelli pseudobezoutiani e fattoriali. *Rend. Sem. Mat. Univ. Padova 37*,273–288.

Boisen, M. B., Jr., and **Larsen, M. D.**

Prüfer and valuation rings with zero divisors. *Pacific J. Math.* (to appear).

Brameret, M. P.

1964a Anneaux et modules de largeur finie. *C. R. Acad. Sci. Paris Ser. A-B 258*, 3605–3608.

1964b Anneaux Noethérian de largeur finie. *C. R. Acad. Sci. Paris Ser. A-B 259*, 3914–3915.

1964c Anneaux d'intégrité de largeur finie. *C. R. Acad. Sci. Paris Ser. A-B 259*, 2047–2049.

Brewer, J. W.

1968 The ideal transform and overrings of an integral domain. *Math. Z. 107*, 301–306.

Integral domains of finite character II. *J. Reine Angew. Math.* (to appear).

Brewer, J. W., and **Gilmer, R. W.**

Integral domains whose overrings are ideal transforms. *Math. Nachr.* (to appear).

Brewer, J. W., and **Mott, J. L.**

1970 Integral domains of finite character. *J. Reine Angew. Math. 241*, 34–41.

Brown, B., and **McCoy, N. H.**

1946 Rings with unit element which contain a given ring. *Duke Math. J. 13*, 9–20.

Brungs, H. H.

1969 Generalized discrete valuation rings. *Canad. J. Math. 6*, 1404–1408.

Buchsbaum, D. A.

1961 Some remarks on factorization in power series rings. *J. Math. Mech. 10*, 749–753.

Budach, L.

1963a Quotalringe und ihre Anwendungen. *Math. Nachr. 27*, 29–66.

1963b Verallgemeinerte lokale Ringe. *Math. Nachr. 25*, 65–81.

1963c Aufbau der ganzen Abschleissung einartiger Noetherscher Integritätsvereicher I, II. *Math. Nachr. 25*, 5–17, 129–149.

Budach, L., and **Kerstan, J.**

1963/64 Über eine Charakterisierung der Grellschen Schemata. *Math. Nachr. 27*, 253–264.

Burgess, W. D.

1969 Rings of quotients of group rings. *Canad. J. Math. 21*, 865–875.

Butts, H. S.

1964 Unique factorization of ideals into nonfactorable ideals. *Proc. Amer. Math. Soc. 15*, 21.

1965 Quasi-invertible prime ideals. *Proc. Amer. Math. Soc. 16*, 291–292.

Butts, H. S., and **Cranford, R. H.**

1965 Some containment relations between classes of ideals of an integral domain. *J. Sci. Hiroshima Univ. Ser. A-I 29*, 1–10.

Butts, H. S., and **Estes, D.**
1968 Modules and binary quadratic forms over integral domains. *Linear Algebra and Appl. 1*, 153–180.
Butts, H. S., and **Gilbert, J. R., Jr.**
1968 Rings satisfying the three Noether axioms. *J. Sci. Hiroshima Univ. Ser. A-I 32*, 211–224.
Butts, H. S., and **Gilmer, R. W.**
1966 Primary ideals and prime power ideals. *Canad. J. Math. 18*, 1183–1195.
Butts, H. S., and **Phillips, R. C.**
1965 Almost multiplication rings. *Canad. J. Math. 17*, 267–277.
Butts, H. S., and **Smith, W. W.**
1966 On the integral closure of a domain. *J. Sci. Hiroshima Univ. Ser. A-I 30*, 117–122.
1967 Prüfer rings. *Math. Z. 95*, 196–211.
Butts, H. S., and **Vaughan, N.**
1969 On overrings of a domain. *J. Sci. Hiroshima Univ. Ser. A-I 33*, 95–104.
Butts, H. S., and **Wade, L. I.**
1966 Two criteria for Dedekind domains. *Amer. Math. Monthly 73*, 14–21.
Butts, H. S., Mann, H. B., and **Hall, M., Jr.**
1954 On integral closure. *Canad. J. Math. 6*, 471–473.
Cashwell, E. D., and **Everett, C. J.**
1963 Formal power series. *Pacific J. Math. 13*, 45–64.
Cateforis, V. C.
Flat regular quotient rings. *Trans. Amer. Math. Soc.* (to appear).
Chadeyras, M.
1960 Sur les anneaux semi-principaux ou de Bezout. *C. R. Acad. Sci. Paris Ser. A-B 251*, 2116–2117.
1961 Modules sur les anneaux semi-principaux. *C. R. Acad. Sci. Paris Ser. A-B 252*, 3179–3180.
Chatland, H., and **Mann, H. B.**
1949 Integral extensions of a ring. *Bull. Amer. Math. Soc. 55*, 592–594.
Chevalley, C.
1943 On the theory of local rings. *Ann. of Math.* (2) *44*, 690–708.
1944a On the notion of the ring of quotients of a prime ideal. *Bull. Amer. Math. Soc. 50*, 93–97.
1944b Some properties of ideals in rings of power series. *Trans. Amer. Math. Soc. 55*, 68–84.
1954 La notion d'anneau de décomposition. *Nagoya Math. J. 7*, 21–33.
Chew, K. L.
1968 On a conjecture of D. C. Murdoch concerning primary decompositions of an ideal. *Proc. Amer. Math. Soc. 19*, 925–932.
Claborn, L.
1965a Note generalizing a result of Samuel's. *Pacific J. Math. 15*, 805–808.
1965b Dedekind domains and rings of quotients. *Pacific J. Math. 15*, 59–64.
1965c Dedekind domains: Overrings and semi-prime elements. *Pacific J. Math. 15*, 799–804.

1965d The dimension of $A[X]$. *Duke Math. J. 32*, 233–236.
1966 Every Abelian group is a class group. *Pacific J. Math. 18*, 219–222.
1967 A generalized approximation theorem for Dedekind domains. *Proc. Amer. Math. Soc. 18*, 378–380.
1968 Specified relations in the ideal class group. *Michigan Math. J. 15*, 249–255.

Cohen, I. S.
1946 On the structure and ideal theory of complete local rings. *Trans. Amer. Math. Soc. 59*, 54–106.
1950 Commutative rings with restricted minimum condition. *Duke Math. J. 17*, 27–42.
1954 Lengths of prime ideal chains. *Amer. J. Math. 76*, 654–668.

Cohen, I. S., and **Kaplansky, I.**
1946 Rings with a finite number of primes I. *Trans. Amer. Math. Soc. 60*, 468–477.
1951 Rings for which every module is a direct sum of cyclic modules. *Math. Z. 54*, 97–101.

Cohen, I. S., and **Seidenberg, A.**
1946 Prime ideals and integral dependence. *Bull. Amer. Math. Soc. 52*, 252-261.

Cohen, I. S., and **Zariski, O.**
1957 A fundamental inequality in the theory of extensions of valuations. *Illinois J. Math. 1*, 1–8.

Cohn, P. M.
1958 Rings of zero divisors. *Proc. Amer. Math. Soc. 9*, 909–914.
1968 Bezout rings and their subrings. *Proc. Cambridge Philos. Soc. 64*, 251–264.

Cohn, R. M.
1969 Systems of ideals. *Canad. J. Math. 21*, 783–807.

Corbas, B.
1969 Rings with few zero divisors. *Math. Ann. 181*, 1–7.

Cunnea, W. M.
1964 Unique factorization in algebraic function fields. *Illinois J. Math. 8*, 425–438.

Curtis, C. W.
1952 On additive ideal theory in general rings, *Amer. J. Math. 74*, 687–700.

Dade, E. C.
1962 Rings in which no fixed power of ideal classes becomes invertible. *Math. Ann. 148*, 65–66.

Dade, E. C., Taussky, O., and **Zassenhaus, H.**
1962 On the theory of orders, in particular on the semigroup of ideal classes and genera of an order in an algebraic number field. *Math. Ann. 148*, 31–64.

Davis, E. D.
1962 Overrings of commutative rings I. Noetherian overrings. *Trans. Amer. Math. Soc. 104*, 52–61.
1964 Overrings of commutative rings II. Integrally closed overrings. *Trans. Amer. Math. Soc. 110*, 196–212.
1965 Rings of algebraic numbers and functions. *Math. Nachr. 29*, 1–7.
1967 Ideals of the principal class, R-sequences, and a certain monoidal transformation. *Pacific J. Math. 20*, 197–205.

1968 Further results on ideals of the principal class. *Pacific J. Math. 27*, 49–51.
1969 A remark on Prüfer rings. *Proc. Amer. Math. Soc. 20*, 235–237.

Deckard, D., and **Durst, L. K.**
1966 Unique factorization in power series rings and semi-groups. *Pacific J. Math. 16*, 239–242.

Dieudonne, M. J.
1941 Sur la théorie de la divisibilité. *Bull. Soc. Math. France 69*, 133–144.
1942 Sur le nombre de dimensions d'un module. *C. R. Acad. Sci. Paris Ser. A-B 215*, 563–565.

Dilworth, R. P.
1962 Abstract commutative ideal theory. *Pacific J. Math. 12*, 481–498.

Dodo, T.
1938 Bemerkungen zur Zerlegung der Hauptideale in allgemeinen dutegritätsbereichen. *J. Sci. Hiroshima Univ. Ser. A 8*, 1–5.

Dribin, D. M.
1938 Prüfer ideals in commutative rings. *Duke Math. J. 4*, 737–751.

Eakin, P. M.
1968 The converse to a well-known theorem on Noetherian rings. *Math. Ann. 177*, 278–282.

Eakin, P. M., and **Heinzer, W. J.**
1968 Some open questions on minimal primes of a Krull domain. *Canad. J. Math. 20*, 1261–1264.
1970 Non-finiteness in finite dimensional Krull domains. *J. Algebra. 14*, 333–340.

Eggert, N., and **Rutherford, H.**
A local characterization of Prüfer rings. *J. Reine Angew. Math.* (to appear).

Endler, O.
1959 Modules and rings of fractions. *Summa Brasil. Math. 4*, 149–182.
1963 Über einen Existenzsatz der Bewertungstheorie. *Math. Ann. 150*, 54–65.

Endo, S.
1959 On regular rings. *J. Math. Soc. Japan 11*, 159–170.

Estes, D., and **Ohm, J.**
1967 Stable range in commutative rings. *J. Algebra 7*, 343–362.

Everett, C. J.
1942 Vector spaces over rings. *Bull. Amer. Math. Soc. 48*, 312–316.

Fields, K. L.
1969 Examples of orders over discrete valuation rings. *Math. Z. 111*, 126–130.

Fitting, H.
1935 Primärkomponentenzelegung in nicht kommutativen Ringen. *Math. Ann. 111*, 19–41.

Flanders, H.
1954 A remark on Hilbert's Nullstellensatz. *J. Math. Soc. Japan 6*, 160–161.
1960 The meaning of the form calculus in classical ideal theory. *Trans. Amer. Math. Soc. 95*, 92–100.

Fleischer, I.
1957 Modules of finite rank over Prüfer rings. *Ann. of Math. 65*, 250–254.

Fletcher, C. R.
1969 Unique factorization rings. *Proc. Cambridge Philos. Soc. 65*, 579–583.

1970 The structure of unique factorization rings. *Proc. Cambridge Philos. Soc. 67*, 535–540.

Forster, O.

1964 Über die Anzahl der Erzeugenden eines Ideals in einem Noetherschen Ring. *Math. Z. 84*, 80–87.

Forsythe, A.

1943 Divisors of zero in polynomial rings. *Amer. Math. Monthly 50*, 7–8.

Fossum, R. M.

1968 Maximal orders over Krull domains. *J. Algebra 10*, 321–332.

Fuchs, L.

1947 On quasi-primary ideals. *Acta. Sci. Math. (Szeged) 11*, 174–183.

1948a Domains d'intégrité ou tout ideal est quasi-primaire. *C. R. Acad. Sci. Paris Ser. A-B 226*, 1660–1662.

1948b A note on semi-prime ideals. *Norske Vid. Selsk. Forh. (Trondhjem) 20*, 112–114.

1948c A condition under which an irreducible ideal is primary. *Quart. J. Math. Oxford Ser. 19*, 235–237.

1949 Über die Ideale arithmetische Ringe. *Comment. Math. Helv. 23*, 334–341.

1950 A note on the idealizer of a subring. *Publ. Math. Debrecen 1*, 160–161.

1951 A generalization of the valuation theory. *Duke Math. J. 18*, 19–26.

1954 On the fundamental theorem of commutative ideal theory. *Acta. Math. Acad. Sci. Hungar. 5*, 95–99.

Fuelberth, J. D.

1970 On commutative splitting ring. *Proc. London Math. Soc. (3) 20*, 393–408.

Fukawa, M.

1965a On the theory of valuations. *J. Fac. Sci. Univ. Tokyo Sect. I 12*, 57–79.

1965b An extention theorem on valuations. *J. Math. Soc. Japan 17*, 67–71.

Ganesan, N.

1964 Properties of rings with a finite number of zero divisors. *Math. Ann. 157*, 215–218.

1965 Properties of rings with a finite number of zero divisors II. *Math. Ann. 161*, 241–246.

Gilmer, R. W.

1962 Rings in which semi-primary ideals are primary. *Pacific J. Math. 12*, 1273–1276.

1963a Finite rings having a cyclic multiplicative group of units. *Amer. J. Math. 85*, 447–452.

1963b Commutative rings containing at most two prime ideals. *Michigan Math. J. 10*, 263–268.

1963c Rings in which the unique primary decomposition theorem holds. *Proc. Amer. Math. Soc. 14*, 777–781.

1963d On a classical theorem of Noether in ideal theory. *Pacific J. Math. 13*, 579–583.

1964a Integral domains which are almost Dedekind, *Proc. Amer. Math. Soc. 15*, 813–818.

1964b Extension of results concerning rings in which semi-primary ideals are primary, *Duke Math. J. 31*, 73–78.

1965a A class of domains in which primary ideals are valuation ideals, *Math. Ann. 161*, 247–254.

1965b Some containment relations between classes of ideals of a commutative ring, *Pacific J. Math. 15*, 497–502.

1965c The cancellation law for ideals in a commutative ring, *Canad. J. Math. 17*, 281–287.

1966a Overrings of Prüfer domains, *J. Algebra 4*, 331–340.

1966b Eleven nonequivalent conditions on a commutative ring, *Nagoya Math. J. 26*, 183–194.

1966c Domains in which valuation ideals are prime powers, *Arch. Math. (Basel) 17*, 210–215.

1966d The pseudo-radical of a commutative ring, *Pacific J. Math. 19*, 275–284.

1967a A class of domains in which primary ideals are valuation ideals II, *Math. Ann. 171*, 93–96.

1967b A note on the quotient field of the domain $D[[X]]$, *Proc. Amer. Math. Soc. 18*, 1138–1140.

1967c Contracted ideals with respect to integral extensions. *Duke Math. J. 34*, 561–572.

1967d A counterexample to two conjectures in ideal theory. *Amer. Math. Monthly 74*, 195–197.

1967e If $R[X]$ is Noetherian, R contains an identity. *Amer. Math. Monthly 74*, 700.

1967f A note on two criteria for Dedekind domains. *Enseignment Math. 13*, 253–256.

1967g Some applications of the Hilfssatz von Dedekind-Mertens. *Math. Scand. 20*, 240–244.

1968a On a condition of J. Ohm for integral domains. *Canad. J. Math. 20*, 970–983.

1968b R-automorphisms of $R[X]$. *Proc. London Math. Soc. (3) 18*, 328–336.

1969a A note on generating sets for invertible ideals. *Proc. Amer. Math. Soc. 22*, 426–427.

1969b Power series rings over a Krull domain. *Pacific J. Math. 29*, 543, 549.

1969c Commutative rings in which each prime ideal is principal. *Math. Ann. 183*, 151–158.

1969d Two constructions of Prüfer domains. *J. Reine Angew. Math. 239/240*, 153–162.

1969e The unique primary decomposition theorem in commutative rings without identity. *Duke Math. J. 36*, 737–747.

1970a Integral domains with Noetherian subrings. *Comment. Math. Helv. 45*, 129–134.

1970b R-automorphisms of $R[[X]]$. *Michigan Math. J. 17*, 15–21.

1970c Integral dependence in power series ring. *J. Algebra 11*, 488–502.

1970d An embedding theory for HCF-rings. *Proc. Cambridge Philos. Soc. 68*, 583–587.

1970e Contracted ideals in Krull domains. *Duke Math. J. 37*, 769–774.

Gilmer, R. W., and **Heinzer, W. J.**

1966 On the complete integral closure of an integral domain. *J. Austral. Math. Soc. 6*, 351–361.

1967a Overrings of Prüfer domains II. *J. Algebra 7*, 281–302.

1967b Intersections of quotient rings of an integral domain. *J. Math. Kyoto Univ. 7*, 133–150.

1968a Irredundant intersections of valuation rings. *Math. Z. 103*, 306–317.

1968b Primary ideals and valuation ideals II. *Trans. Amer. Math. Soc. 131*, 149–162.

1970 On the number of generators of an invertible ideal. *J. Algebra 14*, 139–151.

Gilmer, R. W., and **Husch, L. S.**

1966 On the uniqueness of the $*$-representation of an ideal. *Amer. Math. Monthly 73*, 876–877.

Gilmer, R. W., and **Mott, J. L.**

1965 Multiplication rings as rings in which ideals with prime radical are primary. *Trans. Amer. Math. Soc. 114*, 40–52.

1968 On proper overring of integral domains. *Monatsh. Math. 72*, 61–71.

1970 Some results on contracted ideals. *Duke Math. J. 37*, 751–767.

Integrally closed subrings of an integral domain. *Trans. Amer. Math. Soc.* (to appear).

Gilmer, R. W., and **Ohm, J.**

1964 Integral domains with quotient overrings. *Math. Ann. 153*, 97–103.

1965 Primary ideals and valuation ideals. *Trans. Amer. Math. Soc. 117*, 237–250.

Goldman, O.

1951 Hilbert rings and the Hilbert Nullstellensatz. *Math. Z. 54*, 136–140.

1961 Quasi-equality in maximal orders. *J. Math. Soc. Japan 13*, 371–376.

1964 On a special class of Dedekind domains. *Topology* (*Suppl. 1*) *3*, 113–118.

1969 Rings and modules of quotients. *J. Algebra 13*, 10–47.

Gorman, H.

1970 Invertibility of modules over Prüfer rings. *Illinois J. Math. 14*, 283–298.

Greco, S.

1966a Sull' integrita e la fattorialita dei complementi. *Rend. Sem. Mat. Univ. Padova 36*, 50–65.

1966b Sugli ideali frazionari invertibili. *Rend. Sem. Mat. Univ. Padova 36*, 315–333.

Grell, H.

1935 Über die Gültigkeit der gewöhnlichen Idealtheorie in endlichen algebraischen Erweiterungen erster und zweiter Art. *Math. Z. 40*, 503–505.

1947 Über die Erhaltung der Kettensätze der Idealtheorie. *Ber. Math.-Tagung, Tübingen, 1946*, p. 67.

1951 Modulgruppen and inversionen bei primären Integritätsbereichen. *Math. Nachr. 4*, 392–407.

Griffin, M.

1967 Some results on v-multiplication rings. *Canad. J. Math. 19*, 710–722.

1968a Families of finite character and essential valuations. *Trans. Amer. Math. Soc. 130*, 75–85.

1968b Rings of Krull type. *J. Reine Angew. Math. 229*, 1–27.

1969 Prüfer rings with zero divisors. *J. Reine Angew. Math. 239/240*, 55–67.

Gröbner, W.

1951 Ein Irreduzibilitätskriterium für Primärideale in kommutativen Ringen. *Monatsh. Math. 55*, 138–145.

Grundy, P. M.

1942 A generalization of the additive ideal theory. *Proc. Cambridge Philos. Soc. 38*, 241–279.

1947 On integrally dependent integral domains. *Philos. Trans. Roy. Soc. London Ser. A 240*, 295–326.

Guazzoue, S.

1967 Su alcume classi di anelli noetheriani normali. *Rend. Sem. Mat. Univ. Padova 37*, 258–266.

Guerindon, J.

1956 Sur une famille d'équivalences en théorie des idéaux. *C. R. Acad. Sci. Paris Ser. A-B 242*, 2693–2695.

Hacque, M.

1969 Localizations exactes et localizations plates. *Publ. Dép. Math. (Lyon) 6*-2.

Harris, M. E.

1966 Some results on coherent rings. *Proc. Amer. Math. Soc. 17*, 474–479.

1967 Some results on coherent rings II. *Glasgow Math. J. 8*, 123–126.

Harui, H.

1968 Noes on quasi-valuation rings. *J. Sci. Hiroshima Univ. Ser. A-I 32*, 237–240.

Hattori, A.

1957 On Prüfer rings. *J. Math. Soc. Japan 9*, 381–385.

Heinzer, W. J.

1967 Some properties of integral closure. *Proc. Amer. Math. Soc. 18*, 749–753.

1968a Integral domains in which each non-zero ideal is divisorial. *Mathematika 15*, 164–170.

1968b *J*-Noetherian integral domains with 1 in the stable range. *Proc. Amer. Math. Soc. 19*, 1369–1372.

1969a On Krull overrings of a Noetherian domain. *Proc. Amer. Math. Soc. 22*, 217–222.

1969b Some remarks on complete integral closure. *J. Austral. Math. Soc. 9*, 310–314.

1969c On Krull overrings of an affine ring. *Pacific J. Math. 29*, 145–149.

1970a Quotient overrings of integral domains. *Mathematika 17*, 139–148.

1970b A note on rings with Noetherian spectrum. *Duke Math. J. 37*, 573–578.

Heinzer, W. J., Ohm, J., and **Pendleton, R.**

1970 On integral domains of the form $\bigcap D_P$, P minimal. *J. Reine Angew. Math. 241*, 147–159.

Helmer, O.

1940 Divisibility properties of integral functions. *Duke Math. J. 6*, 345–356.

1943 The elementary divisor theorem for rings without chain condition. *Bull. Amer. Math. Soc. 49*, 225–236.

Henriksen, M.

1952 On the ideal structure of the ring of entire functions. *Pacific J. Math. 2,* 179–184.

1953a On rings of entire functions of finite order. *J. Indian Math. Soc. (N.S.) 17,* 59–61.

1953b On the prime ideals of the ring of entire functions. *Pacific J. Math. 3,* 711–720.

Hochster, M.

1969 Prime ideal structure of commutative rings. *Trans. Amer. Math. Soc. 142,* 43–60.

Honig, C. S.

1960 Classification of D-submodules of a quotient field of a Dedekind domain D. *An. Acad. Brasil. Ci. 32,* 329–332.

Hopkins, C.

1939 Rings with minimal condition for left ideals. *Ann. of Math. 40,* 712–730.

Huckaba, J. A.

1969 Extensions of pseudovaluation. *Pacific J. Math. 29,* 295–302.

1970 Some results on pseudovaluations. *Duke Math. J. 37,* 1–9.

Ishikawa, T.

1959 On Dedekind rings. *J. Math. Soc. Japan 1,* 83–84.

Jacob, H. G.

1965 Module finiteness of the integral closure of a domain. *J. London Math. Soc. 40,* 565–567.

Jaffard, Paul

1950 Corps demi-values. *C. R. Acad. Sci. Paris Ser. A-B 231,* 1401–1403.

1952 Théorie arithmétique des anneaux du type de Dedekind. *Bull. Soc. Math. France 80,* 61–100.

1953a Théorie arithmétique des annueaux du type de Dedekind II. *Bull. Soc. Math. France 81,* 41–61.

1953b Contribution à l'étude des groupes ordonnes. *J. Math. Pures Appl. 32,* 203–280.

1954 La notion de valuation. *Enseignement Math. 40,* 5–26.

1958a Dimension des anneaux de polynomes. Comportement asymptotique de la dimension. *C. R. Acad. Sci. Paris Ser. A-B 246,* 3199–3201.

1958b Dimension des anneaux de polynomes. La notion de dimension valuative. *C. R. Acad. Sci. Paris Ser. A-B 246,* 3305–3307.

1961a La théorie des ideaux d'Artin–Prüfer–Lorenzen. *Sem. A. Chatelet et P. Dubreil 5-6,* Nr. 3.

1961b Valuations d'un anneaux Noethérian et théorie de la dimension. *Sem. P. Dubreil–M.-L. Debreil–Jacotin et C. Pisot 13,* Nr. 12.

Jensen, C. U.

1963 On characterizations of Prüfer rings. *Math. Scand. 13,* 90–98.

1964 A remark on arithmetical rings. *Proc. Amer. Math. Soc. 15,* 951–954.

1966a Arithmetical rings. *Acta Math. Acad. Sci. Hungar. 17,* 115–123.

1966b A remark on the distributive law for an ideal of a commutative ring. *Proc. Glasgow Math. Assoc.* 7, 193–198.

1966c A remark on flat and projective modules. *Canad. J. Math. 18,* 943–949.

1966d On homological dimensions of rings with countably generated ideals. *Math. Scand. 18*, 97–105.

1967 Homological dimensions of $\aleph_0$—coherent rings. *Math. Scand. 20*, 55–60.

1968a Some remarks on a change of rings theorem. *Math. Z. 106*, 395–401.

1968b Sur le plongement d'un anneau dans un anneau héréditaire. *C. R. Acad. Sci. Paris Ser. A 267*, 534–535.

1969 Some cardinality questions for flat modules and coherence. *J. Algebra 12*, 231–241.

Johnson, J.

1969 A notion of Krull dimension for differential rings. *Comment. Math. Helv. 44*, 207–216.

Joly, J. R.

1965 Sur les puissances *d*-ièmes des éléments d'un anneau commutatif. *C. R. Acad. Sci. Paris Ser. A-B 261*, 3259–3262.

Kaplansky, I.

1952 Modules over dedekind rings and valuation rings. *Trans. Amer. Math. Soc. 72*, 327–340.

1960 A characterization of Prüfer rings. *J. Indian Math. Soc. (N.S.) 24*, 279–281.

Keller, O. H.

1965 Berechnung eines Primideals aus seiner allgemeinen Nullstelle. *Math. Z. 87*, 160–162.

Kelly, P. H., and **Larsen, M. D.**

Valuation rings with zero divisors. *Proc. Amer. Math. Soc.* (to appear).

Kikuchi, T.

1966 Some remarks on S-domains. *J. Math. Kyoto Univ. 6*, 49–60.

Kirby, D.

1966 Components of ideals in a commutative ring. *Ann. Mat. Pura Appl.* (4) *71*, 109–125.

1969 Closure operations on ideals and submodules. *J. London Math. Soc. 44*, 283–291.

1970 Integral dependence and valuation algebras. *Proc. London Math. Soc.* (3) *20*, 79–100.

Kobayashi, Y., and **Moriya, M.**

1941 Eine hinreichende Bedingung für die eindeutige Primfaktorzerlegung der Ideale in einem kommutativen Ring. *Proc. Imp. Acad. Tokyo 17*, 134–138.

Krull, W.

1935–36 Über allgemeine Multiplikationsringe. *Tohoku Math. J. 41*, 320–326.

1936a Hauptidealzerlegung in Polynomringen. *Math. Z. 41*, 213–217.

1936b Beiträge zur Arithmetik kommutativer Integritätsbereiche. *Math. Z. 41*, 545–577.

1936c Beiträge zur Arithmetik kommutativer Integritätsbereiche II. *v*-Ideale und vollstandig ganz abgeschlossene Integritätsbereiche. *Math. Z. 41*, 665–679.

1936d Über die Entwicklung der Arithmetik kommutativer Integritätschereiche. *Jber. Deutsch. Math.-Verein 46*, 153–171.

1937a BeiträNge zur Arithmetik kommutativer Integritätsbereiche III. Zum Dimensionbegriff der Idealtheorie. *Math. Z. 42*, 745–766.

1937b Beiträge zur Arithmetik kommutativer Integritätsbereiche IV. Unendliche algebraische erweitesungen endlichen diskreet Hauptordrungen. *Math. Z.* *42*, 767–773.

1938a Dimensionstheorie in Stellenringen. *J. Reine Angew. Math.* *179*, 204-226.

1938b Beiträge zur Arithmetik kommutativer Integritätsbereiche V. Potenzreihenringe. *Math. Z.* *43*, 768–782.

1938c Beiträge zur Arithmetik kommutativer Integritätsbereiche, III a. Eine ergänzungen von Beitrag III. *Math. Z.* *43*, 767.

1939 Beiträge zur Arithmetik kommutativer Integritätsbereiche VI. Der allgemeine Diskriminantensatz, Unversweigte Ringerweiterungen. *Math. Z.* *45*, 1–19.

1942–43 Beiträge zur Arithmetik kommutativer Integritätsbereiche. Eine bemerkung zu den Beitrag VI Und VII. *Math. Z.* *48*, 530–531.

1943 Beiträge zur Arithmetik kommutativer Integritätsbereiche VIII. Muliplikativ abgeschlossene Systeme von endichen Idealen. *Math. Z.* *48*, 533–552.

1950a Subdirekte Sumemendarstellungen von Integritätsbereichen. *Math. Z.* *52*, 810–826.

1950b Die Verzweigungsgroppen in der Galoisschen Theorie Beliebiger arithmetischer Körper. *Math. Ann.* *121*, 446–466.

1951a Jacobsonsche Ringe, Hilbertscher Nullstellensatz, Dimension theorie. *Math. Z.* *54*, 354–387.

1951b Zur Arithmetik der endlichen diskreten Hauptordnungen. *J. Reine Angew. Math.* *189*, 118–128.

1953 Zur Theorie der kommutativen Integritätsbereiche. *J. Reine Angew. Math.* *192*, 230–252.

1956 Eine Bermerkung über primäre Integritätsbereiche. *Math. Ann.* *130*, 394–398.

1957/58 Zur Idealtheorie der unendlichen algebraischen Zahlkörper. *J. Sci. Hiroshima Univ. Ser. A* *21*, 79–88.

1958 Über Laskersche Ringe. *Rend. Circ. Mat. Palermo Ser. 2* 7, 155–166.

1959 Über einen Existenzsatz der Bewertungstheorie. *Abh. Math. Sem. Univ. Hamburg* *23*, 29–35.

1960 Einbettungsfreie, fast-Noethersche Ringe und ihre Oberringe. *Math. Nachr.* *21*, 319–338.

1961 Ordnungsfunktionen und Berwertungen von Körpern. *Math. Z.* 77, 135–148.

Kubo, K.

1940 Über die Noetherschen fünf Axiome in kommutativen Ringen. *J. Sci. Hiroshima Univ. Ser. A* *10*, 77–84.

Kurata, Y.

1965 On an additive ideal theory in a non-associative ring. *Math. Z.* *88*, 129–135.

Lafon, J. P.

1965a Remarques sur lemme de preparation de Wierstrass. *C. R. Acad. Sci. Paris Ser. A-B* *260*, 2660–2663.

1965b Series formelles algebriques. *C. R. Acad. Sci. Paris Ser. A-B* *260*, 3238–3241.

Laplaza, M. L.

1963 Some properties of immersion with respect to a unit. *Rev. Mat. Hisp.-Amer.* (*4*) *23*, 242–252.

Larsen, M. D.

1967 Equivalent conditions for a ring to be a P-ring and a note on flat overrings. *Duke Math. J. 34*, 273–280.

1968 Containment relations between classes of regular ideals in a ring with few zero divisors. *J. Sci. Hiroshima Univ. Ser. A-I 32*, 241–246.

1969 Harrison primes in a ring with few zero divisors. *Proc. Amer. Math. Soc. 22*, 111–116.

1970 A generalization of almost Dedekind domains. *J. Reine Angew. Math. 245*, 119–123.

Prüfer rings of finite character. *J. Reine Angew. Math.* (to appear).

Ideal theory in Prüfer rings with zero divisors. *J. Reine Angew. Math.* (to appear).

Larsen, M. D., and **McCarthy, P. J.**

1967 Overrings of almost multiplication rings. *J. Sci. Hiroshima Univ. Ser. A-I 31*, 123–129.

Lazard, D.

1967/68 Epimorphismes plats. *Sém. P. Samuel* (*algèbre commutative*). *Exposé No.* 4. (Secrétariat mathématique, Paris 1968.)

1968 Epimorphismes plats d'anneau. *C. R. Acad. Sci. Paris Ser. A-B 266*, 314–317.

1969 Autour de la platitude. *Bull. Soc. Math. France 97*, 81–128.

Lazard, M.

1961 Théorie des idéaux dans les anneaux de Dedekind. *Sem. A. Chatelet et P. Dubreil 5-6*, Nr. 2.

Leger, G. F.

1955 A note on some properties of finite rings. *Proc. Amer. Math. Soc. 6*, 968–969.

Lesieur, L.

1949 Sur les domaines d'intégrité intégralement fermés. *C. R. Acad. Sci. Paris Ser. A-B 229*, 691–693.

1950a Un théorème de transfert d'un anneau abstrait à l'anneau des polynomes. *Canad. J. Math. 2*, 50–65.

1950b Le transfert de certaines propriétés d'un anneau A à l'anneau des polynomes $A[X]$. *Colloq. Internationaux du Centre National de la Recherche Scientifique No. 24*, 99–101.

1967 Divers aspects de la théorie des idéaux d'un anneau commutatif. *Enseignement Math. 13*, 75–87.

Lesieur, L., and **Croisot, R.**

1956 Théorie Noethérieune des anneaux des demi-groupes et des modules dans le cas non commutatif, I. *Colloq. d'Algèbre Sup., C.B.R.M., Bruxelles*, pp. 79–121.

1958a Théorie Noethérienne des anneaux, des demi-groupes et des modules dans le cas non commutatif, II. *Math. Ann. 134*, 458–476.

1958b Théorie Noethérienne des anneaux, des demi-groupes et des modules dans le cas non commutatif, III, Sur la notion de radical. *Acad. Roy. Belg. Bull. Cl. Sci. 44*, 75–93.

1960 Extension au cas non commutatif d'un théorème de Krull et d'un lemme d'Artin-Rees. *J. Reine Angew. Math. 204*, 216–220.

Lorenzen, P.

1939 Abstrakte Begründung der multiplikativen Idealtheorie. *Math. Z. 45*, 533–553.

1952 Teilbarkeitstheorie in Bereichen. *Math. Z. 55*, 269–275.

MacLane, S.

1939 The universality of formal power series fields. *Bull. Amer. Math. Soc. 45*, 888–890.

MacLane, S., and **Schilling, O. F. G.**

1939 Infinite number fields with Noether ideal theories. *Amer. J. Math. 61*, 771–782.

Macaulay, F. S.

1933/34 Modern algebra and polynomial ideals. *Proc. Cambridge Philos. Soc. 30*, 27–46.

Manganey, M.

1966 Sur la finitude de la fermeture integrale. *Bull. Soc. Math. France 94*, 277–286.

1967 *Errantum ibid. 95*, 311.

Manis, M. E.

1967 Extension of valuation theory. *Bull. Amer. Math. Soc. 73*, 735–736.

1969 Valuations on a commutative ring. *Proc. Amer. Math. Soc. 20*, 193–198.

Mann, H. B.

1958 On integral bases. *Proc. Amer. Math. Soc. 9*, 167–172.

Maranda, J. M.

1957 Factorization rings. *Canad. J. Math. 9*, 597–623.

Marot, J.

1967 Une propriété des sur-anneaux plats d'un anneau. *C. R. Acad. Sci. Paris Ser. A-B 265*, 8–10.

1968 Extension de la notion d'anneau de valuation. *Publ. Dép. Math. (Brest)*.

1969a Une généralisation de la notion d'anneau de valuation. *C. R. Acad. Sci. Paris Ser. A-B 268*, 1451–1454.

1969b Anneaux héréditaires commutatifs. *C. R. Acad. Sci. Paris Ser. A-B 269*, 58–61.

Matlis, E.

1959 Injective modules over Prüfer rings. *Nagoya Math. J. 15*, 57–69.

1961 Some properties of Noetherian domains of dimension one. *Canad. J. Math. 13*, 569–586.

Matusita, K.

1941 Über die Idealtheorie im Integritätschereich mit dem eingeschränkten Vielfachkettensatz. *Proc. Phys.-Math. Soc. Japan (3) 23*, 8–12.

1944 Über ein bewertungstheoretische Axiomensystem für die Dedekind-Noethersche Idealtheorie. *Japan. J. Math. 19*, 97–110.

Maury, G.

1961 La condition "intégralement clos" dans quelques structures algebriques. *Ann. Sci. Ecole Norm. Sup. 78*, 31–100.

1963 Théorème de transfert de propriétés de l'anneau A à l'anneau $A[\theta]$ extension simple entière de A. *C. R. Acad. Sci. Paris Ser. A-B 256*, 5024–5027.

McCarthy, P. J.

1966 Note on primary ideal decompositions. *Canad. J. Math. 18*, 950–952.

1967 Note on abstract commutative ideal theory. *Amer. Math. Monthly 74*, 706–707.

1969 Arithmetical rings and multiplicative lattices. *Ann. Mat. Pura Appl. 82*, 267–274.

McCoy, N. H.

1938 Subrings of infinite direct sums. *Duke Math. J. 4*, 486–494.

1942 Remarks on divisors of zero. *Amer. Math. Monthly 49*, 286–295.

1947 Subdirect sums of rings. *Bull. Amer. Math. Soc. 53*, 856–877.

1948 Prime ideals in general rings. *Amer. J. Math. 71*, 823–833.

1953 Factorization of certain polynomials over a commutative ring. *Duke Math. J. 20*, 113–118.

1957 A note on finite unions of ideals and subgroups. *Proc. Amer. Math. Soc. 8*, 633–637.

Merker, J.

1969 Idéaux faiblement associés. *Bull. Sci. Math. 93*, 15–21.

Mollier, D.

1970 Descente de la propriété Noethérienne. *Bull. Sci. Math. 94*, 25–31.

Mori, S.

1934a Über allgemeine Multiplikationsringe I. *J. Sci. Hiroshima Univ. Ser. A 4*, 1–26.

1934b Über allgemeine Multiplikationsringe II. *J. Sci. Hiroshima Univ. Ser. A 4*, 99–109.

1935 Über primäre Ringe. *J. Sci. Hiroshima Univ. Ser. A 5*, 131–139.

1936a Über Totalnullteiler kommutativer Ringe mit abgeschwachten U-Satz I. *J. Sci. Hiroshima Univ. Ser. A 6*, 139–146.

1936b Über Totalnullteiler kommutativer Ringe mit abgeschwachtem U-Satz II. *J. Sci. Hiroshima Univ. Ser. A 6*, 257–269.

1938 Über die Produktzerlegung der Hauptideale. *J. Sci. Hiroshima Univ. Ser. A 8*, 7–13.

1939a Zerlegung der Hauptideale aus Polynomringen in minimale Primideale II. *J. Sci. Hiroshima Univ. Ser. A 9*, 1–6.

1939b Über die Produktzerlegung der Hauptideale II. *J. Sci. Hiroshima Univ. Ser. A 9*, 145–155.

1940a Zerlegung der Hauptideale aus Polynomringen in minimale Primidealen III. *J. Sci. Hiroshima Univ. Ser. A 10*, 1–6.

1940b Über die Produktzerlegung der Hauptideale III. *J. Sci. Hiroshima Univ. Ser. A 10*, 85–94.

1940c Allgemeine Z.P.I.-Ringe. *J. Sci. Hiroshima Univ. Ser. A 10*, 117–136.

1941 Über die Produktzerlegung der Hauptideale IV. *J. Sci. Hiroshima Univ. Ser. A 11*, 7–14.

1943 Rings with the property of intersection decomposition I. *J. Sci. Hiroshima Univ. Ser. A 12*, 205–215.
1944 Rings with the property of intersection decomposition II. *J. Sci. Hiroshima Univ. Ser. A 13*, 1–10.
1951 Über die Symmetrie des Prädokates "relativ prim". *J. Sci. Hiroshima Univ. Ser. A 15*, 79–85.
1952 Struktur der Multiplikationsringe. *J. Sci. Hiroshima Univ. Ser. A 16*, 1–11.
1956a Über Idealtheorie der Multiplikations ringe. *J. Sci. Hiroshima Univ. Ser. A 19*, 429–437.
1956b Über die eindentige Darstellung der Ideals als Durchschmitt swacher Primärideale. *Proc. Japan Acad. 32*, 83–85.

Mori, S., and **Dodo, T.**
1937a Bedingungen für ganze Abgeschlossenheit in Integritätsbereichen. *J. Sci. Hiroshima Univ. Ser. A 7*, 15–27.
1937b Bemerkungen zur Zerlegung der Hauptideale. *J. Sci. Hiroshima Univ. Ser. A 7*, 131–140.
1938 Zerlegung der Hauptideale aus Polynomringen in minimale Primideale. *J. Sci. Hiroshima Univ. Ser. A 8*, 135–144.

Mori, Y.
1953 On the integral closure of an integral domain. *Mem. Coll. Sci. Univ. Kyoto 27*, 249–256, *28*, 327–328.
1955 On the integral closure of an integral domain II. *Bull. Kyoto Gakugei Univ. 7*, 19–30.
1959 On the integral closure of an integral domain IV. *Bull. Kyoto Gakugei Univ. Ser. B 15*, 14–16.
1964a On the integral closure of an integral domain VII. *Bull. Kyoto Gakugei Univ. Ser. B 23*, 1–4.
1964b On the integral closure of an integral domain VIII. Generalized F-ideals. *Bull. Kyoto Gakugei Univ. Ser. B 25*, 1–3.

Moriya, M.
1940 Bewertungstheoretischer Aufban der multiplikativen Idealtheorie. *J. Fac. Sci. Hokkaido Univ. Ser. I 8*, 109–144.

Moriya, M., and **Kobayashi, Y.**
1941a Eine notwendige Bedingung für die eindeutige Primfaktorzerlegung der Ideale in einem kommutativen Ring. *Proc. Imp. Acad. Tokyo 17*, 129–133.
1941b Eine notwendige Bedingung für die eindeutige Primfaktorzerlegung der Hauptideale IV. *J. Sci. Hiroshima Univ. Ser. A 11*, 7–14.

Mott, J. L.
1964 Equivalent conditions for a ring to be a multiplication ring. *Canad. J. Math. 16*, 429–434.
1965 On irredundant components of the kernel of an ideal. *Mathematika 12*, 65–72.
1966 Integral domains with quotient overrings. *Math. Ann. 166*, 239–232.
1967 On the complete integral closure of an integral domain of Krull type. *Math. Ann. 173*, 238–240.
1969 Multiplication rings containing only finitely many minimal prime ideals. *J. Sci. Hiroshima Univ. Ser. A-I 33*, 73–83.

Motzkin, T.

1949 The Euclidean algorithm. *Bull. Amer. Math. Soc. 55*, 1142–1146.

Moyls, B. N.

1951 The structure of the valuations of the rational function field $K(X)$. *Trans. Amer. Math. Soc. 71*, 102–112.

Muhly, H. T., and **Sakuma, M.**

1963 Some multiplicative properties of complete ideals. *Trans. Amer. Math. Soc. 106*, 210–221.

Muller, D.

1961 Verbandsgruppen und Durchschnitte endlich vieler Bewertungsringe. *Math. Z. 77*, 45–62.

Murdoch, D. C.

1952 Contributions to noncommutative ideal theory. *Canad. J. Math. 4*, 43–57.

Nagarjan, K. R.

1968 Groups acting on Noetherian rings. *Nieuw. Arch. Wisk. 16*, 25–29.

Nagata, M.

1950 On the structure of complete local rings. *Nagoya Math. J. 1*, 63–70.

1952 On Krull's conjecture concerning valuation rings. *Nagoya Math. J. 4*, 29–33.

1953 On the theory of Henselian rings. *Nagoya Math. J. 5*, 45–57.

1954a Some remarks on local rings II. *Mem. Coll. Sci. Univ. Kyoto Ser. A Math. 28*, 109–120.

1954b Note on integral closures of Noetherian domains. *Mem. Coll. Sci. Univ. Kyoto Ser. A Math. 28*, 121–124.

1954c On the theory of Henselian rings II. *Nagoya Math. J. 7*, 1–19.

1954d Note on complete local integrity domains. *Mem. Coll. Sci. Univ. Kyoto Ser. A Math. 28*, 271–278.

1955a Corrections to my paper "On Krull's conjecture concerning valuation rings." *Nagoya Math. J. 9*, 209–212.

1955b Basic theorems on commutative rings. *Mem. Coll. Sci. Univ. Kyoto Ser. A Math. 29*, 59–77.

1955c On the derived normal rings of Noetherian integral domains. *Mem. Coll. Sci. Univ. Kyoto Ser. A Math. 29*, 293–303.

1956a A general theory of algebraic geometry over Dedekind domains. *Amer. J. Math. 78*, 76–116.

1956b A treatise on the 14-th problem of Hilbert. *Mem. Coll. Sci. Univ. Kyoto Ser. A Math. 30*, 57–70.

1956c On the chain problem of prime ideals. *Nagoya Math. J. 10*, 51–64.

1958 A general theory of algebraic geometry over Dedekind domains II. Separably generated extensions and regular local rings. *Amer. J. Math. 80*, 382–420.

1959a A general theory of algebraic geometry over Dedekind domains III. *Amer. J. Math. 81*, 401–435.

1959b Note on a chain condition for prime ideals. *Mem. Coll. Sci. Univ. Kyoto Ser. A Math. 32*, 85–90.

1960a On the 14-th problem of Hilbert. *Proc. Int. Cong. Math.*, pp. 459–462. Cambridge Univ. Press, London and New York, 1960.

1960b Some remarks on prime divisors. *Mem. Coll. Sci. Univ. Kyoto Ser. A Math. 33*, 297–299.
1965 Some sufficient conditions for the fourteenth problem of Hilbert. *Actas Coloq. Inter. Geo. Alg.* 107–121.
1966 A theorem on finite generation of a ring. *Nagoya Math. J. 27*, 193–205.
1968 A type of subring of a noetherian ring. *J. Math. Kyoto Univ. 8*, 465–467.
1968 A type of integral extension. *J. Math. Soc. Japan 20*, 266–267.
1969 Flatness of an extension of a commutative ring. *J. Math. Kyoto Univ. 9*, 439–448.

Nagata, M., Nakayama, T., and **Tuzuku, T.**
1953 On an existence lemma in valuation theory. *Nagoya Math. J. 6*, 59–61.

Nakayama, T.
1942 On Krull's conjecture concerning completely integrally closed integrity domains, I, II. *Proc. Imp. Acad. Tokyo 18*, 185–187, 233–236.
1946 On Krull's conjecture concerning completely integrally closed integrity domains III. *Proc. Japan Acad. 22*, 249–250.

Nakano, N.
1943 Über die Umkehrbarkeit der Ideale im Integritätsbereiche. *Proc. Imp. Acad. Tokyo 19*, 230–234.
1952 Über den Fundamentalsatz der Idealtheorie in unendlichen algebraischen Zahlkörpern. *J. Sci. Hiroshima Univ. Ser. A 15*, 171–175.
1953a Idealtheorie in einem speziellen unendlichen algebraischen Zahlkörper. *J. Sci. Hiroshima Univ. Ser. A 16*, 425–439.
1953b Über idempotente Ideale in unendlichen algegraischen Zahlkörpern. *J. Sci. Hiroshima Univ. Ser. A 17*, 11–20.
1953c Über die kürzests Darstellung der Ideale im unendlichen algebraischen Zahlkörper. *J. Sci. Hiroshima Univ. Ser. A 17*, 21–25.
1955a Idealtheorie in unendlichen algebraischen Zahlkorpern. *Proc. Int. Symp. Algebra Number Theory, Tokyo and Nikko, 1955*, pp. 249–251. Science Council of Japan, Tokyo, 1956.
1955b Über den Primäridealquotienten in unendlichen algebraischen Zahlkörper. *J. Sci. Hiroshima Univ. Ser. A 18*, 257–269.
1955c Idealtheorie in Stiemkeschen Körper. *J. Sci. Hiroshima Univ. Ser. A Math. 18*, 271–287.
1956 Über die Multiplikativeigenschaft der Ideale in unendlichen algebraischen Zahlkörper. *J. Sci. Hiroshima Univ. Ser. A 19*, 439–455.

Năstăsescu, C.
1970 Décomposition primaire dans les anneaux semiartiniens. *J. Algebra 14*, 170–181.

Năstăsescu, C. and **Popescu, N.**
1968 Les anneaux semiartiniens. *Bull. Soc. Math. France 96*, 357–368.
1970 On the localization ring of a ring. *J. Algebra 15*, 41–56.

Nishimura, T.
1961 On the *V*-ideal of an integral domain I, II. *Bull. Kyoto Gakugei Univ. Ser. B 17*, 47–50, *18*, 9–12.
1962 On the *V*-ideal of an integral domain IV. *Bull. Kyoto Gakugei Univ. Ser. B 21*, 1–3.

1963 Unique factorization of ideals in the sense of quasi-equality. *J. Math. Kyoto Univ. 3*, 115–125.

1964 On the *V*-ideal of an integral domain V. *Bull. Kyoto Gakugei Univ. Ser. B 25*, 5–11.

Northcott, D. G.

1952 A note on the intersection theorem for ideals. *Proc. Cambridge Philos. Soc. 48*, 366–367.

1955 A note on classical ideal theory. *Proc. Cambridge Philos. Soc. 51*, 766–767.

1959 A generalization of a theorem on the contents of polynomials. *Proc. Cambridge Philos. Soc. 55*, 282–288.

Northcott, D. G., and **Reufel, M.**

1965a Reduction of polynomial modules by means of an arbitrary valuation. *Abh. Math. Sem. Univ. Hamburg 28*, 16–49.

1965b A generalization of the concept of length. *Quart. J. Math. Oxford Ser. 16*, 297–321.

Ohm, J.

1966a Primary ideals in Prüfer domain. *Canad. J. Math. 18*, 1024–1030.

1966b Some counterexamples related to integral closure in $D[[X]]$. *Trans. Amer. Math. Soc. 122*, 321–333.

1967 Integral closure and $(x, y)^n = (x^n, y^n)$. *Monatsh. Math. 71*, 32–39.

1969 Semi-valuations and groups of divisibility. *Canad. J. Math. 21*, 576–591.

Ohm, J., and **Pendleton, R. L.**

1968 Rings with Noetherian spectrum. *Duke Math. J. 35*, 631–639, 875.

O'Malley, M. J.

1970 R-automorphisms of $R[[X]]$. *Proc. London Math. Soc. 20*, 60–78.

O'Malley, M. J., and **Wood, C.**

1970 R-endomorphisms of $R[[X]]$. *J. Algebra 15*, 314–327.

O'Meara, O. T.

1956 Basis structure of modules. *Proc. Amer. Math. Soc. 7*, 965–974.

Osofsky, B. L.

1969 A commutative local ring with finite global dimension and zero divisors. *Trans. Amer. Math. Soc. 141*, 377–385.

Ostmann, H. H.

1950 Euklidsche Ringe mit eindeutiger Partialbruchzerlegung. *J. Reine Angew. Math. 188*, 150–161.

Pendleton, R. L.

1966 A characterization of Q-domains. *Bull. Amer. Math. Soc. 72*, 499–500.

Perić, V.

1964 Eine Bemerkung zu den invertierbaren und fastinvertierbaren Idealen. *Arch. Math. 15*, 415–417.

1966 Zu den fastinvertierbaren Idealen in kommutativen Ringen und Halbgruppen. *Glasnik Mat. Ser. III 1*, 139–146.

1967a Ein Charakterisierung der S-Komponenten der Ideale und der Halbgruppe $(IR_S, \cap)$. *Publ. Inst. Math. (Beograd)* 7, 93–98.

1967b Fastinvertierbaren Primideale der kommutativen Ringe. *Publ. Inst. Math. (Beograd)* 7, 99–109.

Petro, J. W.

1964 Some results on the asymptotic completion of an ideal. *Proc. Amer. Math. Soc. 15*, 519–524.

Pirtle, E. M.

1968 Integral domains which are almost Krull. *J. Sci. Hiroshima Univ. Ser. A-I 32*, 441–44.

1969 Families of valuations and semigroups of fractionary ideal classes. *Trans. Amer. Math. Soc. 144*, 427–440.

1970 On a generalization of Krull domains. *J. Algebra 14*, 485–492.

Pollak, G.

1959 On types of Euclidean norm. *Acta Sci. Math. (Szeged) 20*, 252–269.

1961 Über die Struktur kommutativer Hauptidealringe. *Acta Sci. Math. (Szeged) 22*, 62–74.

Rabin, M.

1953 Sur la représentation des idéaux par des idéaux primaires. *C. R. Acad. Sci. Paris Ser. A-B 237*, 544–545.

Radu, N.

1960 Lema lui Krull si structura inelelor. *Acad. R. P. Romine Stud. Cerc. Mat. 11*, 183–191.

1964 Intersectii de ideale in uncle ivele. *Acad. R. P. Romine Stud. Cerc. Mat. 15*, 209–215.

1966 Sur les anneaux cohérents Laskériens. *Revue Roumaine Math. Pures Appl. 11*, 865–867.

Rees, D.

1955a A note on valuations associated with a local domain. *Proc. Cambridge Philos. Soc. 51*, 252–253.

1955b Valuations associated with a local ring I. *Proc. London Math. Soc. 5*, 107–128.

1956a Valuations associated with a local ring II. *J. London Math. Soc. 31*, 228–235.

1956b Valuations associated with ideals. *J. London Math. Soc. 31*, 221–228.

1956c Valuations associated with ideals II. *Proc. London Math. Soc. (3) 6*, 161–174.

1956d Two classical theorems of ideal theory. *Proc. Cambridge Philos. Soc. 52*, 155–157.

1957 Filtrations as limits of valuations. *J. London Math. Soc. 32*, 97–102.

Reis, C. M., and **Viswanathan, T. M.**

1970 A compactness property for prime ideals in Noetherian rings. *Proc. Amer. Math. Soc. 25*, 353–356.

Ribenboim, P.

1952 Modules sur un anneau de Dedekind. *Summa Brasil. Math. 3*, 21–36.

1955 Sur une conjecture de Krull en theorie des valuations. *Nagoya Math. J. 9*, 87–97.

1956a Sur une note de Nagata relative à un problème de Krull. *Math. Z. 64*, 159–168.

1956b Sur la theorie des idéaux dans certains anneaux de type infini. *An. Acad. Brasil. Ci. 28*, 21–39.

1956c Anneaux normaux réels à caractère fini. *Summa Brasil. Math. 3*, 213–253.

1956d Un théorème sur les anneaux primaires et complètement intégralement clos. *Math. Ann. 130*, 399–404.

1957 Le théorème d'approximation pour les valuations de Krull. *Math. Z. 68*, 1–18.

1958 Sur les groupes totalement ordornnés et l'arithmétique des anneaux de valuation. *Summa Brasil. Math. 4*, 1–64.

1959a On the theory of Krull valuations. *Bol. Soc. Mat. São Paulo 11*, 1–106.

1959b Remarques sur le prolongement des valuations de Krull. *Rend. Circ. Mat. Palermo (2) 8*, 152–159.

1960 Anneaux de Rees intégralement clos. *J. Reine Angew. Math. 204*, 99–107.

1961 Sur la théorie du prolongement des valuation de Krull. *Math. Z. 75*, 449–466.

1967 On ordered modules. *J. Reine Angew. Math. 225*, 120–146.

1969 Corrections to my paper "On ordered modules." *J. Reine Angew. Math. 238*, 132–134.

Richman, F.

1965 Generalized quotient rings. *Proc. Amer. Math. Soc. 16*, 794–799.

Riley, J. A.

1962 Axiomatic primary and tertiary decomposition theory, *Trans. Amer. Math. Soc. 105*, 177–201.

1965 Reflexive ideals in maximal orders. *J. Algebra 2*, 451–465.

Robinson, A.

1967 Non-standard theory of Dedekind rings. *Proc. Kon. Nederl. Akad. Weten. 70*, 444–452.

Rotman, J. J.

1960 Mixed modules over valuation rings. *Pacific J. Math. 10*, 607–623.

Rotman, J. J., and **Yen, T.**

1961 Modules over a complete discrete valuation ring. *Trans. Amer. Math. Soc. 98*, 242–254.

Sakuma, M.

1956/57 On prime operations in the theory of ideals. *J. Sci. Hiroshima Univ. Ser. A 20*, 101–106.

1957/58 Existence theorems of valuations centered in a local domain with preassigned dimension and rank. *J. Sci. Hiroshima Univ. Ser. A 2*, 61–67.

Salmon, P.

1962 Sur les séries formelles restreintes. *C. R. Acad. Sci. Paris Ser. A-B 255*, 227–228.

1964 Sur les séries formelles restreintes. *Bull. Soc. Math. France 92*, 385–410.

Samuel, P.

1952 Some asymptotic properties of powers of ideals. *Ann. of Math. 56*, 11–21.

1957 La notion de place dans un anneau. *Bull. Soc. Math. France 85*, 123–133.

1958 Sur les anneaux gradués. *An. Acad. Brasil. Ci. 30*, 447–450.

1960 Un example d'anneau factoriel. *Bol. Soc. Mat. São Paulo 15*, 1–4.

1961a On the unique factorization rings. *Illinois J. Math. 5*, 1–17.

1961b Sur les anneaux factoriels. *Bull. Soc. Math. France 89*, 155–173.

1964a Classes de diviseurs et dérivées logarithmiques. *Topology* (*Suppl. 1*) *3*, 81–96.

1964b Anneaux gradués factoriels et modules réflexifs. *Bull. Soc. Math. France 92*, 237–249.

1968 On a construction due to P. M. Cohn. *Proc. Cambridge Philos. Soc. 64*, 249–250.

Sato, H.

1953 Zum Teilerkettensatz in kommutativen Ringen. *Proc. Japan Acad. 29*, 10–12.

1956/57 Some remarks on Zariski rings. *J. Sci. Hiroshima Univ. Ser. A 20*, 93–99.

1957/58a On splitting of valuations in extensions of local domains. *J. Sci. Hiroshima Univ. Ser. A 21*, 69–75.

1957/58b A note on principal ideals. *J. Sci. Hiroshima Univ. Ser. A 21*, 77–78.

Satyanarayana, M.

1969 Generalized primary rings. *Math. Ann. 179*, 109–114.

Scheja, G.

1965 Über Primfactorzerlegung in zweidimensionalen lokalen Ringen. *Math. Ann. 159*, 252–258.

Schilling, O. F. G.

1944 Automorphisms of fields of formal power series. *Bull. Amer. Math. Soc. 50*, 892–901.

Schmidt, F. K.

1936 Über die Erhaltung der Kettensätze der Idealtheorie bei beliebigen endlichen Körpererweiterungen. *Math. Z. 41*, 443–450.

Schneider, J. E.

1969 A note on the theory of primes. *Pacific J. Math. 30*, 805–810.

Scholz, A.

1943 Zu Idealtheorie in unendlichen algebraischen Zahlkorpern. *J. Reine Angew. Math. 185*, 113–126.

Scott, W. R.

1954 Divisors of zero in polynomial rings. *Amer. Math. Monthly 61*, 336.

Seidenberg, A.

1945 Valuation ideals in polynomial rings. *Trans. Amer. Math. Soc. 57*, 387–425.

1953 A note on the dimension theory of rings. *Pacific J. Math. 3*, 505–512.

1954 On the dimension theory of rings II. *Pacific J. Math. 4*, 603–614.

1966 Derivations and integral closure. *Pacific J. Math. 16*, 167–173.

1967 Differential ideals in rings of finitely generated type. *Amer. J. Math. 89*, 22–42.

Shannon, R. T.

1970 The rank of a flat module. *Proc. Amer. Math. Soc. 24*, 452–456.

Skolem, T.

1939 Eine Bemerkung uber gewisse Ringe mit Anwendung auf die Produktzerlegung von Polynom. *Nordisk Mat. Tidskr. 21*, 99–107.

Smith, W. W.

1969 Projective ideals of finite type. *Canad. J. Math. 21*, 1057–1061.

Snapper, E.

1950 Completely primary rings I. *Ann. of Math.* (*2*) *52*, 666–693.

1951 Completely primary rings II. Algebraic and transcendental extensions. *Ann. of Math.* (*2*) *53*, 125–142.

1951b Completely primary rings III. Imbedding and isomorphism theorems. *Ann. of Math.* (*2*) *53*, 207–234.

1952 Completely primary rings IV. Chain conditions. *Ann. of Math.* (*2*) *55*, 46–64.

Storrer, H. H.

1968 Epimorphismen von kommutativen Ringen. *Comment. Math. Helv. 43*, 378–401.

1969 A characterization of Prüfer rings. *Canad. Math. Bull. 12*, 809–812.

Strooker, J. R.

1966 A remark on Artin-van der Waerden equivalence of ideals. *Math. Z. 93*, 241–242.

Szpiro, L.

1966 Fonctions d'ordre et valuations. *Bull. Soc. Math. France 94*, 301–311.

Tirvari, K.

1967 On complete rings of quotients. *J. Indian Math. Soc. 31*, 95–99.

Underwood, D. H.

1969 On some uniqueness question in primary representations of ideals. *J. Math. Kyoto Univ. 9*, 69–94.

Uzkov, A. I.

1939 Zur Idealtheorie der kommutativen Ringe I. *Rec. Math.* (*Moscou*) [*Mat. Sb.*] *N.S. 5* (*47*), 513–520.

1948 On rings of quotients of commutative rings. *Mat. Sb.* (*N.S.*) *22* (*64*), 439–441.

1963 On the decomposition of modules over a commutative ring into direct sums of cyclic submodules. *Mat. Sb.* (*N.S.*) *62*, (104), 469–475.

van der Welt, A. P. J.

1964 Contributions to ideal theory in general rings. *Nederl. Akad. Wetensch. Proc. Ser. A 67*, 26, 68–77.

Vasconcelos, W. V.

1969a On finitely generated flat modules. *Trans. Amer. Math. Soc. 138*, 505–512.

1969b On projective modules of finite rank. *Proc. Amer. Math. Soc. 22*, 430–433.

1970 Quasi-normal rings. *Illinois J. Math. 14*, 268–273.

Viswanathan, T. M.

1969a Ordered modules of fractions. *J. Reine Angew. Math. 235*, 78–107.

1969b Generalization of Hölders' theorem to ordered modules. *Canad. J. Math. 21*, 149–157.

Wajnryb, B., and **Zaks, A.**

On the flat overrings of an integral domain. *Glasglow Math. J.* (to appear).

Ward, M.

1937 Residuation in structures over which a multiplication is defined. *Duke Math. J. 3*, 627–636.

1940 Residuated distributive lattices. *Duke Math. J. 6*, 641–651.

Ward, M., and **Dilworth, R. P.**

1939 Residuated lattices. *Trans. Amer. Math. Soc. 45*, 335–354.

Warner, S.
1960 Compact Noetherian rings. *Math. Ann. 141*, 161–170.
Weinert, H. J.
1961 Über die Einbettung von Ringen in Oberringe mit Einselement. *Acta Sci. Math. (Szeged) 22*, 91–105.
Wichman, M.
1970 The width of a module. *Canad. J. Math. 22*, 102–115.
Wolfisz, A.
1941 Über primare Ideale. *Bull. Acad. Sci. Georgian SSR 2*, 383–388.
Wood, C. A.
1969a Commutative rings for which each proper homomorphic image is a multiplication ring. *J. Sci. Hiroshima Univ. Ser. A-I 33*, 85–94.
1969b On general Z.P.I.-rings. *Pacific J. Math. 30*, 837–846.
Yoshida, M.
1956a On polynomial extensions of rings. *Canad. J. Math. 8*, 1–2.
1956b A theorem on Zariski rings. *Canad. J. Math. 8*, 3–4.
Yoshida, M., and **Sakuma, M.**
1953 A note on semilocal rings. *J. Sci. Hiroshima Univ. Ser. A 17*, 181–184.
1954a On integrally closed Noetherian rings. *J. Sci. Hiroshima Univ. Ser. A 17*, 311–315.
1954b The intersection theorem on Noetherian rings. *J. Sci. Hiroshima Univ. Ser. A 17*, 317–320.
Zariski, O.
1938 Polynomial ideals defined by infinitely near base points. *Amer. J. Math. 60*, 151–204.
Zelinsky, D.
1948 Topological characterization of fields with valuations. *Bull. Amer. Math. Soc. 54*, 1145–1150.

Subject Index

P

Q

R

S

Pure and Applied Mathematics

A Series of Monographs and Textbooks

Editors

Paul A. Smith and Samuel Eilenberg

Columbia University, New York

1: Arnold Sommerfeld. Partial Differential Equations in Physics. 1949 (Lectures on Theoretical Physics, Volume VI)
2: Reinhold Baer. Linear Algebra and Projective Geometry. 1952
3: Herbert Busemann and Paul Kelly. Projective Geometry and Projective Metrics. 1953
4: Stefan Bergman and M. Schiffer. Kernel Functions and Elliptic Differential Equations in Mathematical Physics. 1953
5: Ralph Philip Boas, Jr. Entire Functions. 1954
6: Herbert Busemann. The Geometry of Geodesics. 1955
7: Claude Chevalley. Fundamental Concepts of Algebra. 1956
8: Sze-Tsen Hu. Homotopy Theory. 1959
9: A. M. Ostrowski. Solution of Equations and Systems of Equations. Second Edition. 1966
10: J. Dieudonné. Treatise on Analysis. Volume I, Foundations of Modern Analysis, enlarged and corrected printing, 1969. Volume II, 1970.
11: S. I. Goldberg. Curvature and Homology. 1962.
12: Sigurdur Helgason. Differential Geometry and Symmetric Spaces. 1962
13: T. H. Hildebrandt. Introduction to the Theory of Integration. 1963.
14: Shreeram Abhyankar. Local Analytic Geometry. 1964
15: Richard L. Bishop and Richard J. Crittenden. Geometry of Manifolds. 1964
16: Steven A. Gaal. Point Set Topology. 1964
17: Barry Mitchell. Theory of Categories. 1965
18: Anthony P. Morse. A Theory of Sets. 1965

Pure and Applied Mathematics

A Series of Monographs and Textbooks

19: Gustave Choquet. Topology. 1966

20: Z. I. Borevich and I. R. Shafarevich. Number Theory. 1966

21: José Luis Massera and Juan Jorge Schaffer. Linear Differential Equations and Function Spaces. 1966

22: Richard D. Schafer. An Introduction to Nonassociative Algebras. 1966

23: Martin Eichler. Introduction to the Theory of Algebraic Numbers and Functions. 1966

24: Shreeram Abhyankar. Resolution of Singularities of Embedded Algebraic Surfaces. 1966

25: François Treves. Topological Vector Spaces, Distributions, and Kernels. 1967

26: Peter D. Lax and Ralph S. Phillips. Scattering Theory. 1967.

27: Oystein Ore. The Four Color Problem. 1967

28: Maurice Heins. Complex Function Theory. 1968

29: R. M. Blumenthal and R. K. Getoor. Markov Processes and Potential Theory. 1968

30: L. J. Mordell. Diophantine Equations. 1969

31: J. Barkley Rosser. Simplified Independence Proofs: Boolean Valued Models of Set Theory. 1969

32: William F. Donoghue, Jr. Distributions and Fourier Transforms. 1969

33: Marston Morse and Stewart S. Cairns. Critical Point Theory in Global Analysis and Differential Topology. 1969

34: Edwin Weiss. Cohomology of Groups. 1969

35: Hans Freudenthal and H. De Vries. Linear Lie Groups. 1969

36: Laszlo Fuchs. Infinite Abelian Groups: Volume I. 1970

37: Keio Nagami. Dimension Theory. 1970

38: Peter L. Duren. Theory of H^p Spaces. 1970

39: Bodo Pareigis. Categories and Functors. 1970

40: Paul L. Butzer and Rolf J. Nessel. Fourier Analysis and Approximation: Volume 1, One-Dimensional Theory. 1971

41: Eduard Prugovečki. Quantum Mechanics in Hilbert Space. 1971

42: D. V. Widder: An Introduction to Transform Theory. 1971

43: Max D. Larsen and Paul J. McCarthy. Multiplicative Theory of Ideals. 1971

44: Ernst-August Behrens. Ring Theory. 1971

In preparation

Werner Greub, Steve Halperin, and James Vanstone. De Rham Cohomology of Fibre Bundles: Volume 1; Manifolds, Sphere Bundles, Vector Bundles.